Springer Handbook of Crystal Growth

Govindhan Dhanaraj, Kullaiah Byrappa,
Vishwanath Prasad, Michael Dudley (Eds.)

Springer Handbook of Crystal Growth
Organization of the Handbook

使 用 说 明

1.《晶体生长手册》原版为一册，分为A～H部分。考虑到使用方便以及内容一致，影印版分为6册：第1册—Part A，第2册—Part B，第3册—Part C，第4册—Part D、E，第5册—Part F、G，第6册—Part H。

2.各册在页脚重新编排页码，该页码对应中文目录。保留了原书页眉及页码，其页码对应原书目录及主题索引。

3.各册均给出完整6册书的章目录。

4.作者及其联系方式、缩略语表各册均完整呈现。

5.主题索引安排在第6册。

6.文前介绍基本采用中英文对照形式，方便读者快速浏览。

材料科学与工程图书工作室

联系电话　0451-86412421

　　　　　0451-86414559

邮　　箱　yh_bj@yahoo.com.cn

　　　　　xuyaying81823@gmail.com

　　　　　zhxh6414559@yahoo.com.cn

Springer 手册精选系列

晶体生长手册

晶体生长及缺陷形成概论

【第1册】

Springer Handbook of Crystal Growth

〔美〕Govindhan Dhanaraj 等主编

（影印版）

黑版贸审字08-2012-047号

Reprint from English language edition:
Springer Handbook of Crystal Growth
by Govindhan Dhanaraj, Kullaiah Byrappa, Vishwanath Prasad
and Michael Dudley

图书在版编目（CIP）数据

晶体生长手册. 1, 晶体生长及缺陷形成概论 =Handbook of Crystal Growth. 1, Fundamentals of Crystal Growth and Defect Formation : 英文 / (美)德哈纳拉(Dhanaraj,G.) 等主编. —影印本. —哈尔滨：哈尔滨工业大学出版社, 2013.1
（Springer手册精选系列）
ISBN 978-7-5603-3386-1

Ⅰ. ①晶… Ⅱ. ①德… Ⅲ. ①晶体生长 – 手册 – 英文②晶体缺陷 – 手册 – 英文 Ⅳ. ①O78-62②O77-62

中国版本图书馆CIP数据核字(2012)第292355号

材料科学与工程
图书工作室

责任编辑 杨　桦　许雅莹　张秀华
出版发行 哈尔滨工业大学出版社
社　　址 哈尔滨市南岗区复华四道街10号 邮编 150006
传　　真 0451-86414749
网　　址 http://hitpress.hit.edu.cn
印　　刷 哈尔滨市石桥印务有限公司
开　　本 787mm×960mm 1/16 印张 14.75
版　　次 2013年1月第1版 2013年1月第1次印刷
书　　号 ISBN 978-7-5603-3386-1
定　　价 48.00元

序 言

多年以来，有很多探索研究已经成功地描述了晶体生长的生长工艺和科学，有许多文章、专著、会议文集和手册对这一领域的前沿成果做了综合评述。这些出版物反映了人们对体材料晶体和薄膜晶体的兴趣日益增长，这是由于它们的电子、光学、机械、微结构以及不同的科学和技术应用引起的。实际上，大部分半导体和光器件的现代成果，如果没有基本的、二元的、三元的及其他不同特性和大尺寸的化合物晶体的发展则是不可能的。这些文章致力于生长机制的基本理解、缺陷形成、生长工艺和生长系统的设计，因此数量是庞大的。

本手册针对目前备受关注的体材料晶体和薄膜晶体的生长技术水平进行阐述。我们的目的是使读者了解经常使用的生长工艺、材料生产和缺陷产生的基本知识。为完成这一任务，我们精选了50多位顶尖科学家、学者和工程师，他们的合作者来自于22个不同国家。这些作者根据他们的专业所长，编写了关于晶体生长和缺陷形成共计52章内容：从熔体、溶液到气相体材料生长；外延生长；生长工艺和缺陷的模型；缺陷特性的技术以及一些现代的特别课题。

本手册分为七部分。Part A介绍基础理论：生长和表征技术综述，表面成核工艺，溶液生长晶体的形态，生长过程中成核的层错，缺陷形成的形态。

Part B介绍体材料晶体的熔体生长，一种生长大尺寸晶体的关键方法。这一部分阐述了直拉单晶工艺、泡生法、布里兹曼法、浮区熔融等工艺，以及这些方法的最新进展，例如应用磁场的晶体生长、生长轴的取向、增加底基和形状控制。本部分涉及材料从硅和Ⅲ–Ⅴ族化合物到氧化物和氟化物的广泛内容。

第三部分，本书的Part C关注了溶液生长法。在前两章里讨论了水热生长法的不同方面，随后的三章介绍了非线性和激光晶体、KTP和KDP。通过在地球上和微重力环境下生长的比较给出了重力对溶液生长法的影响的知识。

Part D的主题是气相生长。这一部分提供了碳化硅、氮化镓、氮化铝和有机半导体的气相生长的内容。随后的Part E是关于外延生长和薄膜的，主要包括从液相的化学气相淀积到脉冲激光和脉冲电子淀积。

Part F介绍了生长工艺和缺陷形成的模型。这些章节验证了工艺参数和产生晶体质量问题包括缺陷形成的直接相互作用关系。随后的Part G展示了结晶材料特性和分析的发展。Part F和G说明了预测工具和分析技术在帮助高质量的大尺寸晶体生长工艺的设计和控制方面是非常好用的。

最后的Part H致力于精选这一领域的部分现代课题，例如蛋白质晶体生长、凝胶结晶、原位结构、单晶闪烁材料的生长、光电材料和线切割大晶体薄膜。

我们希望这本施普林格手册对那些学习晶体生长的研究生，那些从事或即将从事这一领域研究的来自学术界和工业领域的研究人员、科学家和工程师以及那些制备晶体的人是有帮助的。

我们对施普林格的Dr. Claus Acheron，Dr. Werner Skolaut和le-tex的Ms Anne Strobach的特别努力表示真诚的感谢，没有他们本书将无法呈现。

我们感谢我们的作者编写了详尽的章节内容和在本书出版期间对我们的耐心。一位编者（GD）感谢他的家庭成员和Dr. Kedar Gupta(ARC Energy 的CEO)，感谢他们在本书编写期间的大力支持和鼓励。还对Peter Rudolf, David Bliss,Ishwara Bhat和Partha Dutta在A、B、E部分的编写中所给予的帮助表示感谢。

Nashua, New Hampshire, April 2010 G. Dhanaraj
Mysore, India K. Byrappa
Denton, Texas V. Prasad
Stony Brook, New York M. Dudley

Preface

Over the years, many successful attempts have been made to describe the art and science of crystal growth, and many review articles, monographs, symposium volumes, and handbooks have been published to present comprehensive reviews of the advances made in this field. These publications are testament to the growing interest in both bulk and thin-film crystals because of their electronic, optical, mechanical, microstructural, and other properties, and their diverse scientific and technological applications. Indeed, most modern advances in semiconductor and optical devices would not have been possible without the development of many elemental, binary, ternary, and other compound crystals of varying properties and large sizes. The literature devoted to basic understanding of growth mechanisms, defect formation, and growth processes as well as the design of growth systems is therefore vast.

The objective of this Springer Handbook is to present the state of the art of selected topical areas of both bulk and thin-film crystal growth. Our goal is to make readers understand the basics of the commonly employed growth processes, materials produced, and defects generated. To accomplish this, we have selected more than 50 leading scientists, researchers, and engineers, and their many collaborators from 22 different countries, to write chapters on the topics of their expertise. These authors have written 52 chapters on the fundamentals of crystal growth and defect formation; bulk growth from the melt, solution, and vapor; epitaxial growth; modeling of growth processes and defects; and techniques of defect characterization, as well as some contemporary special topics.

This Springer Handbook is divided into seven parts. Part A presents the fundamentals: an overview of the growth and characterization techniques, followed by the state of the art of nucleation at surfaces, morphology of crystals grown from solutions, nucleation of dislocation during growth, and defect formation and morphology.

Part B is devoted to bulk growth from the melt, a method critical to producing large-size crystals. The chapters in this part describe the well-known processes such as Czochralski, Kyropoulos, Bridgman, and floating zone, and focus specifically on recent advances in improving these methodologies such as application of magnetic fields, orientation of the growth axis, introduction of a pedestal, and shaped growth. They also cover a wide range of materials from silicon and III–V compounds to oxides and fluorides.

The third part, Part C of the book, focuses on solution growth. The various aspects of hydrothermal growth are discussed in two chapters, while three other chapters present an overview of the nonlinear and laser crystals, KTP and KDP. The knowledge on the effect of gravity on solution growth is presented through a comparison of growth on Earth versus in a microgravity environment.

The topic of Part D is vapor growth. In addition to presenting an overview of vapor growth, this part also provides details on vapor growth of silicon carbide, gallium nitride, aluminum nitride, and organic semiconductors. This is followed by chapters on epitaxial growth and thin films in Part E. The topics range from chemical vapor deposition to liquid-phase epitaxy to pulsed laser and pulsed electron deposition.

Modeling of both growth processes and defect formation is presented in Part F. These chapters demonstrate the direct correlation between the process parameters and quality of the crystal produced, including the formation of defects. The subsequent Part G presents the techniques that have been developed for crystalline material characterization and analysis. The chapters in Parts F and G demonstrate how well predictive tools and analytical techniques have helped the design and control of growth processes for better-quality crystals of large sizes.

The final Part H is devoted to some selected contemporary topics in this field, such as protein crystal growth, crystallization from gels, in situ structural studies, growth of single-crystal scintillation materials, photovoltaic materials, and wire-saw slicing of large crystals to produce wafers.

We hope this Springer Handbook will be useful to graduate students studying crystal growth and to researchers, scientists, and engineers from academia and industry who are conducting or intend to conduct research in this field as well as those who grow crystals.

We would like to express our sincere thanks to Dr. Claus Acheron and Dr. Werner Skolaut of Springer and Ms Anne Strohbach of le-tex for their extraordinary efforts without which this handbook would not have taken its final shape.

We thank our authors for writing comprehensive chapters and having patience with us during the publication of this Handbook. One of the editors (GD) would like to thank his family members and Dr. Kedar Gupta (CEO of ARC Energy) for their generous support and encouragement during the entire course of editing this handbook. Acknowledgements are also due to Peter Rudolf, David Bliss, Ishwara Bhat, and Partha Dutta for their help in editing Parts A, B, E, and H, respectively.

Nashua, New Hampshire, April 2010 — G. Dhanaraj
Mysore, India — K. Byrappa
Denton, Texas — V. Prasad
Stony Brook, New York — M. Dudley

About the Editors

Govindhan Dhanaraj is the Manager of Crystal Growth Technologies at Advanced Renewable Energy Company (ARC Energy) at Nashua, New Hampshire (USA) focusing on the growth of large size sapphire crystals for LED lighting applications, characterization and related crystal growth furnace development. He received his PhD from the Indian Institute of Science, Bangalore and his Master of Science from Anna University (India). Immediately after his doctoral degree, Dr. Dhanaraj joined a National Laboratory, presently known as Rajaramanna Center for Advanced Technology in India, where he established an advanced Crystal Growth Laboratory for the growth of optical and laser crystals. Prior to joining ARC Energy, Dr. Dhanaraj served as a Research Professor at the Department of Materials Science and Engineering, Stony Brook University, NY, and also held a position of Research Assistant Professor at Hampton University, VA. During his 25 years of focused expertise in crystal growth research, he has developed optical, laser and semiconductor bulk crystals and SiC epitaxial films using solution, flux, Czochralski, Bridgeman, gel and vapor methods, and characterized them using x-ray topography, synchrotron topography, chemical etching and optical and atomic force microscopic techniques. He co-organized a symposium on Industrial Crystal Growth under the 17th American Conference on Crystal Growth and Epitaxy in conjunction with the 14th US Biennial Workshop on Organometallic Vapor Phase Epitaxy held at Lake Geneva, WI in 2009. Dr. Dhanaraj has delivered invited lectures and also served as session chairman in many crystal growth and materials science meetings. He has published over 100 papers and his research articles have attracted over 250 rich citations.

Kullaiah Byrappa received his Doctor's degree in Crystal Growth from the Moscow State University, Moscow in 1981. He is Professor of Materials Science, Head of the Crystal Growth Laboratory, and Director of the Internal Quality Assurance Cell of the University of Mysore, India. His current research is in crystal engineering of polyscale materials through novel solution processing routes, particularly covering hydrothermal, solvothermal and supercritical methods. Professor Byrappa has co-authored the Handbook of Hydrothermal Technology, and edited 4 books as well as two special editions of Journal of Materials Science, and published 180 research papers including 26 invited reviews and book chapters on various aspects of novel routes of solution processing. Professor Byrappa has delivered over 60 keynote and invited lectures at International Conferences, and several hundreds of colloquia and seminars at various institutions around the world. He has also served as chair and co-chair for numerous international conferences. He is a Fellow of the World Academy of Ceramics. Professor Byrappa is serving in several international committees and commissions related to crystallography, crystal growth, and materials science. He is the Founder Secretary of the International Solvothermal and Hydrothermal Association. Professor Byrappa is a recipient of several awards such as the Sir C.V. Raman Award, Materials Research Society of India Medal, and the Golden Jubilee Award of the University of Mysore.

Vishwanath "Vish" Prasad is the Vice President for Research and Economic Development and Professor of Mechanical and Energy Engineering at the University of North Texas (UNT), one of the largest university in the state of Texas. He received his PhD from the University of Delaware (USA), his Masters of Technology from the Indian Institute of Technology, Kanpur, and his bachelor's from Patna University in India all in Mechanical Engineering. Prior to joining UNT in 2007, Dr. Prasad served as the Dean at Florida International University (FIU) in Miami, where he also held the position of Distinguished Professor of Engineering. Previously, he has served as a Leading Professor of Mechanical Engineering at Stony Brook University, New York, as an Associate Professor and Assistant Professor at Columbia University. He has received many special recognitions for his contributions to engineering education. Dr. Prasad's research interests include thermo-fluid sciences, energy systems, electronic materials, and computational materials processing. He has published over 200 articles, edited/co-edited several books and organized numerous conferences, symposia, and workshops. He serves as the lead editor of the Annual Review of Heat Transfer. In the past, he has served as an Associate Editor of the ASME Journal of Heat. Dr. Prasad is an elected Fellow of the American Society of Mechanical Engineers (ASME), and has served as a member of the USRA Microgravity Research Council. Dr. Prasad's research has focused on bulk growth of silicon, III-V compounds, and silicon carbide; growth of large diameter Si tube; design of crystal growth systems; and sputtering and chemical vapor deposition of thin films. He is also credited to initiate research on wire saw cutting of large crystals to produce wafers with much reduced material loss. Dr. Prasad's research has been well funded by US National Science Foundation (NSF), US Department of Defense, US Department of Energy, and industry.

Michael Dudley received his Doctoral Degree in Engineering from Warwick University, UK, in 1982. He is Professor and Chair of the Materials Science and Engineering Department at Stony Brook University, New York, USA. He is director of the Stony Brook Synchrotron Topography Facility at the National Synchrotron Light Source at Brookhaven National Laboratory, Upton New York. His current research focuses on crystal growth and characterization of defect structures in single crystals with a view to determining their origins. The primary technique used is synchrotron topography which enables analysis of defects and generalized strain fields in single crystals in general, with particular emphasis on semiconductor, optoelectronic, and optical crystals. Establishing the relationship between crystal growth conditions and resulting defect distributions is a particular thrust area of interest to Dudley, as is the correlation between electronic/optoelectronic device performance and defect distribution. Other techniques routinely used in such analysis include transmission electron microscopy, high resolution triple-axis x-ray diffraction, atomic force microscopy, scanning electron microscopy, Nomarski optical microscopy, conventional optical microscopy, IR microscopy and fluorescent laser scanning confocal microscopy. Dudley's group has played a prominent role in the development of SiC and AlN growth, characterizing crystals grown by many of the academic and commercial entities involved enabling optimization of crystal quality. He has co-authored some 315 refereed articles and 12 book chapters, and has edited 5 books. He is currently a member of the Editorial Board of Journal of Applied Physics and Applied Physics Letters and has served as Chair or Co-Chair for numerous international conferences.

List of Authors

Francesco Abbona
Università degli Studi di Torino
Dipartimento di Scienze Mineralogiche
e Petrologiche
via Valperga Caluso 35
10125 Torino, Italy
e-mail: *francesco.abbona@unito.it*

Mohan D. Aggarwal
Alabama A&M University
Department of Physics
Normal, AL 35762, USA
e-mail: *mohan.aggarwal@aamu.edu*

Marcello R.B. Andreeta
University of São Paulo
Crystal Growth and Ceramic Materials Laboratory,
Institute of Physics of São Carlos
Av. Trabalhador Sãocarlense, 400
São Carlos, SP 13560-970, Brazil
e-mail: *marcello@if.sc.usp.br*

Dino Aquilano
Università degli Studi di Torino
Facoltà di Scienze Matematiche, Fisiche e Naturali
via P. Giuria, 15
Torino, 10126, Italy
e-mail: *dino.aquilano@unito.it*

Roberto Arreguín-Espinosa
Universidad Nacional Autónoma de México
Instituto de Química
Circuito Exterior, C.U. s/n
Mexico City, 04510, Mexico
e-mail: *arrespin@unam.mx*

Jie Bai
Intel Corporation
RA3-402, 5200 NE Elam Young Parkway
Hillsboro, OR 97124-6497, USA
e-mail: *jie.bai@intel.com*

Stefan Balint
West University of Timisoara
Department of Computer Science
Blvd. V. Parvan 4
Timisoara, 300223, Romania
e-mail: *balint@math.uvt.ro*

Ashok K. Batra
Alabama A&M University
Department of Physics
4900 Meridian Street
Normal, AL 35762, USA
e-mail: *ashok.batra@aamu.edu*

Handady L. Bhat
Indian Institute of Science
Department of Physics
CV Raman Avenue
Bangalore, 560012, India
e-mail: *hlbhat@physics.iisc.ernet.in*

Ishwara B. Bhat
Rensselaer Polytechnic Institute
Electrical Computer
and Systems Engineering Department
110 8th Street, JEC 6031
Troy, NY 12180, USA
e-mail: *bhati@rpi.edu*

David F. Bliss
US Air Force Research Laboratory
Sensors Directorate Optoelectronic Technology
Branch
80 Scott Drive
Hanscom AFB, MA 01731, USA
e-mail: *david.bliss@hanscom.af.mil*

Mikhail A. Borik
Russian Academy of Sciences
Laser Materials and Technology Research Center,
A.M. Prokhorov General Physics Institute
Vavilov 38
Moscow, 119991, Russia
e-mail: *borik@lst.gpi.ru*

Liliana Braescu
West University of Timisoara
Department of Computer Science
Blvd. V. Parvan 4
Timisoara, 300223, Romania
e-mail: *lilianabraescu@balint1.math.uvt.ro*

Kullaiah Byrappa
University of Mysore
Department of Geology
Manasagangotri
Mysore, 570 006, India
e-mail: *kbyrappa@gmail.com*

Dang Cai
CVD Equipment Corporation
1860 Smithtown Ave.
Ronkonkoma, NY 11779, USA
e-mail: *dcai@cvdequipment.com*

Michael J. Callahan
GreenTech Solutions
92 Old Pine Drive
Hanson, MA 02341, USA
e-mail: *mjcal37@yahoo.com*

Joan J. Carvajal
Universitat Rovira i Virgili (URV)
Department of Physics and Crystallography
of Materials and Nanomaterials (FiCMA-FiCNA)
Campus Sescelades, C/ Marcel·lí Domingo, s/n
Tarragona 43007, Spain
e-mail: *joanjosep.carvajal@urv.cat*

Aaron J. Celestian
Western Kentucky University
Department of Geography and Geology
1906 College Heights Blvd.
Bowling Green, KY 42101, USA
e-mail: *aaron.celestian@wku.edu*

Qi-Sheng Chen
Chinese Academy of Sciences
Institute of Mechanics
15 Bei Si Huan Xi Road
Beijing, 100190, China
e-mail: *qschen@imech.ac.cn*

Chunhui Chung
Stony Brook University
Department of Mechanical Engineering
Stony Brook, NY 11794-2300, USA
e-mail: *chuchung@ic.sunysb.edu*

Ted Ciszek
Geolite/Siliconsultant
31843 Miwok Trl.
Evergreen, CO 80437, USA
e-mail: *ted_ciszek@siliconsultant.com*

Abraham Clearfield
Texas A&M University
Distinguished Professor of Chemistry
College Station, TX 77843-3255, USA
e-mail: *clearfield@chem.tamu.edu*

Hanna A. Dabkowska
Brockhouse Institute for Materials Research
Department of Physics and Astronomy
1280 Main Str W.
Hamilton, Ontario L8S 4M1, Canada
e-mail: *dabkoh@mcmaster.ca*

Antoni B. Dabkowski
McMaster University, BIMR
Brockhouse Institute for Materials Research,
Department of Physics and Astronomy
1280 Main Str W.
Hamilton, Ontario L8S 4M1, Canada
e-mail: *dabko@mcmaster.ca*

Rafael Dalmau
HexaTech Inc.
991 Aviation Pkwy Ste 800
Morrisville, NC 27560, USA
e-mail: *rdalmau@hexatechinc.com*

Govindhan Dhanaraj
ARC Energy
18 Celina Avenue, Unit 77
Nashua, NH 03063, USA
e-mail: *dhanaraj@arc-energy.com*

Ramasamy Dhanasekaran
Anna University Chennai
Crystal Growth Centre
Chennai, 600 025, India
e-mail: *rdhanasekaran@annauniv.edu; rdcgc@yahoo.com*

Ernesto Diéguez
Universidad Autónoma de Madrid
Department Física de Materiales
Madrid 28049, Spain
e-mail: *ernesto.dieguez@uam.es*

Vijay K. Dixit
Raja Ramanna Center for Advance Technology
Semiconductor Laser Section,
Solid State Laser Division
Rajendra Nagar, RRCAT.
Indore, 452013, India
e-mail: *dixit@rrcat.gov.in*

Sadik Dost
University of Victoria
Crystal Growth Laboratory
Victoria, BC V8W 3P6, Canada
e-mail: *sdost@me.uvic.ca*

Michael Dudley
Stony Brook University
Department of Materials Science and Engineering
Stony Brook, NY 11794-2275, USA
e-mail: *mdudley@notes.cc.sunysb.edu*

Partha S. Dutta
Rensselaer Polytechnic Institute
Department of Electrical, Computer
and Systems Engineering
110 Eighth Street
Troy, NY 12180, USA
e-mail: *duttap@rpi.edu*

Francesc Díaz
Universitat Rovira i Virgili (URV)
Department of Physics and Crystallography
of Materials and Nanomaterials (FiCMA-FiCNA)
Campus Sescelades, C/ Marcel·lí Domingo, s/n
Tarragona 43007, Spain
e-mail: *f.diaz@urv.cat*

Paul F. Fewster
PANalytical Research Centre,
The Sussex Innovation Centre
Research Department
Falmer
Brighton, BN1 9SB, UK
e-mail: *paul.fewster@panalytical.com*

Donald O. Frazier
NASA Marshall Space Flight Center
Engineering Technology Management Office
Huntsville, AL 35812, USA
e-mail: *donald.o.frazier@nasa.gov*

James W. Garland
EPIR Technologies, Inc.
509 Territorial Drive, Ste. B
Bolingbrook, IL 60440, USA
e-mail: *jgarland@epir.com*

Thomas F. George
University of Missouri-St. Louis
Center for Nanoscience,
Department of Chemistry and Biochemistry,
Department of Physics and Astronomy
One University Boulevard
St. Louis, MO 63121, USA
e-mail: *tfgeorge@umsl.edu*

Andrea E. Gutiérrez-Quezada
Universidad Nacional Autónoma de México
Instituto de Química
Circuito Exterior, C.U. s/n
Mexico City, 04510, Mexico
e-mail: *30111390@escolar.unam.mx*

Carl Hemmingsson
Linköping University
Department of Physics, Chemistry
and Biology (IFM)
581 83 Linköping, Sweden
e-mail: *cah@ifm.liu.se*

Antonio Carlos Hernandes
University of São Paulo
Crystal Growth and Ceramic Materials Laboratory,
Institute of Physics of São Carlos
Av. Trabalhador Sãocarlense
São Carlos, SP 13560-970, Brazil
e-mail: *hernandes@if.sc.usp.br*

Koichi Kakimoto
Kyushu University
Research Institute for Applied Mechanics
6-1 Kasuga-kouen, Kasuga
816-8580 Fukuoka, Japan
e-mail: *kakimoto@riam.kyushu-u.ac.jp*

Imin Kao
State University of New York at Stony Brook
Department of Mechanical Engineering
Stony Brook, NY 11794-2300, USA
e-mail: *imin.kao@stonybrook.edu*

John J. Kelly
Utrecht University,
Debye Institute for Nanomaterials Science
Department of Chemistry
Princetonplein 5
3584 CC, Utrecht, The Netherlands
e-mail: *j.j.kelly@uu.nl*

Jeonggoo Kim
Neocera, LLC
10000 Virginia Manor Road #300
Beltsville, MD, USA
e-mail: *kim@neocera.com*

Helmut Klapper
Institut für Kristallographie
RWTH Aachen University
Aachen, Germany
e-mail: *klapper@xtal.rwth-aachen.de; helmut-klapper@web.de*

Christine F. Klemenz Rivenbark
Krystal Engineering LLC
General Manager and Technical Director
1429 Chaffee Drive
Titusville, FL 32780, USA
e-mail: *ckr@krystalengineering.com*

Christian Kloc
Nanyang Technological University
School of Materials Science and Engineering
50 Nanyang Avenue
639798 Singapore
e-mail: *ckloc@ntu.edu.sg*

Solomon H. Kolagani
Neocera LLC
10000 Virginia Manor Road
Beltsville, MD 20705, USA
e-mail: *harsh@neocera.com*

Akinori Koukitu
Tokyo University of Agriculture and Technology (TUAT)
Department of Applied Chemistry
2-24-16 Naka-cho, Koganei
184-8588 Tokyo, Japan
e-mail: *koukitu@cc.tuat.ac.jp*

Milind S. Kulkarni
MEMC Electronic Materials
Polysilicon and Quantitative Silicon Research
501 Pearl Drive
St. Peters, MO 63376, USA
e-mail: *mkulkarni@memc.com*

Yoshinao Kumagai
Tokyo University of Agriculture and Technology
Department of Applied Chemistry
2-24-16 Naka-cho, Koganei
184-8588 Tokyo, Japan
e-mail: *4470kuma@cc.tuat.ac.jp*

Valentin V. Laguta
Institute of Physics of the ASCR
Department of Optical Materials
Cukrovarnicka 10
Prague, 162 53, Czech Republic
e-mail: *laguta@fzu.cz*

Ravindra B. Lal
Alabama Agricultural and Mechanical University
Physics Department
4900 Meridian Street
Normal, AL 35763, USA
e-mail: *rblal@comcast.net*

Chung-Wen Lan
National Taiwan University
Department of Chemical Engineering
No. 1, Sec. 4, Roosevelt Rd.
Taipei, 106, Taiwan
e-mail: *cwlan@ntu.edu.tw*

Hongjun Li
Chinese Academy of Sciences
R & D Center of Synthetic Crystals,
Shanghai Institute of Ceramics
215 Chengbei Rd., Jiading District
Shanghai, 201800, China
e-mail: *lh_li@mail.sic.ac.cn*

Elena E. Lomonova
Russian Academy of Sciences
Laser Materials and Technology Research Center,
A.M. Prokhorov General Physics Institute
Vavilov 38
Moscow, 119991, Russia
e-mail: *lomonova@lst.gpi.ru*

Ivan V. Markov
Bulgarian Academy of Sciences
Institute of Physical Chemistry
Sofia, 1113, Bulgaria
e-mail: *imarkov@ipc.bas.bg*

Bo Monemar
Linköping University
Department of Physics, Chemistry and Biology
58183 Linköping, Sweden
e-mail: *bom@ifm.liu.se*

Abel Moreno
Universidad Nacional Autónoma de México
Instituto de Química
Circuito Exterior, C.U. s/n
Mexico City, 04510, Mexico
e-mail: *carcamo@unam.mx*

Roosevelt Moreno Rodriguez
State University of New York at Stony Brook
Department of Mechanical Engineering
Stony Brook, NY 11794-2300, USA
e-mail: *roosevelt@dove.eng.sunysb.edu*

S. Narayana Kalkura
Anna University Chennai
Crystal Growth Centre
Sardar Patel Road
Chennai, 600025, India
e-mail: *kalkura@annauniv.edu*

Mohan Narayanan
Reliance Industries Limited
1, Rich Branch court
Gaithersburg, MD 20878, USA
e-mail: *mohan.narayanan@ril.com*

Subramanian Natarajan
Madurai Kamaraj University
School of Physics
Palkalai Nagar
Madurai, India
e-mail: *s_natarajan50@yahoo.com*

Martin Nikl
Academy of Sciences of the Czech Republic (ASCR)
Department of Optical Crystals, Institute of Physics
Cukrovarnicka 10
Prague, 162 53, Czech Republic
e-mail: *nikl@fzu.cz*

Vyacheslav V. Osiko
Russian Academy of Sciences
Laser Materials and Technology Research Center,
A.M. Prokhorov General Physics Institute
Vavilov 38
Moscow, 119991, Russia
e-mail: *osiko@lst.gpi.ru*

John B. Parise
Stony Brook University
Chemistry Department
and Department of Geosciences
ESS Building
Stony Brook, NY 11794-2100, USA
e-mail: *john.parise@stonybrook.edu*

Srinivas Pendurti
ASE Technologies Inc.
11499, Chester Road
Cincinnati, OH 45246, USA
e-mail: *spendurti@asetech.com*

Benjamin G. Penn
NASA/George C. Marshall Space Flight Center
ISHM and Sensors Branch
Huntsville, AL 35812, USA
e-mail: *benjamin.g.penndr@nasa.gov*

Jens Pflaum
Julius-Maximilians Universität Würzburg
Institute of Experimental Physics VI
Am Hubland
97078 Würzburg, Germany
e-mail: *jpflaum@physik.uni-wuerzburg.de*

Jose Luis Plaza
Universidad Autónoma de Madrid
Facultad de Ciencias,
Departamento de Física de Materiales
Madrid 28049, Spain
e-mail: *joseluis.plaza@uam.es*

Udo W. Pohl
Technische Universität Berlin
Institut für Festkörperphysik EW5-1
Hardenbergstr. 36
10623 Berlin, Germany
e-mail: *pohl@physik.tu-berlin.de*

Vishwanath (Vish) Prasad
University of North Texas
1155 Union Circle
Denton, TX 76203-5017, USA
e-mail: *vish.prasad@unt.edu*

Maria Cinta Pujol
Universitat Rovira i Virgili
Department of Physics and Crystallography
of Materials and Nanomaterials (FiCMA-FiCNA)
Campus Sescelades, C/ Marcel·lí Domingo
Tarragona 43007, Spain
e-mail: *mariacinta.pujol@urv.cat*

Balaji Raghothamachar
Stony Brook University
Department of Materials Science and Engineering
310 Engineering Building
Stony Brook, NY 11794-2275, USA
e-mail: *braghoth@notes.cc.sunysb.edu*

Michael Roth
The Hebrew University of Jerusalem
Department of Applied Physics
Bergman Bld., Rm 206, Givat Ram Campus
Jerusalem 91904, Israel
e-mail: *mroth@vms.huji.ac.il*

Peter Rudolph
Leibniz Institute for Crystal Growth
Technology Development
Max-Born-Str. 2
Berlin, 12489, Germany
e-mail: *rudolph@ikz-berlin.de*

Akira Sakai
Osaka University
Department of Systems Innovation
1-3 Machikaneyama-cho, Toyonaka-shi
560-8531 Osaka, Japan
e-mail: *sakai@ee.es.osaka-u.ac.jp*

Yasuhiro Shiraki
Tokyo City University
Advanced Research Laboratories,
Musashi Institute of Technology
8-15-1 Todoroki, Setagaya-ku
158-0082 Tokyo, Japan
e-mail: *yshiraki@tcu.ac.jp*

Theo Siegrist
Florida State University
Department of Chemical
and Biomedical Engineering
2525 Pottsdamer Street
Tallahassee, FL 32310, USA
e-mail: *siegrist@eng.fsu.edu*

Zlatko Sitar
North Carolina State University
Materials Science and Engineering
1001 Capability Dr.
Raleigh, NC 27695, USA
e-mail: *sitar@ncsu.edu*

Sivalingam Sivananthan
University of Illinois at Chicago
Department of Physics
845 W. Taylor St. M/C 273
Chicago, IL 60607-7059, USA
e-mail: *siva@uic.edu; siva@epir.com*

Mikhail D. Strikovski
Neocera LLC
10000 Virginia Manor Road, suite 300
Beltsville, MD 20705, USA
e-mail: *strikovski@neocera.com*

Xun Sun
Shandong University
Institute of Crystal Materials
Shanda Road
Jinan, 250100, China
e-mail: *sunxun@icm.sdu.edu.cn*

Ichiro Sunagawa
University Tohoku University (Emeritus)
Kashiwa-cho 3-54-2, Tachikawa
Tokyo, 190-0004, Japan
e-mail: *i.sunagawa@nifty.com*

Xu-Tang Tao
Shandong University
State Key Laboratory of Crystal Materials
Shanda Nanlu 27, 250100
Jinan, China
e-mail: *txt@sdu.edu.cn*

Vitali A. Tatartchenko
Saint – Gobain, 23 Rue Louis Pouey
92800 Puteaux, France
e-mail: *vitali.tatartchenko@orange.fr*

Filip Tuomisto
Helsinki University of Technology
Department of Applied Physics
Otakaari 1 M
Espoo TKK 02015, Finland
e-mail: *filip.tuomisto@tkk.fi*

Anna Vedda
University of Milano-Bicocca
Department of Materials Science
Via Cozzi 53
20125 Milano, Italy
e-mail: *anna.vedda@unimib.it*

Lu-Min Wang
University of Michigan
Department of Nuclear Engineering
and Radiological Sciences
2355 Bonisteel Blvd.
Ann Arbor, MI 48109-2104, USA
e-mail: *lmwang@umich.edu*

Sheng-Lai Wang
Shandong University
Institute of Crystal Materials,
State Key Laboratory of Crystal Materials
Shanda Road No. 27
Jinan, Shandong, 250100, China
e-mail: *slwang@icm.sdu.edu.cn*

Shixin Wang
Micron Technology Inc.
TEM Laboratory
8000 S. Federal Way
Boise, ID 83707, USA
e-mail: *shixinwang@micron.com*

Jan L. Weyher
Polish Academy of Sciences Warsaw
Institute of High Pressure Physics
ul. Sokolowska 29/37
01/142 Warsaw, Poland
e-mail: *weyher@unipress.waw.pl*

Jun Xu
Chinese Academy of Sciences
Shanghai Institute of Ceramics
Shanghai, 201800, China
e-mail: *xujun@mail.shcnc.ac.cn*

Hui Zhang
Tsinghua University
Department of Engineering Physics
Beijing, 100084, China
e-mail: *zhhui@tsinghua.edu.cn*

Lili Zheng
Tsinghua University
School of Aerospace
Beijing, 100084, China
e-mail: *zhenglili@tsinghua.edu.cn*

Mary E. Zvanut
University of Alabama at Birmingham
Department of Physics
1530 3rd Ave S
Birmingham, AL 35294-1170, USA
e-mail: *mezvanut@uab.edu*

Zbigniew R. Zytkiewicz
Polish Academy of Sciences
Institute of Physics
Al. Lotnikow 32/46
02668 Warszawa, Poland
e-mail: *zytkie@ifpan.edu.pl*

Acknowledgements

A.1 Crystal Growth Techniques and Characterization: An Overview

by Govindhan Dhanaraj, Kullaiah Byrappa, Vishwanath (Vish) Prasad, Michael Dudley

The authors would like to thank Dr. Kedar Gupta and Dr. Rick Schwerdtfeger for their generous support and encouragement during the preparation this manuscript.

We also wish to acknowledge the help rendered by Ms. K. Namratha, Dept. of Geology, University of Mysore, Mysore, India in preparing this chapter.

A.4 Generation and Propagation of Defects During Crystal Growth

by Helmut Klapper

The author is indebted to J. Thar and R.A. Becker (Institut für Kristallographie, RWTH Aachen University) for the preparation of the figures.

A.6 Defect Formation During Crystal Growth from the Melt

by Peter Rudolph

The author is indebted to his long-term co-workers Dr. M. Neubert, Dr. F.-M. Kießling, Dr. C. Frank-Rotsch, Dr. U. Juda, M. Czupalla, P. Lange, O. Root, U. Kupfer, M. Ziem, T. Wurche, M. Imming, U. Rehse, W. Miller (all from IKZ Berlin), and the director of IKZ Prof. R. Fornari for helpful discussions, and experimental work essentially contributing to the present chapter. He is also grateful for long-term stimulating cooperations with Dr. M. Jurisch, Dr. B. Weinert, and Dr. S. Eichler from Freiberger Compound Materials GmbH. Special thanks go to D. Bliss from the Air Force Res. Lab. (USA) for critical reading and helpful comments for manuscript revision.

目 录

Contents

List of Abbreviations

μ-PD micro-pulling-down
1S-ELO one-step ELO structure
2-D two-dimensional
2-DNG two-dimensional nucleation growth
2S-ELO double layer ELO
3-D three-dimensional
4T quaterthiophene
6T sexithienyl
8MR eight-membered ring
8T hexathiophene

A

a-Si amorphous silicon
A/D analogue-to-digital
AA additional absorption
AANP 2-adamantylamino-5-nitropyridine
AAS atomic absorption spectroscopy
AB Abrahams and Burocchi
ABES absorption-edge spectroscopy
AC alternate current
ACC annular capillary channel
ACRT accelerated crucible rotation technique
ADC analog-to-digital converter
ADC automatic diameter control
ADF annular dark field
ADP ammonium dihydrogen phosphate
AES Auger electron spectroscopy
AFM atomic force microscopy
ALE arbitrary Lagrangian Eulerian
ALE atomic layer epitaxy
ALUM aluminum potassium sulfate
ANN artificial neural network
AO acoustooptic
AP atmospheric pressure
APB antiphase boundaries
APCF advanced protein crystallization facility
APD avalanche photodiode
APPLN aperiodic poled LN
APS Advanced Photon Source
AR antireflection
AR aspect ratio
ART aspect ratio trapping
ATGSP alanine doped triglycine sulfo-phosphate
AVT angular vibration technique

B

BA Born approximation
BAC band anticrossing
BBO BaB_2O_4
BCF Burton–Cabrera–Frank
BCT $Ba_{0.77}Ca_{0.23}TiO_3$
BCTi $Ba_{1-x}Ca_xTiO_3$
BE bound exciton
BF bright field
BFDH Bravais–Friedel–Donnay–Harker
BGO $Bi_{12}GeO_{20}$
BIBO BiB_3O_6
BLIP background-limited performance
BMO $Bi_{12}MO_{20}$
BN boron nitride
BOE buffered oxide etch
BPD basal-plane dislocation
BPS Burton–Prim–Slichter
BPT bipolar transistor
BS Bridgman–Stockbarger
BSCCO Bi−Sr−Ca−Cu−O
BSF bounding stacking fault
BSO $Bi_{20}SiO_{20}$
BTO $Bi_{12}TiO_{20}$
BU building unit
BaREF barium rare-earth fluoride
BiSCCO $Bi_2Sr_2CaCu_2O_n$

C

C–V capacitance–voltage
CALPHAD calculation of phase diagram
CBED convergent-beam electron diffraction
CC cold crucible
CCC central capillary channel
CCD charge-coupled device
CCVT contactless chemical vapor transport
CD convection diffusion
CE counterelectrode
CFD computational fluid dynamics
CFD cumulative failure distribution
CFMO Ca_2FeMoO_6
CFS continuous filtration system
CGG calcium gallium germanate
CIS copper indium diselenide
CL cathode-ray luminescence
CL cathodoluminescence
CMM coordinate measuring machine
CMO $CaMoO_4$
CMOS complementary metal–oxide–semiconductor
CMP chemical–mechanical polishing
CMP chemomechanical polishing

COD	calcium oxalate dihydrate
COM	calcium oxalate-monohydrate
COP	crystal-originated particle
CP	critical point
CPU	central processing unit
CRSS	critical-resolved shear stress
CSMO	$Ca_{1-x}Sr_xMoO_3$
CST	capillary shaping technique
CST	crystalline silico titanate
CT	computer tomography
CTA	$CsTiOAsO_4$
CTE	coefficient of thermal expansion
CTF	contrast transfer function
CTR	crystal truncation rod
CV	Cabrera–Vermilyea
CVD	chemical vapor deposition
CVT	chemical vapor transport
CW	continuous wave
CZ	Czochralski
CZT	Czochralski technique

D

D/A	digital to analog
DBR	distributed Bragg reflector
DC	direct current
DCAM	diffusion-controlled crystallization apparatus for microgravity
DCCZ	double crucible CZ
DCPD	dicalcium-phosphate dihydrate
DCT	dichlorotetracene
DD	dislocation dynamics
DESY	Deutsches Elektronen Synchrotron
DF	dark field
DFT	density function theory
DFW	defect free width
DGS	diglycine sulfate
DI	deionized
DIA	diamond growth
DIC	differential interference contrast
DICM	differential interference contrast microscopy
DKDP	deuterated potassium dihydrogen phosphate
DLATGS	deuterated L-alanine-doped triglycine sulfate
DLTS	deep-level transient spectroscopy
DMS	discharge mass spectroscopy
DNA	deoxyribonucleic acid
DOE	Department of Energy
DOS	density of states
DPH-BDS	2,6-diphenylbenzo[1,2-*b*:4,5-*b*']diselenophene
DPPH	2,2-diphenyl-1-picrylhydrazyl
DRS	dynamic reflectance spectroscopy
DS	directional solidification
DSC	differential scanning calorimetry
DSE	defect-selective etching
DSL	diluted Sirtl with light
DTA	differential thermal analysis
DTGS	deuterated triglycine sulfate
DVD	digital versatile disk
DWBA	distorted-wave Born approximation
DWELL	dot-in-a-well

E

EADM	extended atomic distance mismatch
EALFZ	electrical-assisted laser floating zone
EB	electron beam
EBIC	electron-beam-induced current
ECE	end chain energy
ECR	electron cyclotron resonance
EDAX	energy-dispersive x-ray analysis
EDMR	electrically detected magnetic resonance
EDS	energy-dispersive x-ray spectroscopy
EDT	ethylene dithiotetrathiafulvalene
EDTA	ethylene diamine tetraacetic acid
EELS	electron energy-loss spectroscopy
EFG	edge-defined film-fed growth
EFTEM	energy-filtered transmission electron microscopy
ELNES	energy-loss near-edge structure
ELO	epitaxial lateral overgrowth
EM	electromagnetic
EMA	effective medium theory
EMC	electromagnetic casting
EMCZ	electromagnetic Czochralski
EMF	electromotive force
ENDOR	electron nuclear double resonance
EO	electrooptic
EP	EaglePicher
EPD	etch pit density
EPMA	electron microprobe analysis
EPR	electron paramagnetic resonance
erfc	error function
ES	equilibrium shape
ESP	edge-supported pulling
ESR	electron spin resonance
EVA	ethyl vinyl acetate

F

F	flat
FAM	free abrasive machining
FAP	$Ca_5(PO_4)_3F$
FCA	free carrier absorption
fcc	face-centered cubic
FEC	full encapsulation Czochralski

FEM finite element method
FES fluid experiment system
FET field-effect transistor
FFT fast Fourier transform
FIB focused ion beam
FOM figure of merit
FPA focal-plane array
FPE Fokker–Planck equation
FSLI femtosecond laser irradiation
FT flux technique
FTIR Fourier-transform infrared
FWHM full width at half-maximum
FZ floating zone
FZT floating zone technique

G

GAME gel acupuncture method
GDMS glow-discharge mass spectrometry
GE General Electric
GGG gadolinium gallium garnet
GNB geometrically necessary boundary
GPIB general purpose interface bus
GPMD geometric partial misfit dislocation
GRI growth interruption
GRIIRA green-radiation-induced infrared absorption
GS growth sector
GSAS general structure analysis software
GSGG $Gd_3Sc_2Ga_3O_{12}$
GSMBE gas-source molecular-beam epitaxy
GSO Gd_2SiO_5
GU growth unit

H

HA hydroxyapatite
HAADF high-angle annular dark field
HAADF-STEM high-angle annular dark field in scanning transmission electron microscope
HAP hydroxyapatite
HB horizontal Bridgman
HBM Hottinger Baldwin Messtechnik GmbH
HBT heterostructure bipolar transistor
HBT horizontal Bridgman technique
HDPCG high-density protein crystal growth
HE high energy
HEM heat-exchanger method
HEMT high-electron-mobility transistor
HF hydrofluoric acid
HGF horizontal gradient freezing
HH heavy-hole
HH-PCAM handheld protein crystallization apparatus for microgravity
HIV human immunodeficiency virus
HIV-AIDS human immunodeficiency virus–acquired immunodeficiency syndrome
HK high potassium content
HLA half-loop array
HLW high-level waste
HMDS hexamethyldisilane
HMT hexamethylene tetramine
HNP high nitrogen pressure
HOE holographic optical element
HOLZ higher-order Laue zone
HOMO highest occupied molecular orbital
HOPG highly oriented pyrolytic graphite
HOT high operating temperature
HP Hartman–Perdok
HPAT high-pressure ammonothermal technique
HPHT high-pressure high-temperature
HRTEM high-resolution transmission electron microscopy
HRXRD high-resolution x-ray diffraction
HSXPD hemispherically scanned x-ray photoelectron diffraction
HT hydrothermal
HTS high-temperature solution
HTSC high-temperature superconductor
HVPE halide vapor-phase epitaxy
HVPE hydride vapor-phase epitaxy
HWC hot-wall Czochralski
HZM horizontal ZM

I

IBAD ion-beam-assisted deposition
IBE ion beam etching
IC integrated circuit
IC ion chamber
ICF inertial confinement fusion
ID inner diameter
ID inversion domain
IDB incidental dislocation boundary
IDB inversion domain boundary
IF identification flat
IG inert gas
IK intermediate potassium content
ILHPG indirect laser-heated pedestal growth
IML-1 International Microgravity Laboratory
IMPATT impact ionization avalanche transit-time
IP image plate
IPA isopropyl alcohol
IR infrared
IRFPA infrared focal plane array
IS interfacial structure
ISS ion-scattering spectroscopy
ITO indium-tin oxide
ITTFA iterative target transform factor analysis
IVPE iodine vapor-phase epitaxy

J

JDS	joint density of states
JFET	junction FET

K

K	kinked
KAP	potassium hydrogen phthalate
KDP	potassium dihydrogen phosphate
KGW	$KY(WO_4)_2$
KGdP	$KGd(PO_3)_4$
KLYF	$KLiYF_5$
KM	Kubota–Mullin
KMC	kinetic Monte Carlo
KN	$KNbO_3$
KNP	$KNd(PO_3)_4$
KPZ	Kardar–Parisi–Zhang
KREW	$KRE(WO_4)_2$
KTA	potassium titanyl arsenate
KTN	potassium niobium tantalate
KTP	potassium titanyl phosphate
KTa	$KTaO_3$
KTaN	$KTa_{1-x}Nb_xO_3$
KYF	KYF_4
KYW	$KY(WO_4)_2$

L

LACBED	large-angle convergent-beam diffraction
LAFB	L-arginine tetrafluoroborate
LAGB	low-angle grain boundary
LAO	$LiAlO_2$
LAP	L-arginine phosphate
LBIC	light-beam induced current
LBIV	light-beam induced voltage
LBO	LiB_3O_5
LBO	$LiBO_3$
LBS	laser-beam scanning
LBSM	laser-beam scanning microscope
LBT	laser-beam tomography
LCD	liquid-crystal display
LD	laser diode
LDT	laser-induced damage threshold
LEC	liquid encapsulation Czochralski
LED	light-emitting diode
LEEBI	low-energy electron-beam irradiation
LEM	laser emission microanalysis
LEO	lateral epitaxial overgrowth
LES	large-eddy simulation
LG	$LiGaO_2$
LGN	$La_3Ga_{5.5}Nb_{0.5}O_{14}$
LGO	$LaGaO_3$
LGS	$La_3Ga_5SiO_{14}$
LGT	$La_3Ga_{5.5}Ta_{0.5}O_{14}$
LH	light hole
LHFB	L-histidine tetrafluoroborate
LHPG	laser-heated pedestal growth
LID	laser-induced damage
LK	low potassium content
LLNL	Lawrence Livermore National Laboratory
LLO	laser lift-off
LLW	low-level waste
LN	$LiNbO_3$
LP	low pressure
LPD	liquid-phase diffusion
LPE	liquid-phase epitaxy
LPEE	liquid-phase electroepitaxy
LPS	$Lu_2Si_2O_7$
LSO	Lu_2SiO_5
LST	laser scattering tomography
LST	local shaping technique
LT	low-temperature
LTa	$LiTaO_3$
LUMO	lowest unoccupied molecular orbital
LVM	local vibrational mode
LWIR	long-wavelength IR
LY	light yield
LiCAF	$LiCaAlF_6$
LiSAF	lithium strontium aluminum fluoride

M

M–S	melt–solid
MAP	magnesium ammonium phosphate
MASTRAPP	multizone adaptive scheme for transport and phase change processes
MBE	molecular-beam epitaxy
MBI	multiple-beam interferometry
MC	multicrystalline
MCD	magnetic circular dichroism
MCT	HgCdTe
MCZ	magnetic Czochralski
MD	misfit dislocation
MD	molecular dynamics
ME	melt epitaxy
ME	microelectronics
MEMS	microelectromechanical system
MESFET	metal-semiconductor field effect transistor
MHP	magnesium hydrogen phosphate-trihydrate
MI	morphological importance
MIT	Massachusetts Institute of Technology
ML	monolayer
MLEC	magnetic liquid-encapsulated Czochralski

MLEK magnetically stabilized liquid-encapsulated Kyropoulos
MMIC monolithic microwave integrated circuit
MNA 2-methyl-4-nitroaniline
MNSM modified nonstationary model
MOCVD metalorganic chemical vapor deposition
MOCVD molecular chemical vapor deposition
MODFET modulation-doped field-effect transistor
MOMBE metalorganic MBE
MOS metal–oxide–semiconductor
MOSFET metal–oxide–semiconductor field-effect transistor
MOVPE metalorganic vapor-phase epitaxy
mp melting point
MPMS mold-pushing melt-supplying
MQSSM modified quasi-steady-state model
MQW multiple quantum well
MR melt replenishment
MRAM magnetoresistive random-access memory
MRM melt replenishment model
MSUM monosodium urate monohydrate
MTDATA metallurgical thermochemistry database
MTS methyltrichlorosilane
MUX multiplexor
MWIR mid-wavelength infrared
MWRM melt without replenishment model
MXRF micro-area x-ray fluorescence

N

N nucleus
N nutrient
NASA National Aeronautics and Space Administration
NBE near-band-edge
NBE near-bandgap emission
NCPM noncritically phase matched
NCS neighboring confinement structure
NGO $NdGaO_3$
NIF National Ignition Facility
NIR near-infrared
NIST National Institute of Standards and Technology
NLO nonlinear optic
NMR nuclear magnetic resonance
NP no-phonon
NPL National Physical Laboratory
NREL National Renewable Energy Laboratory
NS Navier–Stokes
NSF National Science Foundation
nSLN nearly stoichiometric lithium niobate
NSLS National Synchrotron Light Source
NSM nonstationary model
NTRS National Technology Roadmap for Semiconductors
NdBCO $NdBa_2Cu_3O_{7-x}$

O

OCP octacalcium phosphate
ODE ordinary differential equation
ODLN opposite domain LN
ODMR optically detected magnetic resonance
OEIC optoelectronic integrated circuit
OF orientation flat
OFZ optical floating zone
OLED organic light-emitting diode
OMVPE organometallic vapor-phase epitaxy
OPO optical parametric oscillation
OSF oxidation-induced stacking fault

P

PAMBE photo-assisted MBE
PB proportional band
PBC periodic bond chain
pBN pyrolytic boron nitride
PC photoconductivity
PCAM protein crystallization apparatus for microgravity
PCF primary crystallization field
PCF protein crystal growth facility
PCM phase-contrast microscopy
PD Peltier interface demarcation
PD photodiode
PDE partial differential equation
PDP programmed data processor
PDS periodic domain structure
PE pendeo-epitaxy
PEBS pulsed electron beam source
PEC polyimide environmental cell
PECVD plasma-enhanced chemical vapor deposition
PED pulsed electron deposition
PEO polyethylene oxide
PET positron emission tomography
PID proportional–integral–differential
PIN positive intrinsic negative diode
PL photoluminescence
PLD pulsed laser deposition
PMNT $Pb(Mg, Nb)_{1-x}Ti_xO_3$
PPKTP periodically poled KTP
PPLN periodic poled LN
PPLN periodic poling lithium niobate
ppy polypyrrole
PR photorefractive
PSD position-sensitive detector
PSF prismatic stacking fault

PSI	phase-shifting interferometry
PSM	phase-shifting microscopy
PSP	pancreatic stone protein
PSSM	pseudo-steady-state model
PSZ	partly stabilized zirconium dioxide
PT	pressure–temperature
PV	photovoltaic
PVA	polyvinyl alcohol
PVD	physical vapor deposition
PVE	photovoltaic efficiency
PVT	physical vapor transport
PWO	$PbWO_4$
PZNT	$Pb(Zn, Nb)_{1-x}Ti_xO_3$
PZT	lead zirconium titanate

Q

QD	quantum dot
QDT	quantum dielectric theory
QE	quantum efficiency
QPM	quasi-phase-matched
QPMSHG	quasi-phase-matched second-harmonic generation
QSSM	quasi-steady-state model
QW	quantum well
QWIP	quantum-well infrared photodetector

R

RAE	rotating analyzer ellipsometer
RBM	rotatory Bridgman method
RC	reverse current
RCE	rotating compensator ellipsometer
RE	rare earth
RE	reference electrode
REDG	recombination enhanced dislocation glide
RELF	rare-earth lithium fluoride
RF	radiofrequency
RGS	ribbon growth on substrate
RHEED	reflection high-energy electron diffraction
RI	refractive index
RIE	reactive ion etching
RMS	root-mean-square
RNA	ribonucleic acid
ROIC	readout integrated circuit
RP	reduced pressure
RPI	Rensselaer Polytechnic Institute
RSM	reciprocal space map
RSS	resolved shear stress
RT	room temperature
RTA	$RbTiOAsO_4$
RTA	rapid thermal annealing
RTCVD	rapid-thermal chemical vapor deposition
RTP	$RbTiOPO_4$
RTPL	room-temperature photoluminescence
RTR	ribbon-to-ribbon
RTV	room temperature vulcanizing
R&D	research and development

S

S	stepped
SAD	selected area diffraction
SAM	scanning Auger microprobe
SAW	surface acoustical wave
SBN	strontium barium niobate
SC	slow cooling
SCBG	slow-cooling bottom growth
SCC	source-current-controlled
SCF	single-crystal fiber
SCF	supercritical fluid technology
SCN	succinonitrile
SCW	supercritical water
SD	screw dislocation
SE	spectroscopic ellipsometry
SECeRTS	small environmental cell for real-time studies
SEG	selective epitaxial growth
SEM	scanning electron microscope
SEM	scanning electron microscopy
SEMATECH	Semiconductor Manufacturing Technology
SF	stacking fault
SFM	scanning force microscopy
SGOI	SiGe-on-insulator
SH	second harmonic
SHG	second-harmonic generation
SHM	submerged heater method
SI	semi-insulating
SIA	Semiconductor Industry Association
SIMS	secondary-ion mass spectrometry
SIOM	Shanghai Institute of Optics and Fine Mechanics
SL	superlattice
SL-3	Spacelab-3
SLI	solid–liquid interface
SLN	stoichiometric LN
SM	skull melting
SMB	stacking mismatch boundary
SMG	surfactant-mediated growth
SMT	surface-mount technology
SNR	signal-to-noise ratio
SNT	sodium nonatitanate
SOI	silicon-on-insulator
SP	sputtering
sPC	scanning photocurrent
SPC	Scientific Production Company
SPC	statistical process control
SR	spreading resistance
SRH	Shockley–Read–Hall
SRL	strain-reducing layer
SRS	stimulated Raman scattering

SRXRD	spatially resolved XRD
SS	solution-stirring
SSL	solid-state laser
SSM	sublimation sandwich method
ST	synchrotron topography
STC	standard testing condition
STE	self-trapped exciton
STEM	scanning transmission electron microscopy
STM	scanning tunneling microscopy
STOS	sodium titanium oxide silicate
STP	stationary temperature profile
STS	space transportation system
SWBXT	synchrotron white beam x-ray topography
SWIR	short-wavelength IR
SXRT	synchrotron x-ray topography

T

TCE	trichloroethylene
TCNQ	tetracyanoquinodimethane
TCO	thin-film conducting oxide
TCP	tricalcium phosphate
TD	Tokyo Denpa
TD	threading dislocation
TDD	threading dislocation density
TDH	temperature-dependent Hall
TDMA	tridiagonal matrix algorithm
TED	threading edge dislocation
TEM	transmission electron microscopy
TFT-LCD	thin-film transistor liquid-crystal display
TGS	triglycine sulfate
TGT	temperature gradient technique
TGW	Thomson–Gibbs–Wulff
TGZM	temperature gradient zone melting
THM	traveling heater method
TMCZ	transverse magnetic-field-applied Czochralski
TMOS	tetramethoxysilane
TO	transverse optic
TPB	three-phase boundary
TPRE	twin-plane reentrant-edge effect
TPS	technique of pulling from shaper
TQM	total quality management
TRAPATT	trapped plasma avalanche-triggered transit
TRM	temperature-reduction method
TS	titanium silicate
TSC	thermally stimulated conductivity
TSD	threading screw dislocation
TSET	two shaping elements technique
TSFZ	traveling solvent floating zone
TSL	thermally stimulated luminescence
TSSG	top-seeded solution growth
TSSM	Tatarchenko steady-state model
TSZ	traveling solvent zone
TTV	total thickness variation
TV	television
TVM	three-vessel solution circulating method
TVTP	time-varying temperature profile
TWF	transmitted wavefront
TZM	titanium zirconium molybdenum
TZP	tetragonal phase

U

UC	universal compliant
UDLM	uniform-diffusion-layer model
UHPHT	ultrahigh-pressure high-temperature
UHV	ultrahigh-vacuum
ULSI	ultralarge-scale integrated circuit
UV	ultraviolet
UV-vis	ultraviolet–visible
UVB	ultraviolet B

V

VAS	void-assisted separation
VB	valence band
VB	vertical Bridgman
VBT	vertical Bridgman technique
VCA	virtual-crystal approximation
VCSEL	vertical-cavity surface-emitting laser
VCZ	vapor pressure controlled Czochralski
VDA	vapor diffusion apparatus
VGF	vertical gradient freeze
VLS	vapor–liquid–solid
VLSI	very large-scale integrated circuit
VLWIR	very long-wavelength infrared
VMCZ	vertical magnetic-field-applied Czochralski
VP	vapor phase
VPE	vapor-phase epitaxy
VST	variable shaping technique
VT	Verneuil technique
VTGT	vertical temperature gradient technique
VUV	vacuum ultraviolet

W

WBDF	weak-beam dark-field
WE	working electrode

X

XP	x-ray photoemission
XPS	x-ray photoelectron spectroscopy
XPS	x-ray photoemission spectroscopy
XRD	x-ray diffraction
XRPD	x-ray powder diffraction
XRT	x-ray topography

Y

YAB	$YAl_3(BO_3)_4$
YAG	yttrium aluminum garnet
YAP	yttrium aluminum perovskite
YBCO	$YBa_2Cu_3O_{7-x}$
YIG	yttrium iron garnet
YL	yellow luminescence
YLF	$LiYF_4$
YOF	yttrium oxyfluoride
YPS	$(Y_2)Si_2O_7$
YSO	Y_2SiO_5

Z

ZA	Al_2O_3-$ZrO_2(Y_2O_3)$
ZLP	zero-loss peak
ZM	zone-melting
ZNT	ZN-Technologies
ZOLZ	zero-order Laue zone

Part A

Fundame

Part A Fundamentals of Crystal Growth and Defect Formation

1. Crystal Growth Techniques and Characterization: An Overview

Govindhan Dhanaraj, Kullaiah Byrappa, Vishwanath (Vish) Prasad, Michael Dudley

A brief overview of crystal growth techniques and crystal analysis and characterization methods is presented here. This is a prelude to the details in subsequent chapters on fundamentals of growth phenomena, details of growth processes, types of defects, mechanisms of defect formation and distribution, and modeling and characterization tools that are being employed to study as-grown crystals and bring about process improvements for better-quality and large-size crystals.

1.1 Historical Developments

Crystals are the unacknowledged pillars of the world of modern technology. They have attracted human civilization from prehistoric times owing to their beauty and rarity, but their large-scale applications for devices have been realized only in the last six decades. For a long time, crystal growth has been one of the most fascinating areas of research. Although systematic understanding of the subject of crystal growth began during the last quarter of the 19th century with Gibbs' phase equilibrium concept based on a thermodynamical treatment, man practiced crystal growth and or crystallization processes as early as 1500 BC in the form of salt and sugar crystallization. Thus, crystal growth can be treated as an ancient scientific activity. However, the scientific approach to the field of crystal growth started in 1611 when Kepler correlated crystal morphology and structure, followed by Nicolous Steno, who explained the origin of a variety of external forms. Since then crystal growth has evolved steadily to attain its present status. Several theories were proposed from the 1920s onwards. The current impetus in crystal growth started during World War II. Prior to that, applications of crystals and crystal growth technology did not catch the attention of technologists. The growth of small or fine crystals in the early days, which involved uncontrolled or poorly controlled crystal growth parameters without much sophistication in instrumentation or crystal growth equipment, slowly led to the growth of large bulk crystals during World War II. With advancement in instrumentation technology, the attention of crystal growers focused on the quality of the grown crystals and understanding of their formation. Also, tailoring of crystal shape or morphology, size, and properties plays a key role in crystal growth science. In this context it is appropriate to mention nanocrystals, which exhibit desirable physicochemical characteristics. Similarly, the growth of thin films has emerged as a fascinating technology. Further crystal growth research is being carried out in microgravity or space conditions. There are various methods of evaluating the quality of grown crystals. Thus the growth of crystals with tailored physics and chemistry, characterization of crystals with more advanced instrumentation, and their conversion into useful devices play vital roles in science and technology [1.1,2].

Crystal growth is a highly interdisciplinary subject that demands the collaboration of physicists, chemists, biologists, engineers, crystallographers, process engineers, materials scientists, and materials engineers. The significance of the beauty and rarity of crystals is now well knitted with their symmetry, molecular structure, and purity, and the physicochemical environment of their formation. These characteristics endow crystals with unique physical and chemical properties, which have transformed electronic industries for the benefit of human society. Prior to commercial growth or production of crystals, man depended only on the availability of natural crystals for both jewelery and devices. Today the list of uses of artificially grown crystals is growing exponentially for a variety of applications, such as electronics, electrooptics, crystal bubble memories, spintronics, magnetic devices, optics, nonlinear devices, oscillators, polarizers, transducers, radiation detectors, lasers, etc. Besides inorganic crystal growth, the world of organic, semiorganic, biological crystal growth is expanding greatly to make crystal growth activity more cost-effective. Today, the quality, purity, and defect-free nature of crystals is a prerequisite for their technological application. A reader can get useful information on the history of crystal growth from the works of *Scheel* [1.3,4].

Crystal growth is basically a process of arranging atoms, ions, molecules or molecular assemblies into regular three-dimensional periodic arrays. However, real crystals are never perfect, mainly due to the presence of different kinds of local disorder and long-range imperfections such as dislocations. Moreover, they are often polycrystalline in nature. Hence, the ultimate aim of a crystal grower is to produce perfect single crystals of desired shape and size, and to characterize them in order to understand their purity and quality and perfection for end users. Accordingly, crystal growth techniques and characterization tools have advanced greatly in recent years. This has facilitated the growth and characterization of a large variety of technologically important single crystals. Crystal growth can be treated as an important branch of materials science leading to the formation of technologically important materials of different sizes. Hence, it covers crystals from bulk to small and even to fine, ultrafine, and nanoscale sizes. In this respect, crystal growth has a close relationship with crystal engineering, and also polyscale crystal growth is relevant. This concept becomes even more relevant with progress achieved in nanotechnology, wherein the size effect explains changes in the physical properties of crystalline materials with size.

1.2 Theories of Crystal Growth

Growth of single crystals can be regarded as a phase transformation into the solid state from the solid, liquid or vapor state. Solid–solid phase transformations are rarely employed to grow single crystals, except for certain metals and metal alloys, whereas liquid to solid and vapor to solid transformations are most important in crystal growth and have resulted in a great variety of experimental techniques. When a crystal is in dynamic equilibrium with its mother phase, the free energy is at a minimum and no growth can occur. This equilibrium has to be disturbed suitably for growth to occur. This may be done by an appropriate change in temperature, pressure, pH, chemical potential, electrochemical potential, strain, etc. The three basic steps involved in the formation of a crystal from an initially disordered phase are:

1. Achievement of supersaturation or supercooling
2. Nucleation
3. Growth of the nuclei into single crystals of distinct phases

The driving force for crystallization actually derives from supersaturation, supercooling of liquid or gas phase with respect to the component whose growth is required. Therefore steady-state supersaturation/supercooling needs to be maintained during crystal growth to obtain higher-quality results. Nucleation or crystallization centers are an important feature of crystal growth. Nucleation may occur either spontaneously due to the conditions prevailing in the parent phase or it may be induced artificially. Therefore, the study of nucleation forms an integral part of crystal growth process. Several theories to explain nucleation have been proposed from time to time. Perhaps Gibbs was the first to comprehend that the formation of small embryonic clusters of some critical size is a prerequisite for the development of a macroscopic crystal. The

Gibbs–Thomson equation is fundamental for nucleation events [1.5], expressed for a cluster inside a supercooled phase under equilibrium conditions inside a supersaturated/supercooled phase as

$$k_\mathrm{B}T\ln\left(\frac{p}{p^*}\right)=\frac{2\sigma V}{r}\,, \tag{1.1}$$

where r is the radius of the cluster formed inside a vapor at temperature T, k_B is the Boltzmann constant, p is the vapor pressure outside the cluster, p^* is the saturated vapor pressure over a plane liquid surface, σ is the surface energy per unit area, and V is the volume of the growth units.

For nucleation from solution,

$$k_\mathrm{B}T\ln\left(\frac{c}{c^*}\right)=\frac{2\sigma V}{r}\,. \tag{1.2}$$

Here, c is the actual concentration and c^* is the concentration of the solution with a crystal of infinite radius. The condition for nucleation from the melt is

$$\Delta H_\mathrm{m}\left(\frac{T_\mathrm{m}-T_\mathrm{r}}{T_\mathrm{m}}\right)=\frac{2\sigma V}{r}\,. \tag{1.3}$$

Here, T_r is the melting point of a crystal of radius r and T_m is the melting point of a large crystal. ΔH_m is the latent heat of fusion per molecule.

The Gibbs–Thomson equation, which gives the free energy change per unit volume for solution growth, is given by

$$\Delta G_\mathrm{v}=\frac{2\sigma}{r}=-k_\mathrm{B}T\ln\left(\frac{c}{c^*}\right)=-\frac{k_\mathrm{B}}{V}\ln S\,, \tag{1.4}$$

where S is the degree of supersaturation and V is the molecular volume.

There are several theories to explain crystal growth, involving the mechanism and the rate of growth of crystals. The important crystal growth theories are the surface energy theory, diffusion theory, adsorption layer theory, and screw dislocation theory. Gibbs proposed the first theory of crystal growth, in which he assumed growth of crystals to be analogous to the growth of a water droplet from mist. Later *Kossel* and others explained the role of step and kink sites on the growth surface in promoting the growth process [1.6].

1.2.1 Surface Energy Theory

The surface energy theory is based on the thermodynamic treatment of equilibrium states put forward by Gibbs. He pointed out that the growing surface would assume that shape for which the surface energy is lowest. Many researchers later applied this idea. *Curie* [1.7] worked out the shapes and morphologies of crystals in equilibrium with solution or vapor. Later, *Wulff* [1.8] deduced expressions for the growth rate at different faces and the surface free energies. According to him, the equilibrium is such that excess surface free energy $\sigma_{hkl}\mathrm{d}A_{hkl}$ is minimum for crystal with its $\{hkl\}$ faces exposed. The value of σ_{hkl} determines the shape of a small crystal; for example, if σ is isotropic, the form of the crystal is spherical, provided the effect of gravity is negligible. *Marc* and *Ritzel* [1.9] considered the effect of surface tension and solution pressure (solubility) on the growth rate. In their opinion, different faces have different values of solubility. When the difference in solubility is small, growth is mainly under the influence of surface energy, and the change in the surface of one form takes place at the expense of the other. *Bravais* [1.10] proposed that the velocities of growth of the different faces of a crystal depend on the reticular density.

1.2.2 Diffusion Theory

The diffusion theory proposed by *Nernst* [1.11], *Noyes*, and *Whitney* [1.12] is based on the following two basic assumptions:

1. There is a concentration gradient in the neighborhood of the growing surface;
2. Crystal growth is the reverse process of dissolution.

Consequently, the amount of solute that will get deposited on a crystal growing in a supersaturated solution is given by

$$\frac{\mathrm{d}m}{\mathrm{d}t}=\left(\frac{D}{\delta}\right)A(c-c_0)\,, \tag{1.5}$$

where $\mathrm{d}m$ is the mass of solute deposited in a small time interval $\mathrm{d}t$ over an area A of the crystal surface, D is the diffusion coefficient of the solute, c and c_0 are the actual and equilibrium concentrations of the solute, and δ is the thickness of the torpid layer adjacent to the solid surface.

The importance of surface discontinuities in providing nucleation sites during crystal growth was the main consideration of *Kossel* [1.6], *Stranski* [1.13], and *Volmer* [1.14]. Volmer suggested a growth mechanism by assuming the existence of an adsorbed layer of atoms or molecules of the growth units on crystal faces. Later, *Brandes* [1.15], Stranski, and Kossel modified this concept. Volmer's theory was based on thermodynamical reasoning. The units reaching a crystal face are not immediately attached to the lattice but migrate over the

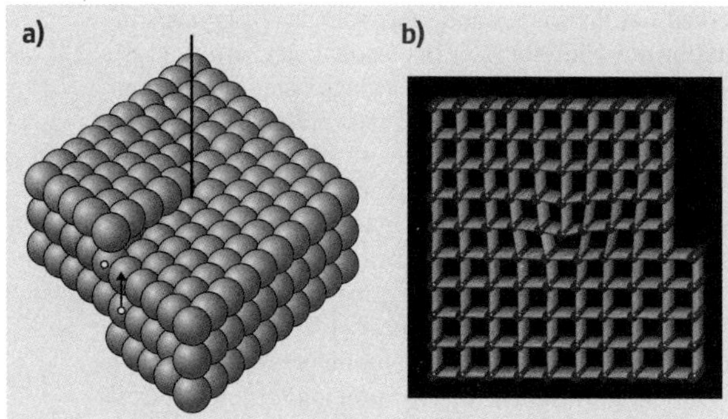

Fig. 1.1a,b Screw dislocation in a crystal (**a**); edge dislocation (**b**)

crystal face to find a suitable site for attachment. They form a loosely adsorbed layer at the interface, and soon a dynamic equilibrium is established between the layer and the bulk solution.

1.2.3 Adsorption Layer Theory

Kossel viewed crystal growth based on atomistic considerations. He assumed that crystal is in equilibrium with its solution when it is just saturated. Also, the attachment energy unit on growing surface is a simple function of distance only. The attachment energy is due to van der Waals forces if the crystal is homopolar, while it is due to electrostatic forces if the crystal is heteropolar (ionic). A growth unit arriving at a crystal surface finds attachment sites such as terraces, ledges, and kinks. The attachment energy of a growth unit can be considered to be the resultant of three mutually perpendicular components. The binding energy or attachment energy of an atom is maximum when it is incorporated into a kink site in a surface ledge, whilst at any point on the ledge it is greater than that for an atom attached to the flat surface (terrace). Hence, a growth unit reaching a crystal surface is not integrated into the lattice immediately. Instead it migrates to a step and moves along it to a kink site, where it is finally incorporated. Based on this consideration of attachment, Kossel was able to determine the most favorable face for growth. According to the Kossel model, growth of a crystal is a discrete process and not continuous. Also, a new layer on a preferably flat face of a homopolar crystal will start growing from the interior of the face. For heteropolar crystals, the corners are the most favorable for growth, while mid-face is least favored. According to Stranski, the critical quantity that determines the growth process is the work necessary to detach a growth unit from its position on the crystal surface. Growth units with the greatest detachment energy are most favored for growth, and vice versa. The greatest attraction of atoms to the corners of ionic and metallic crystals often leads to more rapid growth along these directions, with the result that the crystal grows with many branches called dendrites radiating from a common core.

1.2.4 Screw Dislocation Theory

However, the Kossel, Stranski, and Volmer theory could not explain the moderately high growth rates observed in many cases at relatively low supersaturation, far below those needed to induce surface nucleation. *Frank* [1.16] proposed that a screw dislocation emerging at a point on the crystal surface could act as a continuous source of steps (surface ledges) which can propagate across the surface of the crystal and promote crystal growth. Growth takes place by rotation of the steps around the dislocation point (Fig. 1.1). *Burton* et al. [1.17] proposed the famous screw dislocation theory based on the relative supersaturation as the Burton–Cabrera–Frank (BCF) model determining the absolute value of growth rate depending upon the concentration. Frank's model could explain the experimental observations on the growth rate and spiral pattern mechanism.

1.3 Crystal Growth Techniques

Crystal growth is a heterogeneous or homogeneous chemical process involving solid or liquid or gas, whether individually or together, to form a homogeneous solid substance having three-dimensional atomic arrangement. Various techniques have been employed, depending upon the chemical process involved. All crystal growth processes can be broadly classified according to the scheme presented in Table 1.1. The subject of crystal growth has therefore developed as an interdisciplinary subject covering various branches of science, and it is extremely difficult to discuss the entire subject in this overview chapter. However, the

Table 1.1 Classification of crystal growth processes [1.18]

1. Solid–Solid	Solid	$\xrightarrow{T}$	Solid Devitrification Strain annealing Polymorphic phase change Precipitation from solid solution
2. Liquid–Solid			
i) Melt growth	Molten material	$\xrightarrow{\text{Dec. } T}$	Crystal Bridgman–Stockbarger Kyropoulos Czochralski Zoning Verneuil
ii) Flux growth	Solid(s) + Flux agent(s)	$\xrightarrow{\text{Dec. } T}$	Crystal(s)
iii) Solution growth	Solid(s) + Solvent	$\xrightarrow{\text{Low } T}$	Crystal(s) Evaporation Slow cooling Boiling solutions
iv) Hydrothermal growth	Solid(s) + Solvent	$\xrightarrow[\text{High } p]{\text{High } T}$	Crystal(s) Hydrothermal sintering Hydrothermal reactions Normal temperature gradient Reversed temperature gradient
v) Gel growth	Solution + Gel medium	$\xrightarrow{\text{Low } T}$	Crystal Reaction Complex decomplex Chemical reduction Solubility reduction Counter-flow diffusion
	Solution	$\longrightarrow$	Crystal(s) + products
3. Gas–Solid	Vapor(s)	$\longrightarrow$	Solid Sublimation–condensation Sputtering Epitaxial processes Ion-implantation

present Handbook covers most important techniques adopted in modern crystal growth through the chapters authored by world authorities in their respective fields.

1.3.1 Solid Growth

The solid-state growth technique is basically controlled by atomic diffusion, which is usually very slow except in the case of fast ionic conductors or superionic conductors. The commonly used solid-state growth techniques are annealing or sintering, strain annealing, heat treatment, deformation growth, polymorphic phase transitions, quenching, etc., and most of these are popularly used in metallurgical processes for tailoring material properties. In fact, gel growth is also considered as solid growth by some researchers. Solid growth is not covered in this Handbook.

1.3.2 Solution Growth

This is one of the oldest and most widely used crystal growth techniques compared with vapor-phase melt growth. Solution growth is used not only for growth of technologically important crystals but also for a variety of crystalline products for daily life such as the growth of foods, medicines, fertilizers, pesticides, dye stuffs, etc. Most crystallization processes of ionic salts are conducted in aqueous solutions or in some cases in solvents which are a mixture of miscible and organic solvents. Solution growth is used for substances that melt incongruently, decompose below the melting point, or have several high-temperature polymorphic modifications, and is also often efficient in the absence of such restrictions. The important advantage of solution growth is the control that it provides over the growth temperature, control of viscosity, simplicity of equipment, and the high degree of crystal perfection since the crystals grow at temperatures well below their melting point. We can divide solution growth into three types depending upon the temperature, the nature of the solvent, solute, and the pressure: low-temperature aqueous solution growth, superheated aqueous solution growth, and high-temperature solution growth. Aqueous solution growth has produced the largest crystals known to mankind, such as potassium dihydrogen phosphate (KDP), deuterated potassium dihydrogen phosphate (DKDP), etc. produced at the Lawrence Livermore Laboratory, USA.

For successful growth of a crystal from solution, it is essential to understand certain basic properties (physicochemical features) of the solution. The behavior of water with temperature and pressure; the critical, subcritical, and supercritical conditions; its structure, the variation in pH; viscosity; density; conductivity; dielectric constant; and coefficient of expansion are critical for successful crystal growth. Recently, a rational approach to the growth of a given crystal was carried out in order to: compute the thermodynamic equilibrium as a function of the processing variables, generate equilibrium (yield) diagrams to map the processing variable space for the phases of interest, design experiments to test and validate the computed diagrams, and utilize the results for mass production [1.19]. The change in ionic strength of the solution during crystal growth results in formation of defects, and variation in the crystal habit and even the phases, and therefore has to be maintained constant, often with the help of swamping-electrolyte solutions. Similarly, chelating agents are frequently used to sequester ions and form respective complexes, which are later thermodynamically broken to release their cations very slowly into the solution, which helps in controlling the growth rate and crystal habit.

In the last decade crystal growth from solution under microgravity conditions has been studied extensively to grow a wide variety of crystals such as zeolites, compound semiconductors (InP, GaAs, GaP, AlP, etc.), triglycine sulfate, etc.

Crystal Growth from Low-Temperature Aqueous Solutions

The greatest advantages of crystal growth from low-temperature aqueous solutions are the proximity to ambient temperature, which helps to retain a high degree of control over the growth conditions, especially with reference to thermal shocks, and reduction of both equilibrium and nonequilibrium defects to a minimum (even close to zero). Solution growth can be classified into several groups according to the method by which supersaturation is achieved:

1. Crystallization by changing the solution temperature
2. Crystallization by changing the composition of the solution (solvent evaporation)
3. Crystallization by chemical reaction

Crystal Growth from Superheated Aqueous Solutions

This method is commonly known as the hydrothermal method and is highly suitable for crystal growth of compounds with very low solubility and phase transitions. When nonaqueous solvents are used in the system, it is called the solvothermal method. The largest known single crystal formed in nature (beryl crystal of > 1000 kg) is of hydrothermal origin, and similarly some of the largest quantities of single crystals produced in one experimental run (quartz single crystals of > 1000 kg) are based on the hydrothermal technique. The term "hydrothermal" refers to any heterogenous (usually for bulk crystal growth) or homogeneous (for fine to nanocrystals) chemical reaction in the presence of aqueous solvents or mineralizers under high-pressure and high-temperature conditions to dissolve and recrystallize (recover) materials that are relatively insoluble under ordinary conditions [1.20]. The last decade has witnessed growing popularity of this technique, and a large variety of crystals and crystalline materials starting from native elements to the most complex coordinated compounds such as rare-earth silicates, germinates, phosphates, tungstates, etc. have been obtained. Also, the method is becoming very popular among organic chemists dealing with synthesis of life-forming compounds and problems related to the origin of life. The method is discussed in great detail in Chap. 18.

Crystal Growth from High-Temperature Solutions

This is popularly known as flux growth and gained its importance for growing single crystals of a wide range of materials, especially complex multicomponent systems. In fact, this was one of the earliest methods employed for growing technologically important crystals, for example, single crystals of corundum at the end of the 19th century. The main advantage of this method is that crystals are grown below their melting temperature. If the material melts incongruently, i. e., decomposes before melting or exhibits a phase transition below the melting point or has very high vapor pressure at the melting point, one has indeed to look for growth temperatures lower than these phase transitions. The method is highly versatile for growth of single crystals as well as layers on single-crystal substrates (so-called liquid-phase epitaxy, LPE). The main disadvantages are that the growth rates are smaller than for melt growth or rapid aqueous solution growth, and the unavoidable presence of flux ions as impurities in the final crystals. Some of the important properties to be considered for successful flux growth of crystals are stability and solubility of the crystal to be grown, low melting point and lower vapor pressure of the flux, the lower viscosity of the melt (which should not attack the crucible), and also ease of separation [1.4,21]. The most commonly used fluxes are the basic oxides or fluorides: PbO, PbF_2, BaO, BaF, Bi_2O_3, Li_2O, Na_2O, K_2O, KF, B_2O_3, P_2O_5, V_2O_5, MoO_3, and in most cases a mixture consisting of two or three of them. The prime advantage of this method is that growth can be carried out either through spontaneous nucleation or crystallization on a seed. Supersaturation can be achieved through slow cooling, flux evaporation, and vertical temperature gradient transport methods. Also, during the growth, one can introduce rotation of the seed or crucible, or pulling of the seed, and so on. Accordingly, several versions of flux growth have been developed: slow cooling (SC), slow cooling bottom growth (SCBG), top-seeded solution growth (TSSG), the top-seeded vertical temperature gradient technique (VTGT), bottom growth with a nutrient, growth by traveling solvent zone (TSZ), flux evaporation, LPE, and so on.

The flux method has been popularly used especially for the growth of a large variety of garnets, and recently for a wide range of laser crystals such as rare-earth borates, potassium titanyl phosphates, and so on. The reader can get valuable information from several interesting reviews on flux growth [1.22–24].

1.3.3 Crystal Growth from Melt

Melt growth of crystals is undoubtedly the most popular method of growing large single crystals at relatively high growth rates. In fact, more than half of technological crystals are currently obtained by this technique. The method has been popularly used for growth of elemental semiconductors and metals, oxides, halides, chalcogenides, etc. Melt growth requires that the material melts without decomposition, has no polymorphic transitions, and exhibits low chemical activity (or manageable vapor pressure at its melting point). The thermal decomposition of a substance and also chemical reactions in the melt can disturb the stoichiometry of the crystal and promote formation of physical or chemical defects. Similarly, the interaction between the melt and crucible, or the presence of a third component derived from the crystallization atmosphere, can affect melt growth. Usually, an oxygen-containing atmosphere is used for oxides, a fluorine-containing atmosphere for fluorides, a sulfur-containing atmosphere for sulfides, and so on. In melt growth, crystallization can be carried out in a vacuum, in a neutral atmosphere (helium, argon, nitrogen), or in a reducing atmosphere (air, oxygen). In a large melt volume, convective flows caused by the temperature gradient within the melt lead to several physical and chemical defects. In a small melt volume, transport is affected by diffusion.

Selection of a particular melt growth technique is done on the basis of the physical and chemical characteristics of the crystal to be grown. Metal single crystals with melting point $< 1800\,°C$ are grown by Stockbarger method, and those with melting point $> 1800\,°C$ by zone melting. Semiconducting crystals are grown chiefly by Czochralski method, and by zone melting. Single crystals of dielectrics with melting point $< 1800\,°C$ are usually grown by the Stockbarger or Czochralski methods, while higher-melting materials are produced by flame fusion (Verneuil method). If the physicochemical processes involved in crystallization are taken into account, it is possible to establish optimum growth conditions.

One of the earliest melt techniques used to grow large quantity of high-melting materials was the Verneuil method (flame fusion technique), first described by *Verneuil* in 1902 [1.25]. This marks the beginning of commercial production of large quantities of high-melting crystals, which were essentially used as gems or for various mechanical applications. Today, the technique is popular for growth of a variety of high-quality crystals for laser devices and precision instruments, as well as substrates. The essential features are a seed crystal, the top of which is molten and is fed with molten drops of source material, usually as a powder through a flame or plasma. Following this, the Czochralski method, developed in 1917 and later modified by several researchers, became the most popular technique to grow large-size single crystals which were impossible to obtain by any other techniques in such large quantity. This technique has several advantages over the other related melt-growth technique, viz. the Kyropoulos method, which involves a gradual reduction in the melt temperature. In the Czochralski technique the melt temperature is kept constant and the crystal is slowly pulled out of the melt as it grows. This provides a virtually constant growth rate for the crystal. Several versions of Czochralski crystal pullers are commercially available. A large variety of semiconductor crystals such as Si, Ge, and several III–V compounds are being commercially produced using this technique. Besides, several other crystals of oxides, spinel, garnets, niobates, tantalates, and rare-earth gallates have been obtained by this method. The reader can find more valuable information on this method from the works of *Hurle* and *Cockyane* [1.26].

There are several other popularly used melt growth techniques that are feasible for commercial production of various crystals. Amongst them, the Bridgman–Stockbarger, zone melting, and floating zone methods are the most popular. The Bridgman technique is characterized by the relative translation of the crucible containing the melt to the axial temperature gradient in a vertical furnace. The Stockbarger method is a more sophisticated modification of the Bridgman method. There is a high-temperature zone, an adiabatic loss zone, and a low-temperature zone. The upper and lower temperature zones are generally independently controlled, and the loss zone is either unheated or poorly insulated.

1.3.4 Vapor-Phase Growth

Vapor-phase growth is particularly employed in mass production of crystals for electronic devices because of its proven low cost and high throughput, in addition to its capability to produce advanced epitaxial structures. The technique is especially suitable for growth of semiconductors, despite the rather complex chemistry of the vapor-phase process. The fundamental reason for their success is the ease of dealing with low- and high-vapor-pressure elements. This is achieved by using specific chemical precursors in the form of vapor containing the desired elements. These precursors are introduced into the reactor by a suitable carrier gas and normally mix shortly before reaching the substrate, giving rise to the nutrient phase of the crystal growth process. The release of the elements necessary for construction of the crystalline layer may occur at the solid–gas interface or directly in the gas phase, depending on the type of precursors and on the thermodynamic conditions.

The advantage of vapor growth technique is that crystals tend to have a low concentration of point defects and low dislocation densities compared with crystals grown from the melt, as the temperatures employed are usually considerably lower than the melting temperature. Moreover, if the material undergoes a phase transformation or melts incongruently, vapor growth may be the only choice for the growth of single crystals. Although the method was initially used to grow bulk crystals, with the enormous importance of thin films in electronic and metallurgical applications, vapor growth is now widely used to grow thin films, epitaxial layers, and substrates in the field of semiconductor technology [1.27, 28].

Vapor-phase growth primarily involves three stages: vaporization, transport, and deposition. The vapor is formed by heating a solid or liquid to high temperatures. Transportation of vapor may occur through vacuum, driven by the kinetic energy of vaporization. Deposition of the vapor may occur by condensation or chemical reaction.

Various techniques exist in vapor-phase growth, differentiated by the nature of the source material and the means and mechanism by which it is transported to the growing crystal surface. Conceptually, the simplest technique is that of sublimation, where the source material is placed at one end of a sealed tube and heated so that it sublimates and is then transported to the cooler region of the tube, where it crystallizes.

Among vapor-phase growth techniques, vapor-phase epitaxy is the most popularly used, especially for the growth of p- and n-type semiconductor whose dimers and monomers are difficult to achieve by other methods (e.g., physical evaporation) or too stable to be reduced to the necessary atomic form. Furthermore, there are different variants such as metalorganic vapor-phase epitaxy (MOVPE), plasma-assisted mo-

Table 1.2 Main application fields of vapor-phase epitaxy techniques and the relevant classes of materials

Growth technique	Devices and semiconductor family			
	Si, Ge	II–VI	III–V	III–nitrides
Hydride VPE	SiGe alloys		LEDs and photodetectors (GaP, InGaP, GaAsP)	GaN thick layers
Chloride VPE	Bipolar transistors, MOS			
MOVPE		IR sensors (HgCdTe), LEDs and lasers (ZnCdSe, ZnSSe)	Solar cells (GaAs, AlGaAs, InGaP), transistors (AlGaAs, InGaAs), LEDs (AlGaAs), TC and CD lasers (InGaPAs, AlGaAs), photodetectors, LEDs and lasers (InGaPAs)	LEDs and lasers (GaN, InGaN, GaAlN)

lecular beam epitaxy (MBE), etc. to suit the growth of particular compounds. Table 1.2 summarizes the main application fields of the VPE techniques and the relevant classes of materials [1.29].

1.4 Crystal Defects and Characterization

Characterization of crystals has become an integral part of crystal growth and process development. Crystal defects and their distribution together with composition and elemental purity determine most of their properties such as mechanical strength, electrical conductivity, photoconductivity luminescence, and optical absorption, and these properties influence their performance in applications. Therefore, investigating the origin, concentration, and distribution of imperfections in crystals is critical to controlling them and thereby the crystal properties influenced by these imperfections.

1.4.1 Defects in Crystals

Imperfections or defects can be broadly classified based on their dimensionality.

Point Defects

These zero-dimensional defects are vacancies, interstitials, and impurity atoms deliberately added to control the conductivity of the semiconductor, and impurities that are unintentionally incorporated as contaminants during material growth and processing. Electronic defects such as holes and electrons also constitute point defects. In compounds, point defects form disorders such as Frenkel, Schottky, and antistructure disorders.

Line Defects

Line defects consist of purely geometrical faults called dislocations. The concept of dislocations arose from the crystallographic nature of plastic flow in crystalline materials. A dislocation is characterized by its line direction and Burgers vector $\boldsymbol{b}$, which is, as a rule, one of the shortest lattice translations. Dislocation lines may be straight or follow irregular curves or closed loops. Dislocations whose line segments are parallel to $\boldsymbol{b}$ are called screw dislocations. Edge dislocations have their line segments perpendicular to the $\boldsymbol{b}$ direction. In mixed dislocations, the line direction is inclined to $\boldsymbol{b}$ and hence they have both screw and edge components.

Planar Defects

Planar defects include high- and low-angle boundaries, growth striations, growth-sector boundaries, twin boundaries, stacking faults, and antiphase boundaries. Growth striations are lattice perturbations that arise from local variations of the dopant/impurity concentration created by fluctuations in the growth conditions. Stacking faults are formed when there are errors in the normal stacking arrangement of the lattice planes in the crystal structure. These could be caused by plastic deformation or agglomeration of point defects. High- and low-angle boundaries consist of arrays of dislocations, and they separate regions of different orientations. In crystal growth, high-angle boundaries separate grains that have been nucleated independently, and hence misorientations are generally large. Low-angle grain boundaries are formed during cool down by stress-induced glide and climb of dislocations, leading to these energetically favorable configurations. Misorientations in this case usually do not exceed more than 1°. Twin boundaries are planar defects that separate regions of the crystal whose orientations are related to each other in a definite, symmetrical way.

Volume Defects

Precipitates, inclusions, and voids or bubbles are volume defects, and these are formed when gases dissolved in the melt precipitate out after solidification. For example, in microgravity growth, the absence of buoyancy precludes degassing of the melt, resulting in the formation of voids. While undissolved foreign particles are generally classified as inclusions, a second type of inclusion is formed during growth from nonstoichiometric melt. Compound semiconductors generally sublime incongruently, thereby causing a slight excess of one of the components in a stoichiometric melt. On solidification, the excess component forms inclusions.

1.4.2 Observation of Crystal Defects

Techniques for observing dislocations and their complex structures have been described in detail by *Verma* [1.30] and *Amelincks* [1.31]. The commonly used techniques come under the categories:

1. Optical methods
2. X-ray methods
3. Preferential etching
4. Microscopy techniques
5. Other techniques

All these methods provide almost direct observation of defects. Their merit is limited by the resolution achievable and their versatility. Choice of a suitable technique will depend on several factors, such as:

1. The shape and size of the crystal under investigation
2. Cleaving, cutting, and polishing possibilities
3. Ability to use destructive techniques, and above all
4. The extent of the details required

Optical Methods

A common inspection method for the as-grown optical crystal boule is detailed observation by illuminating the boule using high-intensity white light or a laser beam. Probably, this is the first technique to be applied to assess the quality of as-grown crystal and can reveal bubbles, cavities, growth bands, and seed interfaces which depend on the growth parameters.

The conoscope is a simple optical tool for investigating optical inhomogeneity in very small crystals to large-size boules. Conoscopic patterns are characteristic for every main crystallographic orientation, and this feature is also frequently used for orienting crystals [1.32]. This method shows the overall quality of the crystal. If the whole crystal has low dislocation density without any grain boundaries and block structures, a nice symmetrical circular pattern of dark and bright fringes with four segments and a cross at the center is observed. Figure 1.2a shows the conoscopic pattern of a sapphire ingot with dislocation density $10^2-10^3\ /\mathrm{cm}^2$ and without any low-angle grain boundaries. Figure 1.2b shows the pattern for a sapphire ingot of the same size but with a dislocation density of the order of $10^3-10^4\ /\mathrm{cm}^2$ and a few low-angle grain boundaries. The presence of a few grain boundaries alters the birefringence and distorts the fringes. The fringe thickness and spacing depend on the length of the crystal along the direction of inspection. Even though this technique does not reveal the dislocation density very precisely, it can reveal the presence of grain boundaries and higher-order, complex defects. The crystals are normally sliced perpendicular to the c-axis, polished, and inspected under a polarizer and analyzer. As-cut surfaces without polishing can also be observed with the application of suitable refractive-index-matched fluid. In general, this technique can reveal the misorientations, grain boundaries, block structures, and also the stress levels. Conoscopy can be used under a polarizing microscope to study thinner samples. A custom-made polarizer and analyzer with rotation features for the analyzer and sample support can be used to study large crystal boules. Alternatively, conoscopic fringes can be projected onto a screen using a laser beam, polarizer and analyzer, and beam diffuser. These fringes are more influenced by the birefringence inhomogeneity induced by defect structures than by variation in the thickness distribution of the boule itself.

X-Ray Methods

X-ray methods can be classified into:

1. High-resolution x-ray diffraction
2. X-ray topography
3. Synchrotron x-ray topography

High-Resolution X-Ray Diffraction. Diffraction for a given plane and wavelength takes place over a finite angular range about the exact Bragg condition, known as the rocking-curve width [1.33]. In x-ray diffractometry, the intensity of the diffracted beam and the angle in the vicinity of a Bragg peak are measured and represented as a full-width at half-maxima (FWHM) rocking curve. The double-axis rocking curve is obtained by scanning the specimen in small steps through the exact Bragg condition and recording the diffracted intensity. The peak width of a rocking curve can be affected by tilts and dilations in the sample, and by curvature. Tilts are regions in the sample where grains or subgrains are tilted with respect to each other, although the lattice parameter is the same in each region. Dilations are regions where the lattice planes are still parallel but the spacing is slightly different due to strain. Changes in lattice parameter also occur in alloyed crystals with nonhomogeneous composition distribution. The experimentally obtained rocking-curve width (FWHM) value is a measure of the crystalline quality of the sample, and it can be compared with a theoretically calculated value. It is possible to obtain a rocking-curve width less than 10 arcsec for a good crystalline sample. Additional information that can be obtained from double-axis rocking curves are substrate–epilayer mismatch, epilayer composition, substrate offcut and/or layer tilt, and layer thickness.

A limitation of double-axis diffraction is that it cannot distinguish between tilts and dilations. In triple-axis diffraction, a third axis is introduced in the form of an analyzer crystal, and tilts and strain can be separated; the rocking-curve width is still narrow. Double-axis rocking curve analysis is sufficient for studying substrates and epitaxial films. Triple-axis x-ray diffraction is used for obtaining finer details of the defect structure of the sample.

X-Ray Topography. Localized variations in intensity within any individual diffracted spot arise from structural nonuniformity in the lattice planes causing the spot, and this forms the basis for the x-ray topographic technique. This topographic contrast arises from differences in the intensity of the diffracted beam as a function of position inside the crystal. The difference between the intensities diffracted from one region of the crystal which diffracts kinematically to another which diffracts dynamically is one of the ways that dislocations can be rendered visible in topography [1.34].

Even though the first topographic image of a single crystal was recorded as early as 1931 [1.35], the real potential of the technique was understood only in 1958 when Lang [1.36] demonstrated imaging of individual dislocations in a silicon crystal. In general, there are three main types of x-ray topographic geometries for studying defects:

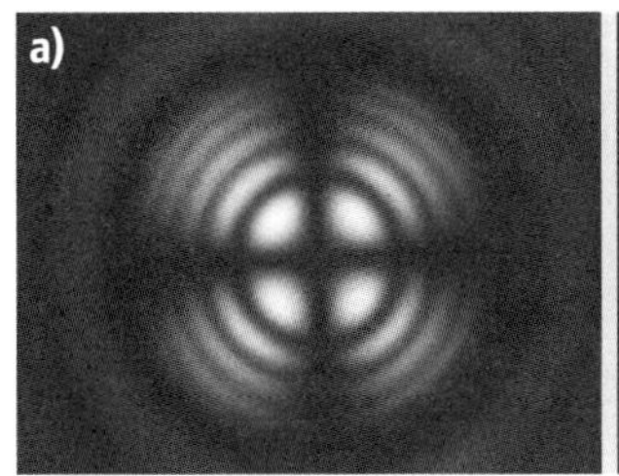

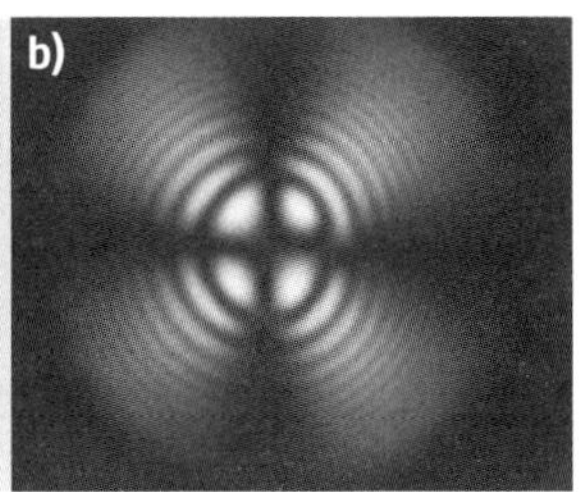

Fig. 1.2 (a) Conoscopic pattern of high-quality sapphire ingot. (b) Conoscopic pattern of sapphire ingot that has a few low-angle boundaries

1. The Berg–Barrett reflection technique [1.37]
2. The double-crystal technique [1.38]
3. The Lang technique [1.36] and its variant – the scanning oscillator technique [1.39]

Following *Lang*'s work [1.36, 40] in imaging of individual dislocations, x-ray topography has become an important quality-control tool for assessment of semiconductor wafers both before and after device fabrication. Using the scanning oscillator technique developed by *Schwuttke* [1.39], it is possible to record transmission topographs of large-size wafers up to 150 mm in diameter, containing appreciable amounts of elastic and/or frozen-in strain.

Synchrotron X-Ray Topography. The advent of dedicated synchrotron radiation sources has enabled the development of a new field of x-ray topography known as synchrotron topography. Synchrotron radiation is especially suitable for x-ray topography because of the high brightness and low divergence of the x-ray beam. Due to the small source dimensions, low divergence angle, as well as the long source–specimen distance, extremely high resolution can be achieved using synchrotron radiation compared with conventional x-ray topography. For example, based on the geometrical factor, the theoretical resolution obtained can be as low as 0.06 μm. Also, it has numerous advantages over laboratory x-ray topography. One of the most important synchrotron topographic techniques developed is white-radiation topography [1.41]. In APS, the white beam is monochromatized by two cooled parallel Si(111) crystals, and the x-ray energy is tunable in the range 2.4–40 keV.

Crystals as large as 150 mm or even 300 mm in diameter can be imaged by using precision translation stages similar to those used in the Lang technique, and the exposure times are much shorter. If a single crystal is oriented in the beam, and the diffracted beams are recorded on a photographic detector, each diffraction spot on the resultant Laue pattern will constitute a map of the diffracting power from a particular set of planes as a function of position in the crystal, with excellent point-to-point resolution. There are three common geometries for synchrotron x-ray topography [1.42]:

1. Transmission geometry, also called Laue geometry: In this mode, the x-ray beam passes through the sample and the topographs recorded reveal the bulk defect information of the crystal. Figure 1.3a shows typical transmission synchrotron topography of a 2 inch LED-grade wafer with a very low dislocation density of 10^2-10^3 /cm^2. The topograph shows the dislocation structure in the entire wafer, which shows the presence of basal dislocations.
2. Gazing-incidence reflection geometry: In this configuration, very small incident angle is used [in the case of SiC, typically 2° used and the $(11\bar{2}8)$ or $(11\bar{2}.12)$ are recorded]. Grazing incidence is used because of the low penetration depth of the x-ray beam, which is more suitable for studying epilayers.
3. Back-reflection geometry: In this mode, a large Bragg angle is used for basal plane reflection $(000l)$ (typically 80° for SiC). Screw dislocations along the c-axis and basal plane dislocations within the x-ray penetration depth can be clearly recorded. The wavelength satisfying Bragg condition is automatically selected in white-beam x-ray topography, while in monochromatic synchrotron x-ray topography (XRT), the energy of the x-ray beam has to be preset to satisfy the diffraction condition. Figure 1.3b shows individual screw dislocations and edge dislocation running almost perpendicular to the wafer.

X-ray topographs are typically recorded on Agfa Structurix D3-SC, Ilford L4 nuclear plate, or VRP-M holographic films, depending on the resolution needed. Exposure time depends on the actual geometry and recording media and varies between a few seconds and 2 h.

Selective Etching

Selective etching is a simple and very sensitive tool for the characterization of single crystals. The usefulness of the etching technique lies in the formation of visible, sharp contrasting etch pits at dislocation sites. The power of etching has been reviewed by several workers [1.31, 43, 44]. The formation of etch pit can be explained as follows. The lattice is distorted for a distance of a few atoms around dislocations. As a result of the stress field generated by the deformation, the lattice elements dissolve more easily at the dislocation sites than in stress-free, undeformed areas. The etch pits are usually straight pyramids with polygonal bases, but other types of pyramids may also be found with various bases and heights. Etch pits can be formed only if certain conditions are satisfied, the most important of these being that the dissolution rate along the surface (V_t) must not greatly exceed the rate of dissolution perpendicular to the surfaces (V_n). The ratio (V_t/V_n) can be increased:

1. By increasing V_n, as has been done in the etchants of several metals
2. By decreasing V_t by adding an inhibitor such as in LiF
3. By varying the temperature to alter the activation energies of V_n and V_t

The etch pits are formed at the dislocation sites, which essentially reveal the emergent point of the dislocations in the surface; they therefore give a direct measure of dislocation density. Since they have certain depths, they also give information on the kind [1.45], configuration, and inclination of dislocations. Etching has also been used to study the stress–velocity rela-

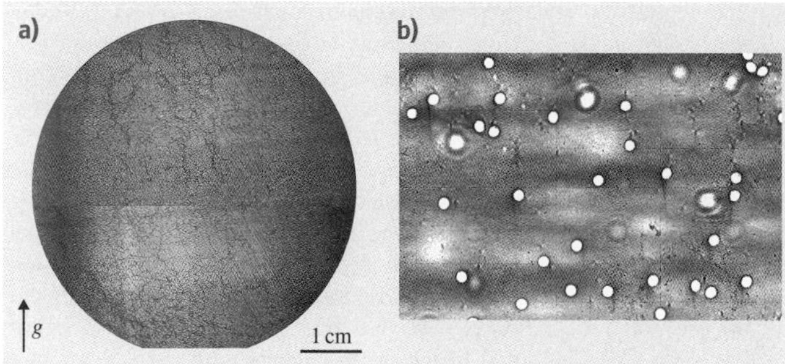

Fig. 1.3 **(a)** Transmission topograph of high-quality sapphire wafer. **(b)** Reflection topograph of SiC revealing individual threading screw dislocations running almost perpendicular to the wafer

tions for individual dislocations [1.46]. Movement of dislocations, deformation patterns like pile-up, origin of dislocations in as-grown crystals, polarity of the crystals, grain boundaries, and distribution of dislocations in crystals can be studied [1.44, 45] (Chap. 43). The greatest advantage of this technique is its simplicity and resolution ($0-10^{12}$ /cm^2). This technique shows the defect density on small areas and hence requires averaging of values taken at a large number of locations. Also, this technique is not a nondestructive method and cannot show the basal plane dislocation when the sample is sliced exactly parallel to the c-axis. Figure 1.4 shows the presence of various defects such as threading edge dislocations, threading screw dislocations, and basal plane dislocations. During the development of SiC crystals, this technique has seen tremendous development and could reveal almost every type of dislocation [1.47].

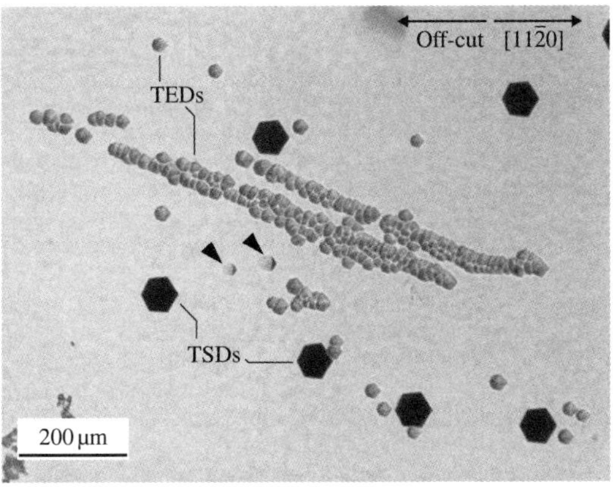

Fig. 1.4 Etch pit pattern of SiC wafer revealing threading edge dislocations (TEDs), threading screw dislocations (TSDs), and basal plane dislocations

Microscopy Techniques

Transmission electron microscopy (TEM) (Chap. 44) is a powerful tool to study dislocations when the sample has higher defect density. It is more commonly used for epitaxial films, where large numbers of dislocations originate due to the lattice misfit between the film and the substrate. This method requires tedious sample preparation and is not considered nondestructive.

Decoration is another important technique, where impurity atoms segregate and settle down along dislocation lines during annealing. The decorated dislocations can be observed easily under an optical microscope in transmission mode [1.31].

Growth spirals, which are true manifestations of screw dislocations, can be observed under optical microscopy, scanning electron microscopy (SEM), and atomic force microscopy (AFM). The presence of growth spirals helps to understand the growth mechanism [1.30].

Infrared (IR) microscopy is similar to optical microscopy except for the fact that IR light is used for illumination, with a wavelength comparable to the bandgap of semiconductor materials. This technique is used to study inclusions, cavities, and even dislocations present in the sample [1.48, 49].

Other Techniques

Photoluminescence (PL) [1.50], electron paramagnetic resonance (EPR) (Chap. 45), positron annihilation (Chap. 46), and micro Raman spectroscopy [1.50] are also used to study semiconductor materials and show electronic defect states and the presence of impurities very successfully.

References

1.1 J. Kepler: *Strena seu de nive sexangula* (Tampach, Frankfurt 1611)

1.2 N. Steno: *De solido intra solidum naturaliter contento dissertationis prodromus* (Stella, Florence 1669), English translation by J.G. Winter (Hafner, New York 1968

1.3 H.J. Scheel: Historical introduction. In: *Handbook of Crystal Growth*, Vol. 1a, ed. by D.T.J. Hurle (Elsevier, Amsterdam 1993) pp. 1–41, Chap. 1

1.4 D. Elwell, H.J. Scheel: *Crystal Growth from High Temperature Solution* (Academic, London 1975)

1.5 J.W. Gibbs: *On the Equilibrium of Heterogeneous Substances*, Collected Works (Longmans Green, New York 1928)

1.6 W. Kossel: Zur Theorie des Kristallwachstums, Nachr. Ges. Wiss. Göttingen **135**, 135–143 (1927)

1.7 P. Curie: Sur la formation des criteaux et sur les constantes capillaires de leurs differentes faces, Bull. Soc. Franc. Mineral. **8**, 145–150 (1885)

1.8 G. Wulff: Zur Frage der Geschwindigkeit des Wachstums und der Auflösung von Krystallflächen, Z. Kristallogr. **34**, 449 (1901)

1.9 R. Marc, A. Ritzel: Über die Faktoren, die den Kristallhabitus bedingen, Z. Phys. Chem. **76**, 584 (1911)

1.10 A. Bravais, A. Etudes: *Crystallographiques* (Gauthier Villers, Paris 1866)

1.11 W. Nernst: Theorie der Reaktionsgeschwindigkeit in heterogenen Systemen, Z. Phys. Chem. **47**(1), 52–55 (1904)
1.12 A.A. Noyes, W.R. Whitney: Über die Auflösungsgeschwindigkeit von festen Stoffen in ihren eigenen Lösungen, Z. Phys. Chem. **23**, 689–692 (1897)
1.13 I.N. Stranski: Zur Theorie des Kristallwachstums, Z. Phys. Chem. **136**, 259–278 (1928)
1.14 M. Volmer, A. Weber: Keimbildung in übersättigten Gebilden, Z. Phys. Chem. **119**, 277–301 (1926)
1.15 E.A. Brandes: *Smithells Reference Book* (Butterworths, London 1983)
1.16 F.C. Frank: The influence of dislocations on crystal growth, Discuss. Faraday Soc. **5**, 48–54 (1949)
1.17 W.K. Burton, N. Cabrera, F.C. Frank: The growth of crystals and the equilibrium structure of their surfaces, Philos. Trans. R. Soc. London A **243**, 299–358 (1951)
1.18 K. Byrappa, D.Y. Pushcharovsky: Crystal chemistry and its significance on the growth of technological materials, Prog. Cryst. Growth Charact. Mater. **24**, 269–350 (1992)
1.19 M.M. Lencka, R.E. Riman: Thermodynamics of the hydrothermal synthesis of calcium titanate with reference to other alkaline-earth titanates, Chem. Mater. **7**(1), 18–25 (1995)
1.20 K. Byrappa, M. Yoshimura: *Handbook of Hydrothermal Technology* (William Andrew Noyes, Norwich 2001)
1.21 W. Tolksdorf: Flux growth. In: *Handbook of Crystal Growth-Bulk Crystal Growth*, Vol. 2, ed. by D.T.J. Hurle (North-Holland, Amsterdam 1994) p. 563, Chap. 10
1.22 R.A. Laudise: *The Growth of Single Crystals* (Prentice Hall, Englewood Cliffs 1970)
1.23 B.M.R. Wanklyn: Practical aspects of flux growth by spontaneous nucleation. In: *Crystal Growth*, Vol. 1, ed. by B.R. Pamplin (Pergamon, Oxford 1974) pp. 217–288
1.24 V.V. Timofeeva: *Growth of Crystals from High Temperature Solutions* (Nauka, Moscow 1975)
1.25 A. Verneuil: Production artificielle du rubis par fusion, C. R. Paris **135**, 791–794 (1902)
1.26 D.T.J. Hurle, B. Cockyane: Czochralski growth. In: *Handbook of Crystal Growth*, Vol. 2a, ed. by D.T.J. Hurle (North Holland, Amsterdam 1994) pp. 99–212, Chap. 3
1.27 D.T.J. Hurle (Ed.): *Handbook of Crystal Growth* (North Holland, Amsterdam 1994)
1.28 G. Stringfellow: *Organometallic Vapor-Phase Epitaxy: Theory and Practice*, 2nd edn. (Academic, New York 1998)
1.29 R. Fornari: Vapor phase epitaxial growth and properties of III-Nitride materials. In: *Crystal Growth of Technologically Important Electronic Materials*, ed. by K. Byrappa, T. Ohachi, H. Klapper, R. Fornari (Allied Publishers, New Delhi 2003)
1.30 A.R. Verma: *Crystal Growth and Dislocations* (Butterworths, London 1953)
1.31 S. Amelinckx: The direct observation of dislocations. In: *Solid State Physics*, ed. by F. Seitz, D. Turnbull (Academic, New York 1964), Suppl. 6
1.32 E.A. Wood: *Crystals and Light* (Dover, New York 1977)
1.33 B.K. Tanne: High resolution x-ray diffraction and topography for crystal characterization, J. Cryst. Growth **99**, 1315 (1990)
1.34 B.K. Tanner: *X-ray Diffraction Topography* (Pergamon, Oxford 1976)
1.35 V.W. Berg: Über eine röntgenographische Methode zur Untersuchung von Gitterstörung an Kristallen, Naturwissenschaften **19**, 391–396 (1931)
1.36 A.R. Lang: Direct observation of individual dislocations, J. Appl. Phys. **29**, 597–598 (1958)
1.37 C.S. Barrett: A new microscopy and its potentialities, Trans. AIME **161**, 15–65 (1945)
1.38 W.L. Bond, J. Andrus: Structural imperfections in quartz crystals, Am. Mineral. **37**, 622–632 (1952)
1.39 G.H. Schwuttke: New x-ray diffraction microscopy technique for study of imperfections in semiconductor crystals, J. Appl. Phys. **36**, 2712–2714 (1965)
1.40 A.R. Lang: Point-by-point x-ray diffraction studies of imperfections in melt-grown crystals, Acta Cryst. **10**, 839 (1957)
1.41 J. Miltat: White beam synchrotron radiation. In: *Characterization of Crystal Growth Defects by X-ray Methods*, NATO ASI Ser. B, Vol. 63, ed. by B.K. Tanner, D.K. Bowen (Plenum, New York 1980) pp. 401–420
1.42 B. Ragothamachar, G. Dhanaraj, M. Dudley: Direct analysis in crystals using x-ray topography, Microsc. Res. Tech. **69**, 343 (2006)
1.43 A.J. Forty: Direct observations of dislocations in crystals, Adv. Phys. **3**, 1–25 (1954)
1.44 W.G. Johnson: Dislocations etchpits in nonmetallic crystals. In: *Progress in Ceramics*, Vol. 2, ed. by J.E. Burke (Pergamon, Oxford 1962) p. 1
1.45 K. Sangawal: *Etching of Crystals* (North-Holland, Amsterdam 1987)
1.46 J.J. Gilman, W.G. Johnston: Behaviour of individual dislocations in strain-hardened LiF crystals, J. Appl. Phys. **31**, 687–692 (1960)
1.47 W.J. Choyke, H. Matsunami, G. Pensl (Eds.): *Silicon Carbide: Recent Major Advances* (Springer, Berlin, Heidelberg 2004)
1.48 A. Hossain, A.E. Bolotnikov, G.S. Camarda, Y. Cui, G. Yang, K-H. Kim, R. Gul, L. Xu, R.B. James: Extended defects in CdZnTe crystals: Effects on device performance, J. Cryst. Growth (2010) in press (doi:10.1016/j.jcrysgro.2010.03.005)
1.49 U.N. Roy, S. Weler, J. Stein, A. Gueorguiev: Unseeded growth of CdZnTe:In by THM technique, Proc. SPIE **7449**, 74490U (2009)
1.50 J. Jimenez (Ed.): *Microprobe Characterization of Optoelectronic Materials* (Taylor Francis, New York 2003)

2. Nucleation at Surfaces

Ivan V. Markov

This chapter deals with the thermodynamics and kinetics of nucleation on surfaces, which is essential to the growth of single crystals and thin epitaxial films. The starting point is the equilibrium of an *infinitely* large crystal and a crystal with a finite size with their ambient phase. When the system deviates from equilibrium density fluctuations or aggregates acquire the tendency to unlimited growth beyond some critical size – the nucleus of the new phase. The Gibbs free energy change of formation of the nuclei is calculated within the framework of the macroscopic thermodynamics and in terms of dangling bonds in the case of small clusters. In the case of nucleation from vapor the nuclei consist as a rule of very small number of atoms. That is why the rate of nucleation is also considered in the limit of high supersaturations. The effect of defect sites and overlapping of nucleation exclusion zones with reduced supersaturation formed around the growing nuclei is accounted for in determining the saturation nucleus density. The latter scales with the ratio of the surface diffusion coefficient and the atom arrival rate. The scaling exponent is a function of the critical nucleus size and depends on the process which controls the frequency of attachment of atoms to the critical nuclei to produce stable clusters, either the surface diffusion or the incorporation of atoms to the critical nuclei. The nucleation on top of two-dimensional (2-D) islands is considered as a reason for roughening in homoepitaxial growth. The mechanism of formation of three-dimensional (3-D) islands in heteroepitaxial growth is also addressed. The effect of surface-active species on the rate of nucleation is explored.

Nucleation at surfaces plays a crucial role in the growth of crystals and epitaxial overlayers for the preparation of advanced materials with potential for technological applications. In homoepitaxy of metal or semiconductor films the instability of planar growth against roughening depends on the kinetics of two-dimensional nucleation [2.1]. The interplay of wetting and strain leads to clustering in overlayers growing under elastic stress in heteroepitaxy and determines the mechanism of growth and in turn the film morphology [2.2–4]. Smooth quantum wells or self-assembled quantum dots can be grown by varying the conditions of growth (temperature or growth rate) or by use of third species which change both the thermodynamics and kinetics of the processes involved [2.5]. The growth of thin epitaxial films in particular by molecular-beam epitaxy (MBE) usually occurs far from equilibrium. Thus, in addition to thermodynamics, one has to account for the kinetic processes taking place on the crystal surface [2.6]. The latter are responsible for the remarkable richness of patterns which are observed during growth [2.7].

This chapter gives the essential physics of the thermodynamics and kinetics of nucleation, both three- and two-dimensional, on like and unlike substrates as well as some later developments such as the Ehrlich–Schwoebel effect on second-layer nucleation and the effect of surface-active species on nucleation rate. The presentation is oriented more to the needs of experimentalists rather than going deeply into theoretical problems. The chapter is organized as follows. We start with problems of equilibrium of crystals and epitaxial overlayers with the parent phase (vapor, solution) in Sect. 2.1 and consider the equilibrium vapor pressure of infinitely large and finite-size crystals, the thermodynamic driving force for nucleation to occur, and the equilibrium shape of three-dimensional (3-D) crystals on unlike surfaces. In Sect. 2.2 we define the work for nucleus formation in the most general way and consider the limiting cases of the classical (capillary) theory of nucleation at low or intermediate values of supersaturation and the atomistic approach at high supersaturations. We derive expressions for the work of formation of three-dimensional nuclei on unlike substrates and two-dimensional nuclei on like and unlike substrates. In Sect. 2.3 we give a general formulation of the nucleation rate and again derive expressions valid for high and low supersaturations. We consider further in Sect. 2.4 the saturation nucleus density accounting for the influence of defect (active) sites stimulating nucleation events and the overlapping of undersaturated nucleation exclusion zones around growing clusters. Making use of the rate equation approach we derive expressions for the saturation nucleus density in thin epitaxial films in diffusion and kinetic regimes of growth. In Sect. 2.5 we consider the effect of the step-edge Ehrlich–Schwoebel barrier on second-layer nucleation as a reason for the formation of mounds and thus roughening of surfaces in homoepitaxy. The mechanism of transformation of monolayer-high two-dimensional (2-D) islands into three-dimensional crystallites in Volmer–Weber and Stranski–Krastanov growth is addressed in Sect. 2.6. In Sect. 2.7 we explore the effect of surface-active species on the kinetics of nucleation. Some conclusions and outlook are given in Sect. 2.8.

2.1 Equilibrium Crystal–Ambient Phase

In treating the title problem we use the atomistic approach developed by *Kaischew* and *Stranski* [2.8]. It is based on the assumption of additivity of bond energies and accounts for the elementary processes taking place during growth and dissolution of the particles of the new phase. Although apparently old fashioned this approach is extremely instructive and informative for understanding the essential physics of the equilibrium of infinitely large phases and phases with finite size as well as of the deviation from equilibrium leading to transitions from one phase to another. Numerical studies of the stability of small clusters performed by making use of modern quantum-mechanical methods lead to the same conclusion that the closed atomic structures are most stable [2.9].

2.1.1 Equilibrium of Infinitely Large Phases

We consider for simplicity one-component system. The equilibrium between infinitely large phases (crystal, liquid or vapor) is determined by the equality of the respective chemical potentials. In 1927 *Kossel* and *Stranski* simultaneously developed an atomistic approach which is in fact identical to the definition of the macroscopic thermodynamics [2.10–12]. They considered the different sites that atoms can occupy on the crystal surface and found that there exists one particular site which plays a crucial role in crystal nucleation and growth. They introduced the concept of the *half-crystal position*, which turned out to be intimately connected with the chemical potential of an infinitely large crystal.

Consider the cubic face of a crystal with a simple cubic lattice (a Kossel crystal) containing a monatomic step (Fig. 2.1). Atoms can be located at different sites on the crystal surface. They can be built in the uppermost lattice plane or into the step edge, be adsorbed at the step edge or on the terrace, or can occupy the corner site (3) which has very peculiar properties. An atom in this position is connected with a half-atomic row, a half-crystal plane, and a half-crystal block. This is the reason the term half-crystal position (*Halbkristalllage* or kink position) was coined for this particular site. Therefore, the work of separation of an atom from this position is exactly equal to the lattice energy of the crystal per building particle. Hence, the work of detachment of an atom from this position is given by

$$\varphi_{1/2} = \tfrac{1}{2}(Z_1\psi_1 + Z_2\psi_2 + Z_3\psi_3 \ldots) ,$$

where Z_i are the numbers of neighbors of the consecutive coordination spheres and ψ_i are the respective bond energies.

Whereas atoms in other positions have different numbers of saturated and unsaturated (dangling) bonds, the atom in the kink position (3) has an equal number of saturated and dangling bonds. Therefore, the separation work from a half-crystal position serves as a specific reference with which the probabilities for elementary processes at other sites to take place can be compared. The detachment of an atom from the half-crystal position gives rise to the same position. It follows that, when an atom is detached from this position, the number of dangling bonds remains unchanged and in turn the surface energy does not change. Hence, the whole crystal (if it is large enough to avoid finite-size effects) can be built up or disintegrated into single atoms by repetitive attachment or detachment of atoms to and from this position.

In equilibrium with its vapor the probability of attachment of atoms to this position must be equal to the probability of their detachment. Hence the work of detachment of atoms from this position will determine the equilibrium vapor pressure and in turn its chemical potential. For simple crystals with monatomic vapor the latter will be given at zero temperature (the change of entropy is equal to zero) by

$$\mu_c^{\infty} = -\varphi_{1/2} , \qquad (2.1)$$

where the superscript ∞ indicates an infinitely large crystal.

As seen the chemical potential of an infinitely large crystal is equal to the work of detachment of atoms from the half-crystal position taken with a negative sign. It is this property which makes this position unique in the theory of crystal nucleation and growth [2.13].

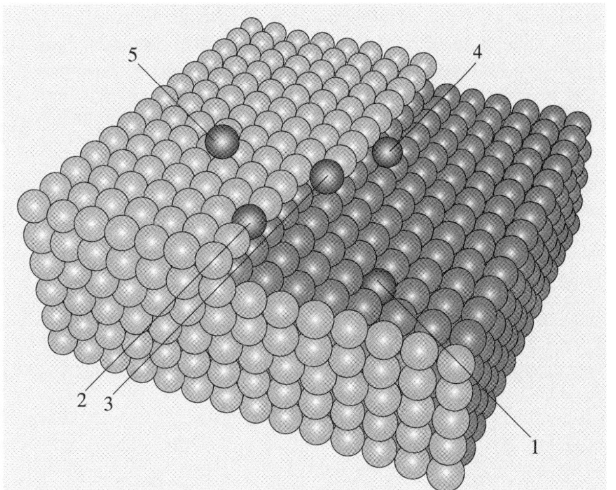

Fig. 2.1 The most important sites an atom can occupy on a crystal surface: 1 – atom embedded into the uppermost crystal plane, 2 – atom embedded into the step edge, 3 – atom in a half-crystal (kink) position, 4 – atom adsorbed at the step, 5 – atom adsorbed on the terrace

There is one more very important property of the half-crystal position. We can divide $\varphi_{1/2}$ into two parts: lateral interaction with the half-atomic row and the half-crystal plane, and the normal interaction with the half-crystal block underneath. If we replace the underlying crystal block by another block of different material and crystal lattice the lateral bonding will remain more or less unchanged if we assume additivity of bond energies. However, the normal bonding will change substantially owing to the difference in both chemical bonding and lattice strain. It is easy to show that the separation work from a kink position in this particular case can be written as

$$\varphi'_{1/2} = \varphi_{1/2} - (\psi - \psi') , \qquad (2.2)$$

where ψ' is the energy of a bond between unlike atoms.

Having in mind (2.1), (2.2) can be written as

$$\mu'_c = \mu_c^{\infty} + (\psi - \psi') . \qquad (2.3)$$

We now define the surface energy of a crystal by the following imaginary process. We cleave isothermally and reversibly the crystal into two halves and produce two surfaces with area S. We count the bonds we break and divide the energy spent by $2S$. If we confine ourselves to nearest-neighbor bonds in the case of Kossel crystal we break one bond per atom and obtain ($S = a^2$)

$$\sigma = \frac{\psi}{2a^2}, \tag{2.4}$$

where a is the atomic diameter.

Using the above definition and the relation of *Dupré* [2.14]

$$\sigma_i = \sigma_A + \sigma_B - \beta, \tag{2.5}$$

which connects the specific interfacial energy σ_i between the unlike crystals A and B with the specific adhesion energy $\beta = \psi'/a^2$, (2.3) can be written as

$$\mu'_c = \mu_c^\infty + a^2(\sigma + \sigma_i - \sigma_s). \tag{2.6}$$

It is immediately seen that the term in the brackets $\Delta\sigma = \sigma + \sigma_i - \sigma_s$ is in fact the parameter that accounts for the wetting of the substrate (the half-crystal block underneath) by the overlayer in epitaxy of one material on the surface of another [2.15]. Thus, when $\Delta\sigma < 0$, or what is the same, $\psi < \psi'$ (complete wetting), the equilibrium vapor pressure of the first monolayer on the unlike substrate will be smaller than the equilibrium vapor pressure of the bulk crystal ($\mu = \mu_0 + k_B T \ln P$), i.e., $P'_\infty < P_\infty$. This means that at least the first monolayer can be deposited at a vapor pressure smaller than the equilibrium vapor pressure of the bulk crystal, or in other words, at undersaturation, $P'_\infty < P < P_\infty$ [2.16]. If the two crystals have different lattice parameters the growth should continue by formation of three-dimensional (3-D) islands. This is the famous Stranski–Krastanov mechanism of growth [2.17], in which the accumulation of strain energy with film thickness makes the planar film unstable against clustering. Obviously, if the lattice misfit is equal to zero the growth will continue layer by layer in the so-called Frank–van der Merwe mechanism of growth [2.18, 19]. In the opposite case of incomplete wetting ($\Delta\sigma > 0$), 3-D islanding will take place from the very beginning of deposition or Volmer–Weber growth, which requires supersaturation, $P > P_\infty$ [2.20]. We thus see that the separation work from a half-crystal position plays a fundamental role in determining the mechanism of epitaxial growth.

The lattice misfit increases the tendency for 3-D islanding by increasing the interfacial energy in (2.6) with the energy per unit area of misfit dislocations or elastic strain. Thus for heteroepitaxial growth the interfacial energy reads [2.21]

$$\sigma_i^* = \sigma_i + \varepsilon_m,$$

where ε_m is either the misfit dislocation energy or the energy of the homogeneous strain.

Thus the interfacial energy between misfitting crystals consists of two parts: a chemical part σ_i accounting for the difference in chemistry and strength of bonding, and a geometrical part ε_m accounting for the difference of lattices and lattice parameters. If the misfit in heteroepitaxy is accidentally or intentionally tailored to be equal to zero (particularly in binary or ternary alloys) $\varepsilon_m = 0$, but the chemical part σ_i remains different from zero and affects the mechanism of growth.

It should be noted that the misfit plays a decisive role for clustering only in Stranski–Krastanov growth, where it changes the sign of $\Delta\sigma$ from negative to positive beyond the so-called wetting layer. In Volmer–Weber growth $\Delta\sigma$ is positive and the strain energy makes a minor contribution with the same sign to it. Frank–van der Merwe growth takes place only in systems with zero misfit [2.22], which is why we will not take into consideration the effect of lattice misfit in nucleation.

2.1.2 Equilibrium of Small Crystal with the Ambient Phase

The separation work from the half-crystal position cannot determine the equilibrium of a crystal with finite size with its surrounding because the role of the crystal edges and corners cannot be ignored. The kink position is no longer a repetitive step for dissolution of the crystal. That is why Stranski and Kaischew suggested that the condition for a small crystal to be in equilibrium with the ambient phase is for the probability of building up a whole crystal plane to be equal to the probability of its dissolution. In this way the effect of the edge and corner atoms are accounted for in addition to the atoms in half-crystal positions. Obviously, the smaller the crystal, the greater will be the role of the corner and edge atoms, and vice versa. Thus they defined the *mean separation work* as the work per atom to disintegrate a whole crystal plane into single atoms. This quantity must have one and the same value for all crystal faces belonging to the equilibrium shape.

Consider for simplicity a small Kossel crystal with a shape of a cube with edge length $l_3 = an_3$, where n_3 is the number of atoms in the edge of the cube. Confining ourselves to nearest-neighbor bond energy ψ the energy for dissolution of a whole lattice plane into single atoms (by counting the bonds we break in the process of disintegration, Fig. 2.2) is $3n_3^2\psi - 2n_3\psi$. Dividing by the number of atoms n_3^2 the mean separation work reads [2.8]

$$\bar{\varphi}_3 = 3\psi - \frac{2\psi}{n_3}\,, \tag{2.7}$$

or, bearing in mind that for a simple cubic lattice $3\psi = \varphi_{1/2}$,

$$\bar{\varphi}_3 = \varphi_{1/2} - \frac{2\psi}{n_3}\,.$$

It follows that the mean work of separation tends asymptotically to the work of separation from a half-crystal position as the crystal size is increased. We conclude that a crystal can be considered as small if $n_3 < 70$, or $l_3 < 2\times10^{-6}$ cm assuming $a \approx 3$ Å.

As $\bar{\varphi}_3$ determines the equilibrium vapor pressure of the small crystal and in turn its chemical potential we can write in analogy with (2.1) for $T = 0$

$$\mu_c = \mu_v = -\bar{\varphi}_3\,.$$

Then

$$\Delta\mu = \mu_v(P) - \mu_c^{\infty}(P) = \varphi_{1/2} - \bar{\varphi}_3 = \frac{2\psi}{n_3} \tag{2.8}$$

is the difference of the chemical potentials of the *infinitely large* vapor and crystal phases which represents the thermodynamic driving force for nucleation to occur, or the *supersaturation*.

The equilibrium of the vapor and the crystal takes place at some vapor pressure P_∞ (to stress the fact that the crystal is infinitely large) so that $\mu_v(P_\infty) = \mu_c(P_\infty)$. Then we can write (2.8) as

$$\Delta\mu = [\mu_v(P) - \mu_v(P_\infty)] - [\mu_c(P) - \mu_c(P_\infty)]\,.$$

For small deviations from equilibrium the differences in the above equation can be replaced by derivatives and

$$\Delta\mu = \int_{P_\infty}^{P} \frac{\partial\mu_v}{\partial P}\,\mathrm{d}P - \int_{P_\infty}^{P} \frac{\partial\mu_c}{\partial P}\,\mathrm{d}P = \int_{P_\infty}^{P} (v_v - v_c)\,\mathrm{d}P\,,$$

where v_v and v_c are the molecular volumes of the vapor and the crystal. As $v_v \gg v_c$ the above equation simplifies to

$$\Delta\mu = \int_{P_\infty}^{P} v_v\,\mathrm{d}P\,.$$

Considering the vapor as an ideal gas ($v_v = k_B T/P$) gives upon integration

$$\Delta\mu = k_B T \ln\left(\frac{P}{P_\infty}\right). \tag{2.9}$$

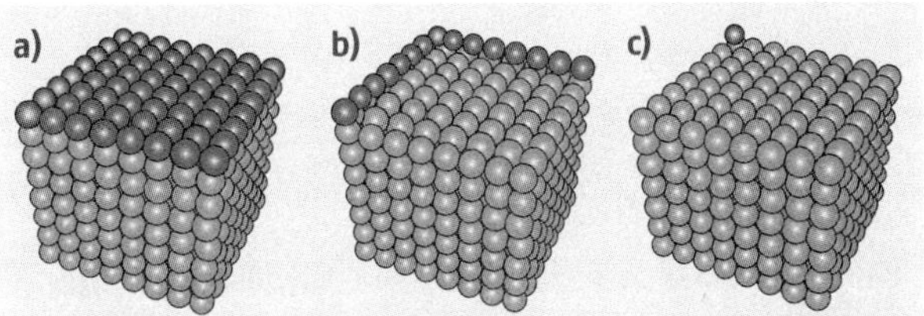

Fig. 2.2a–c Schematic for the evaluation of the mean separation work which determines the equilibrium of a small three-dimensional crystal with the supersaturated vapor phase. In stage (**a**) we detach $(n-1)^2$ atoms, breaking three bonds per atom, in stage (**b**) we detach $2(n-1)$ atoms, breaking two bonds per atom, and finally in (**c**) we detach the last atom, breaking a single bond

The supersaturation $\Delta\mu$ is usually very large in the case of nucleation from vapor, particularly in methods such as MBE. Let us evaluate it for the case of nucleation in MBE growth of Si(111). The supersaturation is given in terms of the ratio of the fluxes R/R_∞, where $R = P/\sqrt{2\pi m k_B T}$, rather than in vapor pressures as in (2.9). Typical growth conditions are $T = 600$ K and $R = 1\times10^{13}$ atom/cm^2 s [2.23]. The equilibrium vapor pressure of Si at 600 K is $P_\infty = 1.3\times10^{-27}$ N/m^2. Then, $R_\infty \cong 6.5\times10^{-8}$ atom/cm^2 s and $\Delta\mu \cong 2.5$ eV. This means that the supersaturation is of the order of the enthalpy of evaporation of Si (≈ 4.5 eV). As we will see below this is why nuclei consist of a number of atoms of the order of unity.

Note that, with the approximation made, (2.9) is valid for very small deviations from equilibrium. If we repeat the above calculations at much higher temperature, say 1300 K, we find $\Delta\mu \cong 0.05$ eV. We can believe this value to be close to the real figure, but for low temperatures we can be sure only of the sign of the supersaturation (growth or evaporation) but not its numerical value.

Equation (2.8) represents the famous Thomson–Gibbs equation which gives the dependence of the equilibrium vapor pressure of a small crystal on its linear size. Using the definition of the specific surface energy (2.4) we obtain the Thomson–Gibbs equation in its form which is well known in the literature

$$\Delta\mu = \frac{4\sigma v_c}{l_3}\,. \tag{2.10}$$

We consider further the equilibrium with the vapor phase (and in turn with the dilute adlayer) of a small two-dimensional crystal with a monolayer height formed on the surface of a large three-dimensional crystal. Such an island grows or dissolves by attach-

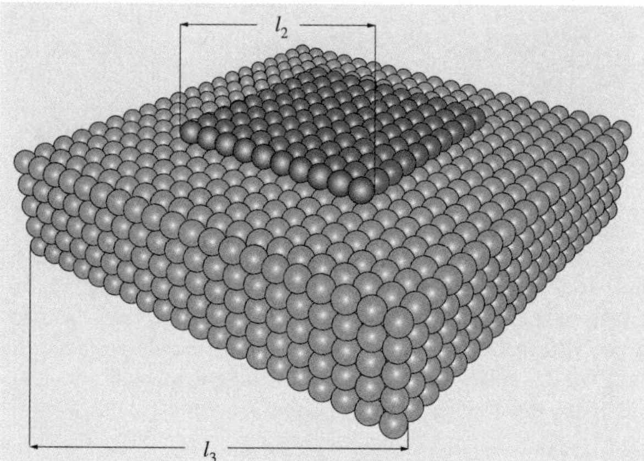

Fig. 2.3 Schematic for the evaluation of the mean separation work which determines the equilibrium of a small two-dimensional crystal with the supersaturated vapor phase. In equilibrium the probabilities of evaporation and building of a whole row of atoms (*black spheres*) are equal

ment or detachment of whole atomic rows. That is why *Kaischew* and *Stranski* suggested that the probability of building of a whole atomic row with length $l_2 = n_2 a$ is equal to the probability of its disintegration into single atoms [2.8]. The equilibrium 2-D island–vapor phase is now determined by the mean separation work $\bar{\varphi}_2$, which is equal to the energy per atom for evaporation of a whole edge row of atoms (Fig. 2.3). Assuming a square-shaped island with n_2 atoms in the edge the mean separation work reads

$$\bar{\varphi}_2 = 3\psi - \frac{\psi}{n_2} = \varphi_{1/2} - \frac{\psi}{n_2} .$$

The supersaturation necessary for the formation of a two-dimensional island with linear size l_2 then reads

$$\Delta\mu = \frac{\psi}{n_2} . \tag{2.11}$$

Note that in nucleation on surfaces the supersaturation can be expressed as a ratio of the real and the equilibrium adatom concentrations (in equilibrium the chemical potential of the vapor is equal to the chemical potential of the adlayer, which in turn depends on the adatom concentration)

$$\Delta\mu = k_B T \ln\left(\frac{N_1}{N_1^e}\right) ,$$

where [2.24]

$$N_1^e = N_0 \exp\left(-\frac{\Delta W}{k_B T}\right) , \tag{2.12}$$

the difference $\Delta W = \varphi_{1/2} - E_{des}$ being the work to transfer an atom from a half-crystal position on the surface of a terrace, and N_0 is the atomic density of the crystal surface.

This is particularly true when the adatom concentration is determined by a dynamic adsorption–desorption equilibrium, i. e., when the atom arrival rate R is equal to the re-evaporation rate N_1/τ_s, where $\tau_s = \nu^{-1} \exp(E_{des}/k_B T)$ is the mean residence time of an atom on the surface before desorption.

We define now the specific edge energy in the same way that we defined the specific surface energy (2.4). We cleave an atomic plane into two halves and produce two edges with length L. We break one bond per atom and for the specific edge energy one obtains

$$\varkappa = \frac{\psi}{2a} . \tag{2.13}$$

Combining (2.11) and (2.13) gives the Thomson–Gibbs equation for the two-dimensional case, or the supersaturation required to form an island with edge length l_2, in its more familiar form [2.24]

$$\Delta\mu = \frac{2\varkappa a^2}{l_2} . \tag{2.14}$$

Equations (2.10) and (2.14) can be derived by using the method of thermodynamic potentials introduced by Gibbs (for a review see [2.21]). However, contrary to the pure thermodynamics, the above *molecular-kinetic* or atomistic approach accounts in addition for the elementary processes of growth and dissolution of crystals. The growth of sufficiently large crystal takes place by attachment of building units to the half-crystal position. Once the atom is incorporated at this position we can say that it has joined the crystal lattice. Small three- and two-dimensional crystals grow and dissolve by building and dissolution of whole crystal planes or atomic rows, respectively.

2.1.3 Equilibrium Shape of Crystals

In 1878 *Gibbs* defined thermodynamically the problem of the equilibrium shape of crystals as the shape at which the crystal has a minimum surface energy at given constant volume [2.25]. This definition later acquired a geometric interpretation in the well-known Gibbs–Wulff theorem [2.26], according to which the distances h_n from an arbitrary (Wulff's) point to the different crystal faces are proportional to the corresponding specific surface energies σ_n of these faces

$$\frac{\sigma_n}{\boldsymbol{h}_n} = \text{const.} \tag{2.15}$$

As a result the equilibrium shape represents a closed polyhedron consisting of the faces with the lowest specific surface energies. The areal extents of the crystal faces belonging to the equilibrium shape have one and the same value of chemical potential.

Half a century later *Kaischew* extended this approach to cover the case of a crystal on a foreign substrate and derived a relation known in the literature as the Wulff–Kaischew theorem [2.27]

$$\frac{\sigma_n}{h_n} = \frac{\sigma_i - \beta}{h_i} = \text{const.}\,, \tag{2.16}$$

where σ_i is the specific surface energy of the crystal face that is in contact with the substrate and h_i is the distance from the Wulff point to the plane of the contact (Fig. 2.4).

It is seen that the distance from the Wulff point to the contact plane is proportional to the difference $\sigma_i - \beta$. Therefore, when the catalytic potency of the substrate β is equal to zero, the distance h_i will have its value in the absence of a substrate. In this case we have *complete nonwetting*. At the other extreme $\beta = \sigma_A + \sigma_B = 2\sigma$ ($\sigma_A = \sigma_B = \sigma$) we have *complete wetting* and the three-dimensional crystal is reduced to a monolayer-high island. In the intermediate case $0 < \beta < 2\sigma$ we have *incomplete wetting* and the crystal height is smaller than its lateral extent.

The introduction of the separation work from half-crystal position and the mean separation works enabled Stranski and Kaischew to provide a new atomistic approach for determination of the equilibrium shape of crystals. The latter is necessary for calculation of the work of nucleus formation as it is assumed that the nuclei preserve the equilibrium shape as the lowest-energy shape. Thus the lowest-energy pathway of the crystallization process is ensured.

The basic idea is that atoms bound more weakly than an atom in the half-crystal position cannot belong to the equilibrium shape. We start from a sufficiently large crystal with a simple crystallographic form and remove in succession from its surface all atoms bound more weakly than in a half-crystal position. Precisely at that process all the faces of the equilibrium shape appear. Then the areas of the faces are varied by removal and addition of whole crystal planes up to the moment when the mean separation works of all crystal faces become equal. As the mean separation works are closely related to the chemical potentials the latter condition is equivalent to the definition of Gibbs. Thus, during the last operation of equating the mean separation works of all crystal faces, those which do not belong to the equilibrium shape disappear [2.28].

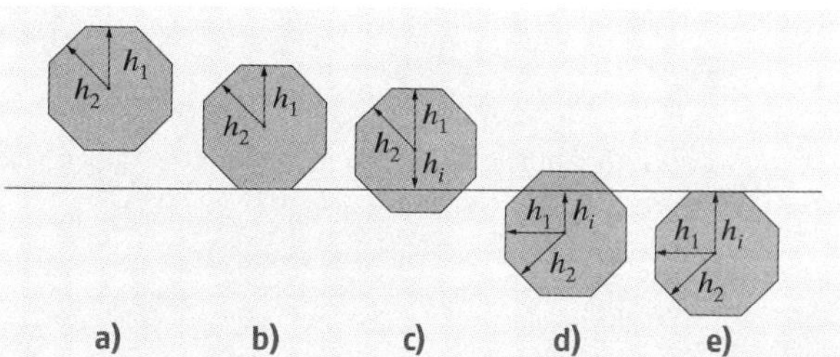

Fig. 2.4a–e Equilibrium shape of a crystal on an unlike substrate. The distances $\boldsymbol{h}_1$ and $\boldsymbol{h}_2$ in the free polyhedron (**a**) are proportional to the specific free energies σ_1 and σ_2 according to the Gibbs–Wulff theorem (2.15). In the presence of unlike substrate the distances to free surfaces remain the same as in the free polyhedron. The distance h_i to the plane of contact is determined by the difference $\sigma_i - \beta$ according to the Wulff–Kaischew theorem (2.16). (**b**) Complete nonwetting ($\beta = 0$); (**c**,**d**) different degrees of incomplete wetting (note that in the latter case the vector h_i is negative); (**e**) complete wetting ($\beta = 2\sigma$)

Therefore, the necessary and sufficient condition for the equilibrium shape of a crystal in the molecular-kinetic approach is equality of the mean separation works, or in other words, of the chemical potentials of all crystal faces. We use this condition to derive the equilibrium aspect ratio of a three-dimensional cubic crystal on the surface of an unlike crystal assuming incomplete wetting ($\Delta\sigma > 0$).

Consider a cubic crystal with a square base with edge length $l = na$ and height $h = n'a$, where n and n' are the number of atoms in the horizontal and vertical edges (Fig. 2.5). The mean separation work calculated

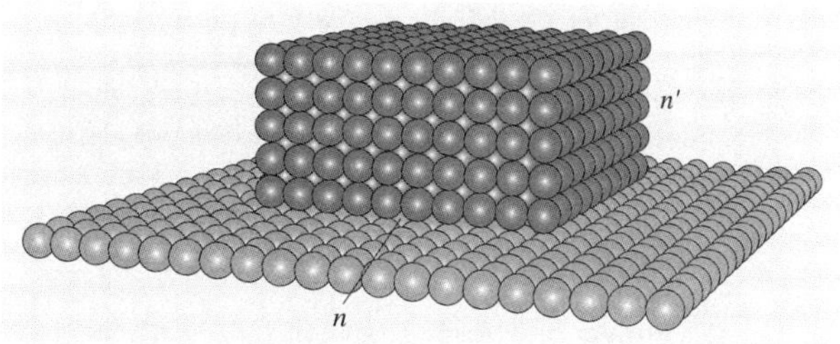

Fig. 2.5 A cubic crystal with n and n' atoms in the base and the height on the surface of an unlike crystal at incomplete wetting

Part A | 2.1

from the side crystal face is

$$\bar{\varphi}'_3 = 3\psi - \frac{\psi - \psi'}{n'} - \frac{\psi}{n},$$

whereas the same quantity calculated for the upper base is given by (2.7). The condition $\bar{\varphi}_3 = \bar{\varphi}'_3$ gives

$$\frac{h}{l} = \frac{n'}{n} = \phi, \quad (2.17)$$

where

$$\phi = 1 - \frac{\psi'}{\psi}. \quad (2.18)$$

Substituting ψ and ψ' by the specific surface and adhesion energies and making use of the relation of Dupré (2.5) gives ϕ in terms of surface energies

$$\phi = \frac{\sigma + \sigma_i - \sigma_s}{2\sigma}. \quad (2.19)$$

As seen, the equilibrium aspect ratio of the crystal is precisely equal to the familiar wetting condition (2.6) relative to 2σ. The parameter ϕ is known in the literature as the *wetting function*; it plays a crucial role in nucleation at surfaces and determines the mechanism of growth of thin epitaxial films [2.15, 29]. It can be shown that (2.19) can be derived by the classical thermodynamic condition of the minimum of the surface energy $\Phi = 4lh\sigma + l^2(\sigma + \sigma_i - \sigma_s)$ at constant volume $V = l^2h$ [2.15].

2.2 Work for Nucleus Formation

2.2.1 General Definition

The nuclei of the new phase represent local fluctuations of the density which can be considered as small molecular aggregates. If the phase is stable the density fluctuations increase the thermodynamic potential of the system. In this sense they are thermodynamically unfavorable. Their concentration is small and they cannot reach considerable size as the probability of decay is greater than the probability of growth. Thus they have no tendency to unlimited growth and can be considered as *lifeless*. *Frenkel* coined for them the term *homophase fluctuations* to emphasize the fact that they are well compatible with the stable state of aggregation [2.30]. As one approaches the phase equilibrium determined by the equality of the chemical potentials, their concentration increases and the maximum of the size distribution shifts to larger sizes. Once the chemical potential of the initial bulk phase (vapor or solution) becomes greater than that of the new, denser phase (liquid or crystal) the probability of growth becomes greater than the probability of decay and the tendency for growth of the density fluctuations prevails after exceeding some critical size. Frenkel referred to these as *heterophase fluctuations* to stress the fact that they are no longer compatible with the old, less dense phase. It is just these density fluctuations or clusters with a critical size which are called the *nuclei* of the new phase. In order to form such nuclei a free energy should be expended.

Consider a volume containing i_v molecules of a vapor with chemical potential μ_v at constant temperature T and pressure P. The thermodynamic potential of this initial state is given by $G_1 = i_v\mu_v$. A small crystal with bulk chemical potential μ_c^∞ is formed from i molecules of the vapor phase and the thermodynamic potential of the final state reads $G_2 = (i_v - i)\mu_v + G(i)$, where $G(i)$ is the thermodynamic potential of a cluster consisting of i molecules. The work of formation of a cluster consisting of i molecules is given by the difference $\Delta G(i) = G_2 - G_1$ and [2.31]

$$\Delta G(i) = G(i) - i\mu_v. \quad (2.20)$$

As seen, the work of formation of the cluster represents the difference between the thermodynamic potential of the cluster and the thermodynamic potential of the same number of molecules but in the ambient phase (vapor, solution or melt). This is the most general definition of the work for nucleation. Taking different expressions for $G(i)$ we can approach different cases of nucleation, such as liquid or crystal nuclei, large or small clusters, clusters with or without equilibrium shape, nuclei on like and unlike surfaces, nuclei formed on small particles or ions, etc.

Equation (2.20) is usually illustrated with the simplest case, when the nucleus is a liquid droplet with the (equilibrium) shape of a sphere with radius r surrounded by its own vapor. We assume that the nucleus is sufficiently large that it can be described by macroscopic thermodynamic quantities. This is in fact the classical or capillary approach introduced by Gibbs. He considered nuclei as small liquid droplets, vapor bubbles or crystallites which, however, are sufficiently large to be described by their bulk properties. Although oversimplified, this approach was a significant step ahead because, when phases with small linear sizes are involved, the surface-to-volume ratio is large.

The thermodynamic potential of the spherical droplet reads

$$G(r) = \frac{4\pi r^3}{3v_l}\mu_l^\infty + 4\pi r^2\sigma ,$$

where $i = 4\pi r^3/3v_l$ is the number of atoms in the nucleus.

Writing the expression for $G(r)$ in this way we suppose that a cluster with radius r has the chemical potential μ_l^∞ of the infinitely large liquid phase. The second term accounts for the excess energy owing to the newly formed interface between the liquid droplet and the ambient vapor phase, to which we ascribe a specific energy σ that is characteristic of the bulk liquid phase.

The thermodynamic potential of a crystalline cluster with a cubic shape and lateral extent l in the capillary approximation is given by a similar expression

$$G(l) = -\frac{l^3}{v_c}\mu_c^\infty + 6l^2\sigma . \quad (2.21)$$

Then for the work of nucleus formation in terms of the size l one obtains

$$\Delta G(l) = \frac{l^3}{v_c}\Delta\mu + 6l^2\sigma , \quad (2.22)$$

where $i = l^3/v_c$, and $\Delta\mu = \mu_v - \mu_c^\infty$ is the supersaturation.

The dependence of $\Delta G(l)$ on the size l is plotted in Fig. 2.6. (Note that the growing cluster preserves its equilibrium shape of a cube with increasing linear size l.) As seen, $\Delta G(l)$ displays a maximum when the ambient phase is supersaturated ($\mu_c^\infty < \mu_v$) at some critical size

$$l^* = \frac{4\sigma v}{\Delta\mu} . \quad (2.23)$$

In the opposite case of undersaturated vapor ($\mu_c^\infty > \mu_v$) both terms in (2.22) are positive and the Gibbs free energy change goes to infinity as the density fluctuations are thermodynamically unfavorable.

Equation (2.23) is in fact the familiar equation (2.10) of Thomson–Gibbs. As discussed above the latter represents the condition of equilibrium of a small particle with its ambient phase. It is important to note that *this equilibrium is unstable*. When more atoms join the nucleus, its size increases and its equilibrium vapor pressure becomes smaller than that of the ambient phase. As a result the probability of growth becomes greater than the probability of decay and the nucleus will continue to grow. If several atoms detach from the nucleus, its equilibrium vapor pressure will increase and become higher than that of the ambient phase. The probability of decay will become dominant and the nucleus will decay further. In other words, any infinitesimal deviation of the size of the nucleus from the critical one leads to a decrease of the thermodynamical potential of the system.

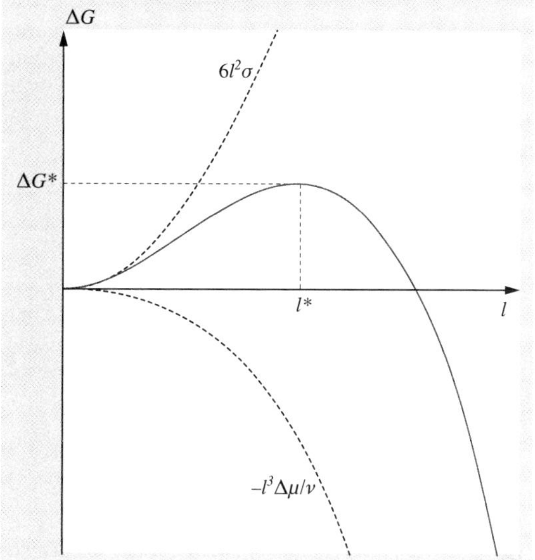

Fig. 2.6 Dependence on the crystal size l (or radius r) of the Gibbs free energy change connected with the formation of a crystalline (liquid) nucleus with a cubic (spherical) shape

Substituting l^* into (2.22) gives the value of the maximum, or in other words, the change of the Gibbs free energy to form the nucleus

$$\Delta G^* = \frac{32\sigma^3 v^2}{\Delta\mu^2} . \quad (2.24)$$

It is inversely proportional to the square of the supersaturation (a result which was obtained for the first time by *Gibbs* in 1878 [2.25]) and increases steeply when approaching the phase equilibrium, thus imposing great difficulties for crystallization to take place.

2.2.2 Formation of 3-D Nuclei on Unlike Substrates

Equation (2.21) gives the thermodynamic potential of a small crystallite with a cubic equilibrium shape whose properties are described in terms of classical macroscopic thermodynamics. In order to relax this restriction Stranski suggested a new approach which can be used for both large crystals and arbitrarily small clusters with arbitrary shape. The thermodynamic potential is given in the more general form

$$G(i) = i\mu_c^\infty + \Phi \,, \tag{2.25}$$

where Φ plays the role of a surface energy.

The work for nucleus formation then reads

$$\Delta G(i) = -i\Delta\mu + \Phi \,. \tag{2.26}$$

According to the definition of *Stranski* the *surface* term is given by [2.32]

$$\Phi = i\varphi_{1/2} - U_i \,, \tag{2.27}$$

where $U_i > 0$ is the energy of disintegration of the whole crystal (or small cluster) into single atoms. In fact $-U_i$ is the potential (binding) energy of the cluster. In the approximation of additivity of bonds energies, U_i is equal to the number of bonds between the atoms of the cluster multiplied by the work ψ to break a single bond.

Equation (2.27) can be easily understood. The first term on the right-hand side gives the energy of the bonds as if all atoms are in the bulk of the crystal (recall that the separation work from the half-crystal position is equal to the lattice energy per atom). The second term gives the energy of the bonds between the atoms of the cluster. Therefore, the difference represents the number of unsaturated (dangling) bonds multiplied by the energy $\psi/2$ of a dangling bond. Obviously, if the cluster is sufficiently large, Φ can be expressed in terms of surface, edge, and apex energies, but as written above it is applicable to arbitrarily small clusters with arbitrary shape.

Combining (2.26) and (2.27) and substituting for $\Delta\mu$ from the Thomson–Gibbs equation (2.8) in atomistic terms in the resulting equation for the Gibbs free energy change for nucleus formation one obtains

$$\Delta G^* = i^*\bar{\varphi}_3 - U_{i^*} \,. \tag{2.28}$$

We can now calculate the work of formation of a nucleus with equilibrium shape shown in Fig. 2.5. In this case $i = n^2n'$ and

$$U_i = 3n^2n'\psi - 2nn'\psi - n^2\psi\phi \,, \tag{2.29}$$

where ϕ is the familiar wetting function (2.17) which determines also the equilibrium shape of a crystal on an unlike substrate.

Combining (2.7), (2.28), and (2.29) gives

$$\Delta G^* = n^{*2}\psi\phi \,, \tag{2.30}$$

where n^* is the number of atoms in the lateral edge of the critical nucleus. Note that $l^* = an^*$ is the length of the edge of the homogeneously formed nucleus in the absence of a substrate or under the condition of complete nonwetting.

We show that (2.30) gives the work of formation of a complete cubic crystallite (2.24) multiplied by the wetting function (2.17), which is positive and smaller than unity in the case of incomplete wetting under study. For this purpose we substitute for n^* and ψ from (2.8) and (2.4), respectively, in (2.30) and obtain ($a^3 = v$)

$$\Delta G^* = \frac{32\sigma^3 v^2}{\Delta\mu^2}\frac{\sigma + \sigma_i - \sigma_s}{2\sigma} \,, \tag{2.31}$$

where the wetting function ϕ is given in terms of surface energies.

It follows that the work for nucleus formation at surfaces (heterogeneous nucleation) is equal to that of the homogeneously formed nuclei in the absence of a surface multiplied by the wetting function. Bearing in mind that

$$\phi = \frac{h}{l} = \frac{l^2h}{l^3} = \frac{V}{V_0} \,,$$

we conclude that the ratio of the works for heterogeneous and homogeneous nucleation is equal to the ratio of the respective volumes in the presence and absence of a substrate

$$\Delta G^*_{\text{het}} = \Delta G^*_{\text{hom}}\frac{V}{V_0} \,.$$

It is interesting to consider the case when a three-dimensional nucleus is formed in the concave edge of a *hill-and-valley* vicinal surface consisting of alternating low-index facets and which is often formed under the effect of adsorbed impurity atoms [2.33,34]. Assuming for simplicity a right angle of the concave edge we find that the nucleus has a prismatic equilibrium shape, having two edges with length $l' = n'a$ and one edge with a length $l = na$. Using the same procedure as before for ΔG^* one obtains

$$\Delta G^* = n^{*2}\psi\phi^2$$

or

$$\Delta G^* = \frac{32\sigma^3 v^2}{\Delta\mu^2}\left(\frac{\sigma + \sigma_i - \sigma_s}{2\sigma}\right)^2 .$$

In the same way we find that the work of formation of a nucleus in a right-angle corner is proportional to the third degree of the wetting function ϕ, etc. As $\phi < 1$ we conclude that a rough surface containing concave edges and corners stimulates nucleation by decreasing the nucleus volume.

2.2.3 Work of Formation of 2-D Crystalline Nuclei on Unlike and Like Substrates

To solve this problem we apply the same procedure, bearing in mind that we have to account for the mean separation work for a two-dimensional square cluster. We consider first the more general case in which the 2-D nucleus is formed on an unlike substrate. Obviously, in order for the 2-D nucleus to be stable the wetting should be complete, although 2-D nuclei can be stable in incomplete wetting but only up to some critical size [2.35]. Beyond this size the monolayer islands become unstable against bilayer islands and should be rearranged into three-dimensional islands as required by the thermodynamics (Sect. 2.6).

The mean separation work calculated for a 2-D square nucleus consisting of $i = n^2$ atoms on unlike substrates reads

$$\bar{\varphi}_2' = 3\psi - \frac{\psi}{n} - \psi\phi$$

and

$$\Delta\mu = \frac{\psi}{n} + \psi\phi \,. \tag{2.32}$$

The binding energy is $U_i = 3n^2\psi - 2n\psi - n^2\psi\phi$ and the Gibbs free energy change reads

$$\Delta G^* = n^*\psi \,. \tag{2.33}$$

Substituting for n^* from (2.32) and ψ from (2.13) in (2.33) gives

$$\Delta G^* = \frac{\psi^2}{\Delta\mu - \psi\phi} = \frac{4\varkappa^2 a^2}{\Delta\mu - a^2(\sigma + \sigma_\mathrm{i} - \sigma_\mathrm{s})} \,. \tag{2.34}$$

In the limiting case of a like substrate (nucleation on the surface of the same crystal) $\Delta\sigma = \sigma + \sigma_\mathrm{i} - \sigma_\mathrm{s} = 0$ and the Gibbs free energy change reads

$$\Delta G^* = \frac{\psi^2}{\Delta\mu} = \frac{4\varkappa^2 a^2}{\Delta\mu} \,. \tag{2.35}$$

Substituting for ψ from the Thomson–Gibbs equation (2.32) in the case of complete wetting, $\phi = 0$, in (2.35) one obtains the very useful result that the work for nucleus formation is precisely equal to the volume part of it

$$\Delta G^* = n^{*2}\Delta\mu = i^*\Delta\mu \,. \tag{2.36}$$

Equations (2.34) and (2.35) lead to some interesting conclusions. In the case of incomplete wetting ($\Delta\sigma > 0$) 2-D nucleation can take place only at supersaturation higher than $\Delta\mu_0 = a^2\Delta\sigma$, because when approaching the latter the work for nucleus formation goes to infinity. In the case of complete wetting ($\Delta\sigma < 0$) both terms in the denominator of (2.34) are positive and 2-D nucleation can take place even at undersaturation. As follows from (2.35) a 2-D nucleation event on the surface of the same crystal ($\Delta\sigma = 0$) can occur only at supersaturations higher than zero.

Equations (2.31) and (2.34) give another critical supersaturation $\Delta\mu_\mathrm{cr} = 2\Delta\mu_0$ at which the 3-D nucleus is reduced to a 2-D nucleus with monolayer height. The reason is that, assuming a constant equilibrium aspect ratio $h/l < 1$, on decreasing the nucleus size with increasing supersaturation a moment comes when the thickness of the 3-D island becomes equal to one monolayer [2.36–38]. As a result three-dimensional nucleation should not take place at supersaturations larger than $\Delta\mu_\mathrm{cr}$. The latter does not contradict the observed layer-by-layer growth of Pb on Ge(001) at 130 K [2.39].

In the end of this subsection we will briefly discuss the very interesting and important question of the existence and formation of one-dimensional nuclei. The latter can be considered as rows of atoms at the edge of a single height step. Using the approach of the mean separation works the equilibrium of a such row of atoms with the ambient phase will be given by the equality of the probabilities of attachment and detachment of atoms to the row's ends. However, the row's ends represent half-crystal positions, so the *mean separation work* reads $\bar{\varphi}_1 = 3\psi = \varphi_{1/2}$ and the supersaturation is $\Delta\mu = \varphi_{1/2} - \bar{\varphi}_1 = 0$. The latter means that a row of atoms has the same chemical potential as the bulk crystal, irrespective of its length. The potential energy of a row consisting of i atoms is $U_i = 3i\psi - \psi$, and the work of formation of a one-dimensional nucleus is $\Delta G_1^* = i\bar{\varphi}_1 - U_i = \psi$. As seen ΔG_1^* does not depend on the row's length, which means that a critical size as in 3-D and 2-D nucleation does not exist. All the above means that one cannot define thermodynamically one-dimensional nuclei. However, as pointed out by several authors, one-dimensional nuclei can be well defined kinetically [2.40–42]. It is in fact the formation of one-dimensional nuclei which allows the propagation of smooth steps, particularly at low temperatures.

We mention here only two cases of great practical importance: the advancement of S_A steps on the surface of Si(001) 2×1 [2.43, 44] and the growth of protein crystals [2.45]. We would like to stress once more that the one-dimensional nucleation is a purely kinetic process and a critical size cannot be defined thermodynamically.

2.3 Rate of Nucleation

As discussed above the equilibrium of a small particle of the new phase with the supersaturated ambient phase is unstable. Accidental detachment of atoms from the critical nucleus can result in a decay of the cluster even to single atoms. Attachment of several atoms could lead to unlimited growth. It is not accidental that the exact solution of the time-dependent problem leads to a diffusion-type equation which reflects the random character of the processes of growth and decay around the critical size [2.30]. We can thus interpret the growth of the clusters as a *diffusion* in the space of the size. We conclude that nucleation is a random process. The steady-state rate of nucleation is a constant quantity which represents an average in time of randomly distributed events.

2.3.1 General Formulation

Becker and *Döring* advanced a purely kinetic approach which allowed them to derive a general expression for the steady-state nucleation rate making the assumptions of: (1) steady-state distribution of the heterophase fluctuations, (2) constant geometrical shape of the growing clusters which coincides with the equilibrium shape, and (3) constant supersaturation which is achieved by removal of clusters which are sufficiently large (much larger than the critical nucleus, $I \gg i^*$) from the system and then are returned back as single atoms [2.46]. The interested reader is referred to the excellent analysis of *Christian* [2.47]. Relaxing assumption 2 did not affect significantly the final result, whereas allowing variable supersaturation changed only the transient character of nucleation but not the steady-state nucleation rate [2.48]. It was in fact the first assumption which played the essential role in solving the problem.

Becker and Döring considered the nucleation process as a series of consecutive bimolecular reactions (a scheme proposed by Leo Szilard)

$$\mathcal{A}_1 + \mathcal{A}_1 \underset{\omega_2^-}{\overset{\omega_1^+}{\rightleftarrows}} \mathcal{A}_2$$

$$\mathcal{A}_2 + \mathcal{A}_1 \underset{\omega_3^-}{\overset{\omega_2^+}{\rightleftarrows}} \mathcal{A}_3$$

$$\dots$$

$$\mathcal{A}_i + \mathcal{A}_1 \underset{\omega_{i+1}^-}{\overset{\omega_i^+}{\rightleftarrows}} \mathcal{A}_{i+1}$$

$$\dots$$

in which the growth and decay of the clusters take place by attachment and detachment of single atoms. Triple and multiple collisions are excluded as less probable. ω_i^+ and ω_i^- denote the rate constants of the direct and reverse reactions. Here $\mathcal{A}$ is used as a chemical symbol.

Clusters consisting of i atoms are formed by the growth of clusters consisting of $i-1$ atoms and the decay of clusters of $i+1$ atoms (birth processes) and disappear by the growth and decay into clusters of $i+1$ and $i-1$ atoms (death processes), respectively. Then the change with time of the concentration $Z_i(t)$ of clusters consisting of i atoms is given by

$$\frac{\mathrm{d}Z_i(t)}{\mathrm{d}t} = J_i(t) - J_{i+1}(t) ,$$

where

$$J_i(t) = \omega_{i-1}^+ Z_{i-1}(t) - \omega_i^- Z_i(t) \tag{2.37}$$

is the net flux of clusters through the size i.

Assuming a steady-state concentration of the clusters in the system, $\mathrm{d}Z_i(t)/\mathrm{d}t = 0$, leads to

$$J_i(t) = J_{i+1}(t) = J_0 ,$$

where we denote by J_0 the time-averaged frequency of formation of clusters of any size. Therefore, J_0 is also equal to the frequency of formation of the clusters with the critical size i^* and thus is equal to the steady-state nucleation rate.

Applying a simple mathematical procedure to the system of rate equations which describe the scheme of Szilard for J_0 one obtains [2.49]

$$J_0 = Z_1 \sum_{i=1}^{I-1} \left(\frac{1}{\omega_i^+} \frac{\omega_2^- \omega_3^- \dots \omega_i^-}{\omega_1^+ \omega_2^+ \dots \omega_{i-1}^+} \right)^{-1} . \tag{2.38}$$

This is the most general expression for the steady-state rate of nucleation. It is applicable to any case of nucleation (homogeneous or heterogeneous, from any ambient phase – vapor, solution or melt, three- or two-dimensional, etc.). It also allows the derivation of equations for the classical as well as the atomistic nucleation rate at small and high supersaturations as limiting cases. The only thing we should know in any particular case are the rate constants ω_i^+ and ω_i^-.

The analysis of (2.38) shows that every term in the sum is equal to $\exp(\Delta G(i)/k_\mathrm{B}T)$, where $\Delta G(i)$ is the work to form a cluster consisting of i atoms [2.50]

$$\frac{\omega_2^-\omega_3^-\ldots\omega_i^-}{\omega_1^+\omega_2^+\ldots\omega_{i-1}^+}=\exp\left(\frac{\Delta G(i)}{k_\mathrm{B}T}\right). \tag{2.39}$$

The condition of an imaginary equilibrium $J_0=0$ applied to (2.37) leads to an equation known in the literature as the equation of *detailed balance*

$$\frac{N_i}{N_{i-1}}=\frac{\omega_{i-1}^+}{\omega_i^-},$$

where N_i denotes the equilibrium concentration of clusters consisting of i atoms. Multiplying the ratios N_i/N_{i-1} from $i=2$ to i gives an expression for the equilibrium concentration of clusters of size i

$$\frac{N_i}{N_1}=\prod_{n=2}^{i}\left(\frac{\omega_{n-1}^+}{\omega_n^-}\right)=\left(\frac{\omega_2^-\omega_3^-\ldots\omega_i^-}{\omega_1^+\omega_2^+\ldots\omega_{i-1}^+}\right)^{-1}. \tag{2.40}$$

Substituting (2.39) into (2.40) gives for the *equilibrium concentration of clusters of size i*

$$N_i=N_1\exp\left(-\frac{\Delta G(i)}{k_\mathrm{B}T}\right). \tag{2.41}$$

We recall that $\Delta G(i)$ displays a maximum at $i=i^*$. It follows that N_i should display a minimum at the critical size.

Substituting (2.39) into (2.38) and replacing the summation by integration valid for large critical nuclei one obtains

$$J_0=\omega^*\Gamma N_{i^*},$$

where $\omega^*\equiv\omega_{i^*}$ is the frequency of attachment of atoms to the critical nucleus, $\Gamma=(\Delta G^*/3\pi k_\mathrm{B}Ti^{*2})^{1/2}$ is the so-called nonequilibrium Zeldovich factor which accounts for neglecting processes taking place far from the critical size, and N_{i^*} is given by (2.41) for the critical nucleus. It is assumed that the equilibrium monomer concentration N_1 is equal to the steady-state concentration Z_1.

In the particular case of nucleation on surfaces we have to account for the configurational entropy of distribution of clusters and single atoms among the adsorption sites of density N_0 ($\approx 1\times10^{15}\,\mathrm{cm}^{-2}$) which should be added to the Gibbs free energy changes (2.31), (2.34) or (2.35) [2.51]. Assuming that the density of clusters is negligible compared with that of single atoms the entropy correction reads

$$\Delta G_\mathrm{conf}\approx -k_\mathrm{B}T\ln\left(\frac{N_0}{N_1}\right).$$

Then for the steady-state nucleation rate on surfaces one obtains

$$J_0=\omega^*\Gamma N_0\exp\left(-\frac{\Delta G^*}{k_\mathrm{B}T}\right), \tag{2.42}$$

where the frequency of attachment of atoms to the critical nucleus ω^* accounts only for the surface diffusion of atoms to the nucleus, the direct impingement from the vapor being neglected [2.52].

As discussed above the capillary nucleation theory is valid at supersaturations which are sufficiently low that the nuclei are large and can be described in terms of the classical thermodynamics. In order to find the limits of validity of (2.42), or in other words, the maximum value of the supersaturation at which the above equation is still valid, we have to find the values of the pre-exponential $K=\omega^*\Gamma N_0$ and ΔG^* and calculate the time τ elapsed from *switching on* the supersaturation to the appearance of the first nucleus. The latter is given by $\tau=1/J_0S$, where S is the area available for nucleation.

Consider for simplicity 2-D nucleation on the surface of the same crystal. The frequency of attachment of atoms to the critical nucleus ω^* is given by the product of the periphery of the nucleus and the flux of adatoms joining the nucleus. We assume that the nucleus consists of at least 49 atoms (a square of 7×7 atoms) in order for the classical theory to be valid. The flux of adatoms to the periphery is $j_\mathrm{s}\approx D_\mathrm{s}N_1/a$, where $D_\mathrm{s}=a^2\nu\exp(-E_\mathrm{sd}/k_\mathrm{B}T)$ is the surface diffusion coefficient, and the adatom concentration is determined by a dynamic adsorption–desorption equilibrium and is given by $N_1=R\tau_\mathrm{s}$. The reason for using this definition is that it is supposed that the temperature is sufficiently high to ensure low supersaturation and the desorption flux N_1/τ_s is significant. Here ν is the attempt frequency and E_sd and E_des are the activation barriers for surface diffusion and desorption, respectively. Taking appropriate values for the parameters involved we find a value for the pre-exponential of the order of $10^{20}-10^{25}\,\mathrm{cm}^{-2}\,\mathrm{s}^{-1}$ for nucleation from vapor. We can further evaluate the supersaturation by using (2.11).

Once we know the supersaturation we can easily evaluate ΔG^* by making use of (2.36).

We consider as an example nucleation on Si(001) at $T = 1500\,\mathrm{K}$ and assume that $S = 1\,\mathrm{cm}^2$, although a more realistic value could be determined from the width of the terraces on the crystal surface. From the enthalpy of evaporation we deduce the bond strength to be of the order of 2–2.2 eV. Then $\Delta\mu \approx 0.3\,\mathrm{eV}$, $\Delta G^* = 15\,\mathrm{eV}$, and $\tau \approx 1\times 10^{15}$ millennia. This behavior of the classical nucleation rate was noticed by *Dash*, who noted that nucleation on defectless crystal surfaces according to the classical theory requires *astronomically* long times [2.53]. The reason for this behavior is that the pre-exponential in J_0 is a very weak function of the supersaturation compared with the exponential $\exp(-\Delta G^*/k_\mathrm{B}T)$, which varies very steeply with the latter. As a result there is a critical supersaturation below which the rate of nucleation is practically equal to zero and beyond which it takes values of many orders of magnitude (Fig. 2.7). We conclude that, in order for a nucleation event to take place on a laboratory scale of time, $\Delta G^*/k_\mathrm{B}T$ should be smaller than ≈ 30 (in the case under consideration it is 4 times larger). This means that, for most materials at working temperatures between 600 and 1000 K, *the number of atoms in the critical nucleus should be of the order of unity*. This is why we will develop in more detail the atomistic theory of nucleation valid for nuclei consisting of very small number of atoms. It is important to note that a small value (usually not larger than ten) of the number of atoms in the critical nucleus should be expected also in the case of three-dimensional nucleation. A value of $i^* = 9$ was obtained in the case of nucleation of $CoSi_2$ from amorphous Co-Si alloy [2.54]. The reason for the comparatively larger size is due to the much greater value of the pre-exponential, which in this particular case is on the order of 10^{35}–$10^{40}\,\mathrm{cm^{-3}\,s^{-1}}$ [2.21].

Fig. 2.7 Plot of the nucleation rate versus the supersaturation. The nucleation rate is practically equal to zero up to a critical supersaturation $\Delta\mu_c$. Beyond this value the rate of nucleation increases sharply by many orders of magnitude

2.3.2 Rate of Nucleation on Single-Crystal Surfaces

Single-crystal surfaces always represent vicinal surfaces consisting of terraces divided by steps due to the tilt of the surface by some small angle with respect to the low-index (singular) crystal face. Numerous processes can take place during deposition on the terraces (Fig. 2.8). We consider first the case of complete wetting. Atoms arrive from the vapor and accommodate thermally with the substrate [2.55], diffuse on the crystal surface, and re-evaporate if the temperature is sufficiently high. The atoms can also join pre-existing steps and diffuse along these steps to incorporate into kink sites. The reverse process of detachment of atoms from kink sites directly to the terrace or through the intermediate state of adsorption at the step edge can also take place. Thus when the temperature is sufficiently high the crystal grows by propagation of the pre-existing steps. If the temperature is low and the atom diffusivity is small the atoms cannot reach the steps and collide with other atoms to produce dimers. The dimers can grow further to produce trimers, tetramers, and finally large islands by attachment of new adatoms, or can decay into single atoms. Arriving atoms will preferably join the islands in a later stage of growth, the formation of new dimers being inhibited. Thus we can distinguish two

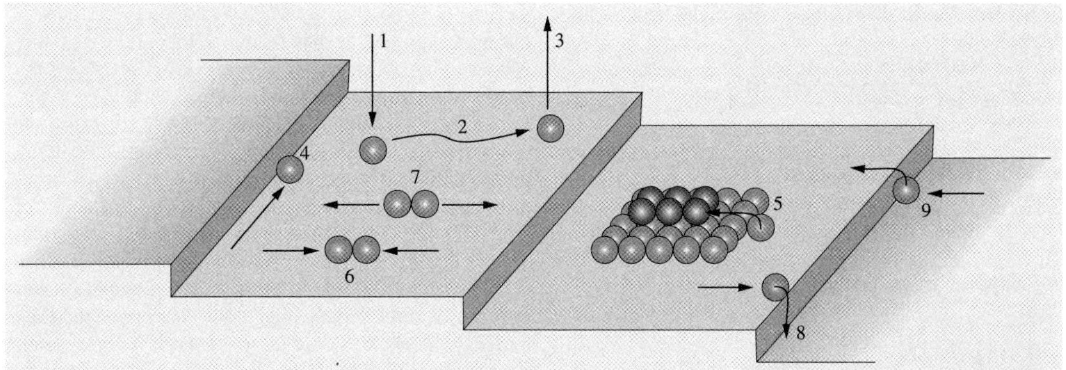

Fig. 2.8 Schematic representation of the different processes which can take place on surfaces during deposition on like and unlike substrates: 1 – adsorption, 2 – surface diffusion, 3 – desorption, 4 – edge diffusion, 5 – transformation of monolayer to bilayer island in heteroepitaxy, 6 – dimer formation, 7 – dimer decay, 8 – step-down hopping, 9 – step-up jump

regimes of growth: step flow growth at high temperatures and growth by two-dimensional nucleation at low temperatures.

In the case of incomplete wetting which favors three-dimensional clustering all the processes listed above remain the same with the exception that step flow growth does not take place (we consider the case of heteroepitaxy with $\psi > \psi'$); nucleation occurs at all temperatures. The mechanism of formation of 3-D clusters depends strongly on the wetting. In the extreme of very weak wetting (metals on alkali halides) visible clustering is observed from the very beginning of deposition. When the wetting is stronger as in the technologically important cases of metals on metals or semiconductors on semiconductors, two-dimensional islands are initially energetically favored but become unstable and transform beyond some critical size into 3-D clusters (Fig. 2.8) [2.35]. The same is observed in Stranski–Krastanov growth beyond the wetting layer [2.56, 57]. Thus in the beginning of deposition the overlayer can be considered as a population of molecules of different size, most of which are one atom high [2.58].

2.3.3 Equilibrium Size Distribution of Clusters

We calculate first the equilibrium concentration of the clusters of size i. The thermodynamic potential of the cluster of size i is given by (2.25), where i is an integer which can be arbitrarily small. Bearing in mind (2.26) and (2.27) the work for nucleus formation reads

$$\Delta G(i) = G(i) - i\mu_\mathrm{v} = i(\varphi_{1/2} - \Delta\mu) - U_i\,. \quad (2.43)$$

Assuming the adlayer consisting of clusters of different size behaves as a two-dimensional ideal gas ($\sum_i N_i \ll N_0$) the thermodynamic potential of the population of clusters of size i will be [2.59]

$$\mathcal{G}(N_i) = N_i G(i) - k_\mathrm{B} T \ln \frac{N_0!}{(N_0 - N_i)! N_i!}\,.$$

Then for the chemical potential of the two-dimensional ideal gas of clusters of size i one obtains

$$\mu_i = \frac{\mathrm{d}\mathcal{G}(N_i)}{\mathrm{d}N_i} = G(i) - k_\mathrm{B} T \ln\left(\frac{N_0}{N_i}\right)\,. \quad (2.44)$$

Suppose now that the pressure of the vapor is precisely equal to the equilibrium vapor pressure of the infinitely large crystal at the given temperature so that $\mu_i = i\mu_\mathrm{c}^\infty$. The system is in a true equilibrium and the nucleation rate is precisely equal to zero. Rearranging (2.44) and inserting the above equality gives for the equilibrium concentration of i-atomic clusters

$$\frac{N_i^\mathrm{e}}{N_0} = \exp\left(-\frac{G(i) - i\mu_\mathrm{c}^\infty}{k_\mathrm{B} T}\right)\,.$$

Assume now that the vapor pressure is higher than the equilibrium vapor pressure so that $\mu_i = i\mu_\mathrm{v} > i\mu_\mathrm{c}^\infty$. The system will be supersaturated and the nucleation rate will differ from zero. We apply as before the artificial condition $J_0 = 0$, which determines a hypothetical equilibrium concentration of clusters of size i

$$\frac{N_i}{N_0} = \exp\left(-\frac{G(i) - i\mu_\mathrm{v}}{k_\mathrm{B} T}\right)\,.$$

Substituting for $G(i)$ from (2.43) in the above equation gives

$$\frac{N_i}{N_0}=\exp\left(-\frac{i\varphi_{1/2}-i\Delta\mu-U_i}{k_B T}\right). \quad (2.45)$$

The condition $i=1$ yields the density of monomers

$$\frac{N_1}{N_0}=\exp\left(-\frac{\varphi_{1/2}-\Delta\mu-U_1}{k_B T}\right),$$

the i-th power of which reads

$$\left(\frac{N_1}{N_0}\right)^i=\exp\left(-\frac{i\varphi_{1/2}-i\Delta\mu-iU_1}{k_B T}\right). \quad (2.46)$$

Dividing (2.45) and (2.46) gives for this hypothetical equilibrium concentration of clusters of size i [2.58]

$$\frac{N_i}{N_0}=\left(\frac{N_1}{N_0}\right)^i\exp\left(\frac{E_i}{k_B T}\right), \quad (2.47)$$

where $E_i=U_i-iU_1$ is the net energy gained to form an i-atom cluster from i single atoms. Bearing in mind that U_1 is, in fact, the adhesion energy ψ', E_i is the potential (binding) energy of the lateral bonds in the cluster. The latter means that the value of E_i does not depend (within the framework of the approximation of the additivity of the bond energies) on the material of the substrate. It should be one and the same on like and unlike substrate crystals. Recall that we defined U_i as a positive quantity. This means that E_i is also positive. As $N_1/N_0 \ll 1$ the pre-exponential decreases whereas the exponential increases with i. It follows that (2.47) should display a minimum at some critical size or, in other words, will have the same qualitative behavior as the classical equilibrium size distribution (2.41).

2.3.4 Rate of Nucleation

An approximate expression for the nucleation rate can be obtained by multiplying (2.47) by the flux of atoms to the critical nucleus. Note, however, that in the case of small clusters the classical definition of a nucleus as a cluster with equal probabilities for growth and decay, each one equal to 0.5, is not valid. The nucleus should be defined as the cluster whose probability of growth is smaller than or equal to 0.5, but which after attachment of one more atom will have a probability of growth greater than or equal to 0.5 [2.58]. The latter is called the *smallest stable cluster*. Thus the nucleation rate is the rate at which clusters of critical size become *supercritical* or smallest stable clusters.

It is clear that for small clusters the requirement of constant geometrical shape required by the classical theory is violated. An analytical expression for i^* cannot be derived and the nucleus structure should be determined by a trial-and-error procedure by estimating the binding energy of the different configurations including the possibility of formation of three-dimensional structures. Let us consider as an example the formation of nuclei on the (111) surface of a face-centered cubic (fcc) metal (Fig. 2.9). At $\Delta\mu=3.25\psi$ the critical nucleus consists of two atoms and the smallest stable cluster consists of three atoms (Fig. 2.10). The work required to decay the nucleus is equal to the work to break a single first-neighbor bond, whereas in order to detach an atom from the smallest stable cluster we have to break simultaneously two first-neighbor bonds. This means that the latter will be much more stable than the nucleus and a higher temperature is required to decay the three-atom cluster. The attachment of additional atoms up to $i=6$ does not change the stability of the respective clusters. Then at $\Delta\mu=2.75\psi$ the nucleus consists of six atoms and the smallest stable cluster represents

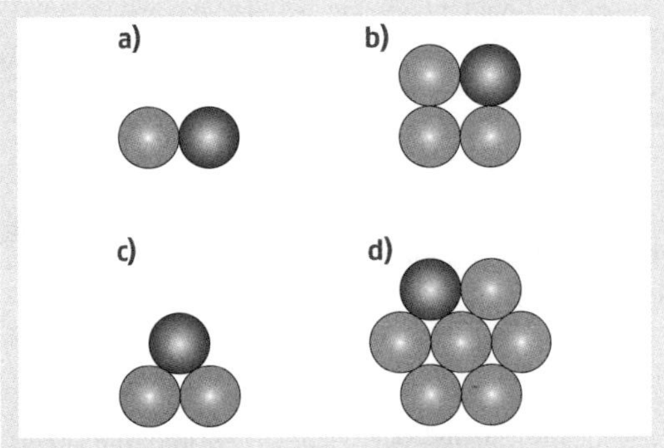

Fig. 2.9a–d Two-dimensional clusters on (001) and (111) surfaces of a crystal with a face-centered cubic lattice. The structure of the nuclei is given by the *gray circles*. The *black circles* denote the atoms that turn the critical nuclei into smallest stable clusters. **(a)** The nucleus consists of a single atom; the stable supercritical cluster is a dimer, which requires a single bond to be broken in order to decay. In **(b)** the nucleus consists of three atoms situated on the apexes of a rectangular triangle on (001) surface; the smallest stable cluster has a square shape. The decay of the latter requires the simultaneous breaking of two bonds. On (111) surface the nuclei consist of **(a)** one, **(c)** two, and **(d)** six atoms. The corresponding stable clusters consist of two, three, and seven atoms, respectively, which require breaking of one, two, and three bonds

a closed structure consisting of a complete ring of six atoms plus an atom in the middle. In order to detach an atom from the smallest stable cluster we have to break simultaneously three first-neighbor bonds. Obviously, such a cluster will be stable at much higher temperatures than a three-atom cluster.

Bearing in mind that every term in the sum of (2.38) is equal to $\exp(\Delta G(i)/k_B T)$ we study the behavior of the latter for small values of i (Fig. 2.10). It is seen that at extremely high supersaturations (low temperatures) $\Delta G(i)$ and $\exp(\Delta G(i)/k_B T)$ are represented by broken curves whereas at low supersaturations (large nuclei) the curve is smooth. Contrary to the classical case where the clusters in the vicinity of the critical size have values of $\exp(\Delta G(i)/k_B T)$ close to that of the nucleus, in the case of small clusters the contribution of $\exp(\Delta G(i^*)/k_B T)$ of the critical nucleus is the largest, all other terms in the sum of the denominator being negligible. Thus, instead of summing all the terms as in the classical theory, we can take the largest term and neglect all the others. For this purpose we write (2.38) in the form

$$J_0 = \omega_1^+ N_1 \left(1 + \frac{\omega_2^-}{\omega_2^+} + \frac{\omega_2^- \omega_3^-}{\omega_2^+ \omega_3^+} + \frac{\omega_2^- \omega_3^- \omega_4^-}{\omega_2^+ \omega_3^+ \omega_4^+} + \ldots\right)^{-1} \quad (2.48)$$

and calculate the rate constants for the birth and death processes.

By analogy with the classical theory, where $\omega_i^+ \approx (P_i/a) D_s N_1$, P_i being the perimeter of the nucleus and P_i/a the number of the dangling bonds, in the atomistic approach [2.60]

$$\omega_i^+ = \alpha_i D_s N_1 ,$$

where α_i is the number of ways of attachment of an atom to a cluster of size i to produce a cluster of size $i+1$. Obviously, this parameter is proportional to the number of dangling bonds.

The decay constant reads

$$\omega_i^- = \beta_i \nu \exp\left(-\frac{E_i - E_{i-1} + E_{sd}}{k_B T}\right) , \quad (2.49)$$

where E_i is the work to disintegrate a cluster of size i into single atoms, and $E_i - E_{i-1}$ is the work required to detach an atom from the cluster of size i. β_i is the number of ways of detachment of an atom from a cluster of size i. It is easy to show that there exists a one-to-one correspondence between the growth ($i \to i+1$) and decay ($i+1 \to i$) processes so that

$$\alpha_i = \beta_{i+1} .$$

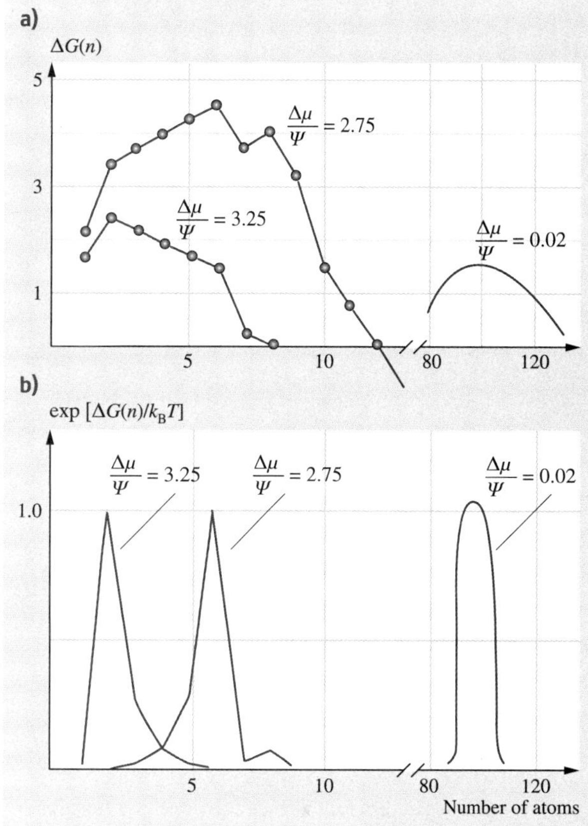

Fig. 2.10a,b Dependence of (**a**) the Gibbs free energy change $\Delta G(i)/\psi$ in units of the work ψ required to break a first-neighbor bond, and (**b**) $\exp(\Delta G(i)/k_B T)$ on the number of atoms i in the cluster at different values of the supersaturation. At small supersaturation ($\Delta\mu = 0.02\psi$) the cluster is large, the respective curves are smooth, and the summation can be replaced by integration. At very large supersaturations the curves are broken and the contribution of the critical nucleus is dominant

Recalling the expression for the diffusion coefficient $D_s = a^2 \nu \exp(-E_{sd}/k_B T)$ we can write (2.49) in the form

$$\omega_i^- = \beta_i D_s N_0 \exp\left(-\frac{E_i - E_{i-1}}{k_B T}\right) ,$$

where $N_0 \cong a^{-2}$.

The assumption that all terms in the denominator in (2.48) are smaller than unity means that $i^* = 1$

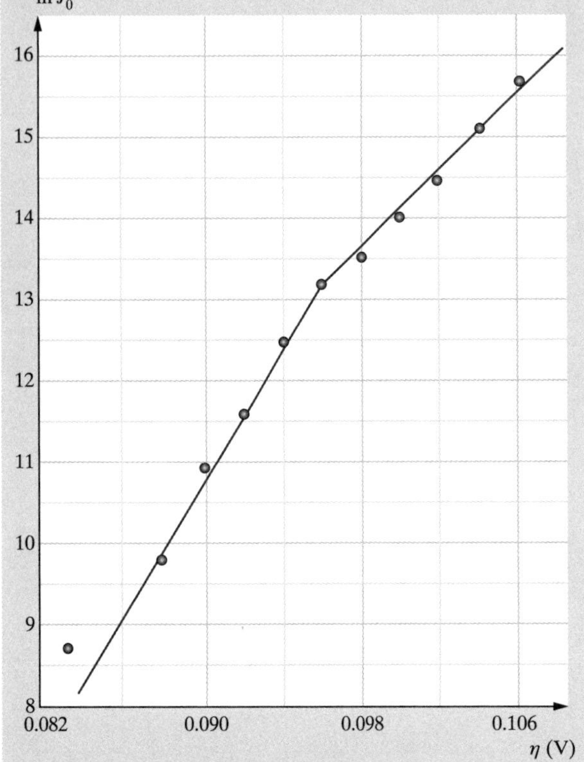

Fig. 2.11 Experimental data for the nucleation rate as a function of the overpotential η in the case of electrochemical nucleation of mercury on platinum single-crystal spheres (after [2.61]), in atomistic coordinates $\ln J_0 - \eta$, according to [2.62]. The number of atoms in the critical nucleus changes at about 0.096 V

($E_1 = 0$) and

$$J_0 = \omega_1^+ N_1 = \alpha_1 D_s N_1^2 .$$

Assuming that the adatom concentration is determined by a dynamic adsorption–desorption equilibrium $N_1 = R\tau_s$ as before, for J_0 one obtains

$$J_0 = \alpha_1 \frac{R^2}{N_0 \nu} \exp\left(\frac{2E_{des} - E_{sd}}{k_B T}\right).$$

When the ratio ω_2^- / ω_2^+ is the largest term in the denominator of (2.48), $i^* = 2$ and

$$J_0 = \omega_1^+ N_1 \frac{\omega_2^+}{\omega_2^-} = \alpha_2 D_s^2 N_1^3 \nu^{-1} \exp\left(\frac{E_2 + E_{sd}}{k_B T}\right)$$

or

$$J_0 = \alpha_2 \frac{R^3}{N_0^2 \nu^2} \exp\left(\frac{E_2 + 3E_{des} - E_{sd}}{k_B T}\right).$$

In the general case

$$J_0 = \alpha^* R \left(\frac{R}{N_0 \nu}\right)^{i^*} \times \exp\left(\frac{E_{i^*} + (i^* + 1)E_{des} - E_{sd}}{k_B T}\right).$$

Very often the process of re-evaporation is negligible (complete condensation) and $N_1 \neq R\tau_s$. Then we can write J_0 in terms of the adatom concentration in the form

$$J_0 = \alpha^* D_s \frac{N_1^{i^*+1}}{N_0^{i^*-1}} \exp\left(\frac{E_{i^*}}{k_B T}\right), \tag{2.50}$$

which is very useful for solving various nucleation problems.

Whereas the attachment or detachment of atoms to and from a comparatively large liquid droplet or crystallite can be considered as a good approximation to a continuous process, this is impossible when the cluster consists of several atoms. In this case the general principles of the thermodynamics are violated, the best example of which is that the Thomson–Gibbs equation is not valid in its familiar form (2.10). The reason becomes obvious if we write it in terms of the number of atoms rather than the linear size of the crystallite

$$\frac{P_i}{P_\infty} = \exp\left(\frac{4\sigma v^{2/3}}{k_B T i^{1/3}}\right).$$

It is immediately seen that the vapor pressure in the left-hand side of the equation can be continuously varied whereas the right-hand side is a discrete function of the cluster size i. The latter means that to any particular size of the cluster corresponds a fixed value of the vapor pressure, but the opposite is not true; an integer number of atoms does not correspond to any arbitrary value of the vapor pressure. It follows that, contrary to the classical concept, a cluster with an integer number of atoms is stable in an interval of supersaturation (or vapor pressure) which becomes larger as the cluster size becomes smaller [2.63]. This interval is equal to $P_i - P_{i+1}$, where P_i is the fixed value of the vapor pressure corresponding to a cluster consisting of i atoms.

Substituting for ΔG^* from (2.26) with $i = i^*$ in (2.42) gives

$$J_0 = \omega^* \Gamma N_0 \exp\left(-\frac{\Phi}{k_B T}\right) \exp\left(i^* \frac{\Delta\mu}{k_B T}\right).$$

As the shape does not change at constant number of atoms the *surface* part Φ remains constant in the interval of stability of a given cluster size. Then the logarithm of the nucleation rate as a function of the supersaturation will represent a broken line when the supersaturation interval is sufficiently wide to cover the intervals of several cluster sizes. The slopes of the consecutive straight line parts will be equal to the respective number of atoms i^* of the critical nuclei. This is shown in Fig. 2.11, which represents experimental data for the nucleation rate in electrodeposition of mercury on platinum single-crystal spheres [2.61], interpreted in terms of the atomistic theory in [2.62] (see also [2.64]). The values $i^* = 6$ and 10 have been found from the slopes of the two parts of the plot. A clear evidence for a transition from $i^* = 1$ to $i^* = 3$ has been reported by *Müller* et al. in the case of nucleation of Cu on Ni(001) [2.65]. Thus a single nucleus size is operative over a temperature (supersaturation) interval. The slopes of the consecutive intervals give a distinct series of consecutive numbers of atoms which depend on the crystallographic orientation of the substrate. Thus in the case of nucleation of (001) surface of fcc metals the numbers are one and three, whereas on (111) surface the numbers are one, two, and six. The corresponding smallest stable clusters ($i^* + 1 = 2, 3, 7$ on the fcc(111) surface) are often referred to as *magic* in the literature. The physics behind this magic is simple. In order to detach an atom from the corresponding smallest stable clusters we have to break simultaneously one, two or three bonds.

2.4 Saturation Nucleus Density

Measurements of the nucleus density as a function of time show that, after sufficiently long time, the nucleus density saturates; this means that the nucleation process ceases. Numerous factors can be responsible for this phenomenon. Preferred nucleation on defect sites, overlapping of zones with reduced supersaturation around growing islands, coalescence of neighboring islands, and growth of larger islands at the expense of smaller ones owing to the Thomson–Gibbs effect (Ostwald ripening) take place most frequently and are most studied [2.66].

Although the preparation of defectless single crystals is already a routine procedure, the complete absence of impurity particles, stacking faults, twin boundaries, emerging points of dislocations, etc. cannot be achieved. It is this presence of defects on the crystal surface which is one of the reasons for the observation of saturation of the nucleus density with time and this was the first to be studied. The defects represents sites on the crystal surface which stimulate nucleation by stronger wetting. Assume for simplicity that they have equal activity (wetting function). Nuclei can form on free active sites whose number is $N_\mathrm{d} - N$ with a frequency J_0' per site, N_d being the total number of active sites. Then the change with time t of the nucleus density reads [2.67]

$$\frac{\mathrm{d}N}{\mathrm{d}t} = J_0'(N_\mathrm{d} - N) .$$

Integration subject to the initial condition $N(0) = 0$ results in a simple exponential function

$$N(t) = N_\mathrm{d}\left[1 - \exp\left(-J_0't\right)\right] ,$$

which tends with time to a saturation value equal to N_d. In the more realistic case of a certain activity distribution of the sites, increasing supersaturation will lead to inclusion of less-active sites in the process and increase of the saturation nucleus density [2.68].

Another reason for saturation of the nucleus density is the appearance of *locally undersaturated* zones around growing nuclei where the nucleation rate is reduced or even equal to zero owing to the consumption of the diffusing adatoms [2.69–71]. *Sigsbee* coined for these zones the term *nucleation exclusion zones* [2.72]. They are also known as *denuded* or *depleted* zones. Nuclei and in turn denuded zones around them are progressively formed and grow during film deposition. When the zones overlap and cover the whole substrate surface the process of nucleation is arrested and saturation of the nucleus density is reached. The radii of the nucleation exclusion zones are defined by the intersection of the gradient of the adatom concentration around the growing island and the critical adatom concentration (or supersaturation) for nucleation to occur (Fig. 2.12). A typical nucleation exclusion zone around a mercury droplet electrodeposited on a platinum single-crystal sphere is shown in Fig. 2.13 [2.73].

The problem of finding the nucleus density when the latter is limited by nucleation exclusion zones has been treated by many authors, such as *Kolmogorov*, *Avrami*, and *Johnson* and *Mehl*, and solutions for different cases have been found [2.74–78] (for a review see [2.47]). The simultaneous influence of both nucleation exclusion zones and active sites has also been

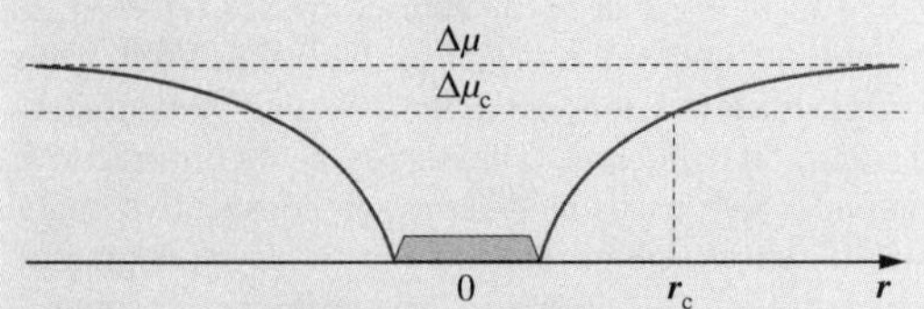

Fig. 2.12 The definition of nucleation exclusion zones. The radius of the latter is determined by the intersection of the gradient of the supersaturation and the critical supersaturation for noticeable nucleation to occur. Because of the very steep dependence of the nucleation rate on the supersaturation (Fig. 2.7) the nucleation rate inside the zone is assumed equal to zero

addressed [2.79,80]. The problem consists of finding the area $\Theta(t)$ uncovered by depleted zones and thus available for nucleation at a moment t. The number of nuclei is then given by

$$N = J_0 \int_0^t \Theta(\tau)\mathrm{d}\tau .$$

The area $1-\Theta(t)$ represents the sum of all nucleation exclusion zones accounting for the area where neighboring zones have overlapped. The latter is equal to the probability of finding an arbitrary point simultaneously in two or more nucleation exclusion zones [2.74]. Assuming that nuclei are formed on randomly distributed sites with a rate J_0 and that the zones grow with a velocity $v(t) = ck(t)$ the area $\Theta(t)$ is given by [2.74]

$$\Theta(t) = \exp\left(-J_0 \int_0^t S'(t')\mathrm{d}t'\right) ,$$

where

$$S'(t', t) = \pi c^2 \left(\int_{t'}^t k(\tau - t')\mathrm{d}\tau\right)^2$$

is the area of a nucleation exclusion zone at a moment t around a nucleus formed at a moment $t' < t$.

Assuming linear growth of the zones ($k(t) = 1$) gives for the nucleus density as a function of time

$$N(t) = J_0 \int_0^t \exp\left(-\frac{\pi}{3} J_0 c^2 t^3\right) \mathrm{d}t . \quad (2.51)$$

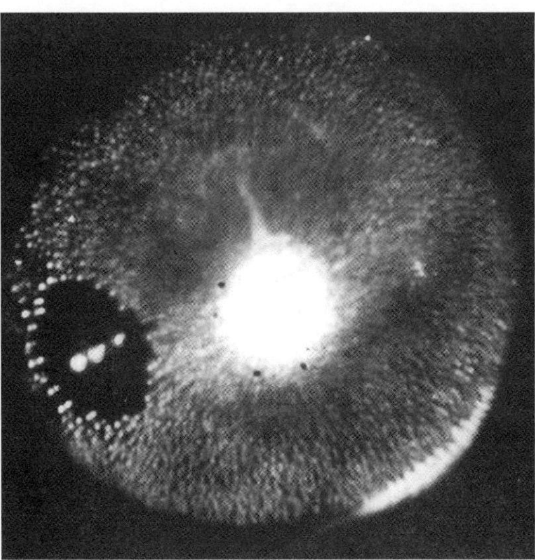

Fig. 2.13 Nucleation exclusion zone around a mercury droplet electrodeposited on a platinum single-crystal sphere. The droplet is practically invisible. Instead, three light reflections from the illuminating lamp are visible. The mercury droplet has been deposited by applying a short electric pulse followed by a lower overpotential in order to grow it to a predetermined size. Then a high electric pulse is applied to cover the whole surface with mercury with the exception of the area around the droplet (after [2.73])

The saturation nucleus density is obtained under the condition $t \to \infty$. Integrating (2.51) from zero to infinity gives

$$N_{\text{sat}} \cong 0.9\left(\frac{J_0}{c}\right)^{2/3} .$$

Another approach was later developed, particularly for nucleation at surfaces, by using a system of kinetic rate equations. It was first introduced by Zinsmeister as a system of equations for the change with time of the concentrations of clusters $\mathrm{d}N_i/\mathrm{d}t$ ($i = 1, 2, 3, \ldots$) for each cluster size, beginning with that of single adatoms [2.81–84]. All birth and death processes were accounted for in $\mathrm{d}N_i/\mathrm{d}t$. In addition, the atom arrival rate and re-evaporation were taken into account in the equation of change of the monomers $\mathrm{d}N_1/\mathrm{d}t$. In order to solve quantitatively the above system of equations the attachment and detachment frequencies had to be determined. As a result a large amount of papers have been devoted to further elaborating the approach [2.85–92]. In the limit $i^* = 1$ (*irreversible aggregation*) the detachment frequencies are equal to zero. The attachment frequencies (capture numbers) were considered by using different approximations, beginning from the mean-field approximation by assuming that the clusters are immersed and grow in a dilute adlayer with an average concentration that does not depend on the location of the clusters, to solutions of diffusion equation around the growing islands in terms of Bessel functions. The system was later greatly simplified by *Venables* et al. to a system of two equations which were sufficient to illustrate the essential physics [2.93].

We consider first the case of irreversible aggregation. The dimers are assumed to be stable (a third atom joins the dimer before the latter to decay) and immobile. The atoms arrive at the crystal surface, diffuse on it, and collide with each other to produce dimers. Atoms join the dimers and larger clusters upon striking without any obstacle of kinetic origin. This means that the growth of clusters is limited only by the surface diffusion. Coalescence of immobile clusters is ruled out. The detachment frequencies are equal to zero and the capture numbers are omitted for simplicity as they represent figures of the order of unity [2.93]. The system of equations is then reduced to

$$\frac{\mathrm{d}N_1}{\mathrm{d}t} = F - 2DN_1^2 - DN_1N_\mathrm{s} , \tag{2.52a}$$

$$\frac{\mathrm{d}N_\mathrm{s}}{\mathrm{d}t} = DN_1^2 , \tag{2.52b}$$

where $F = R/N_0$ is the atom arrival rate in units of number of monolayers, $D = D_\mathrm{s}/a^2 = \nu\exp(-E_\mathrm{sd}/k_\mathrm{B}T)$ is the diffusion (hopping) frequency, and N_s is the sum of all stable clusters

$$N_\mathrm{s} = \sum_{i=2}^{\infty} N_i .$$

Single atoms arrive on the surface with frequency F and are consumed by the formation of dimers (the second term on the right-hand side of (2.52a)) and by incorporation into stable clusters (the third term on the right-hand side of (2.52a)). At the very beginning of deposition most of the adatoms are consumed by the formation of dimers. In a later stage of deposition the density of stable clusters increases and the arriving atoms preferentially join stable clusters rather than colliding with each other to produce dimers. Saturation (or very weak dependence on time) is reached and the consumption of atoms by formation of dimers $2DN_1^2$ is practically arrested and becomes negligible compared with the growth term DN_1N_s. A steady state is reached at this stage ($\mathrm{d}N_1/\mathrm{d}t = 0$) and $N_1 = F/DN_\mathrm{s}$. Substituting the latter into (2.52b) and carrying out the integration gives

$$N_\mathrm{s} \propto \left(\frac{D}{F}\right)^{1/3} .$$

This result is easy to generalize for the case of *reversible aggregation*, assuming the critical nucleus consists of $i^* > 1$ atoms. Then one can write a system of two kinetic equations for the single adatoms and the sum of all clusters larger than i^* [2.93]

$$\frac{\mathrm{d}N_1}{\mathrm{d}t} = F - (i^*+1)DN_1^{i^*+1} - DN_1N_\mathrm{s} , \tag{2.53a}$$

$$\frac{\mathrm{d}N_\mathrm{s}}{\mathrm{d}t} = \omega^* D n_1^{i^*+1} , \tag{2.53b}$$

where $\omega^* = \alpha^* \exp(E^*/k_\mathrm{B}T)$ (see (2.50)).

Following the same procedure as above results in

$$N_\mathrm{s} \propto \left(\frac{D}{F}\right)^{-\chi} , \tag{2.54}$$

where

$$\chi = \frac{i^*}{i^*+2} \tag{2.55}$$

is the scaling exponent valid for the case of diffusion-limited nucleation and growth in the absence of any kinetic barrier inhibiting the attachment of atoms to the critical nucleus.

Later *Kandel* relaxed the condition for diffusion-limited regime of growth, assuming that a barrier exists which inhibits the attachment of atoms to any cluster including the critical nucleus [2.94]. Then the frequency ω^* for collision of atoms with the critical nucleus should contain the term $\exp(-E_\mathrm{b}/k_\mathrm{B}T)$, where E_b is the barrier concerned. He integrated (2.53b) taking for N_1 a value calculated by the solution of a diffusion equation from the radius R of the nucleus to half of the mean distance $L = 1/\sqrt{\pi N_\mathrm{s}}$ between the nuclei and then averaged from R to L. As a result the average adatom concentration included two terms

$$N_1 = A\frac{F}{D}\frac{1}{N_\mathrm{s}} + B\frac{F}{D}\frac{1-\exp(-E_\mathrm{b}/k_\mathrm{B}T)}{\exp(-E_\mathrm{b}/k_\mathrm{B}T)}\frac{1}{\sqrt{N_\mathrm{s}}} ,$$

where A and B are constants.

The first term is inversely proportional to N_s as before and does not include the cluster edge barrier E_b. The second term is inversely proportional to the square root of N_s and includes the barrier E_b. Obviously, when $E_\mathrm{b} = 0$ the second term is equal to zero and the integration of (2.53b) naturally gives the scaling exponent (2.55). In the other extreme of significant cluster edge barrier the second term dominates and the integration of (2.53b) gives the same power-law dependence (2.54) but with a scaling exponent

$$\chi = \frac{2i^*}{i^*+3}\,, \tag{2.56}$$

which is valid for a kinetic regime of growth.

Equation (2.54) shows a simple power-law dependence of N_s on the ratio D/F of the frequency of surface diffusion to the frequency of atom arrival. While F represents the increase of atoms with time, D introduces the fluxes of disappearance of atoms due either to formation of nuclei or to the further growth of these nuclei. Physically this is the ratio of the flux of consumption of atoms on the crystal surface to the flux of their arrival. A constant ratio D/F means a constant adatom concentration or a constant supersaturation. The increase of D/F can be performed by either increasing the temperature or decreasing the atom arrival rate. The fact that the island density scales with D/F simply means that it depends on the supersaturation. The island density should have one and the same value at a given value of D/F, irrespective of whether it is a result of increasing (decreasing) of temperature or decreasing (increasing) of the atom arrival rate. Increasing D/F means decreasing the supersaturation, which in turn leads to an increase of the nucleus size i^*. Thus, at sufficiently low values of D/F of the order of 10^4–10^5, i^* is expected to be equal to one, whereas at D/F of the order of 10^7–10^8, i^* is expected to be equal to three on a square lattice [2.95]. Assuming a constant atom arrival rate of the order of 10^{-2} monolayers per second, attempt frequency of the order of $1\times10^{13}\,\mathrm{s}^{-1}$, and a surface diffusion barrier of 0.75 eV an increase of D/F by four orders of magnitude is equivalent to a temperature increase of 200 K.

It should be noted that considering the size of the critical nucleus as an integer above which all clusters are stable is an approximation which strongly simplifies the mathematical treatment of the problem [2.95]. In fact there are never fully stable clusters. Atoms can always detach from them, particularly at high values of D/F or high temperatures. Things look better at low temperatures when bond breaking is strongly inhibited.

The scaling exponent (2.55) varies with i^* from 1/3 to 1, whereas (2.56) has values larger than unity already at $i^* > 2$. Thus, one can distinguish between diffusion and kinetic regimes of growth if χ is smaller or greater than unity. Examples of the scaling exponent (2.56) have been reported in surfactant-mediated epitaxial growth: homoepitaxy of Si on Sn-precovered surface of Si(111) [2.96], and of Ge on Pb-precovered surface of Si(111) [2.97]. In the former paper a value of $\chi = 1.76$ has been found from the plot of $\ln N_s$ versus $\ln F$. In the case of homoepitaxial growth of Si(111) under clean conditions a value of $\chi = 0.85$ has been obtained from the same plot of $\ln N$ versus $\ln F$ [2.98]. It could be concluded that the nucleation process takes place either in a diffusion regime with $i^* = 6$ or in a kinetic regime with $i^* = 2$. The latter seems more reasonable, bearing in mind the comparatively low temperature of growth (< 700 K) and that Si is a very strongly bonded material.

2.5 Second-Layer Nucleation in Homoepitaxy

Growth of defectless low-index crystal surfaces takes place by formation and growth of 2-D nuclei with monolayer height. When the linear size L of the crystal face is small, in fact, smaller than $L_c = (v/J_0)^{1/3}$ [2.99], where v is the rate of lateral growth and J_0 is the nucleation rate, the growth proceeds by a periodic process of formation of a single nucleus followed by its growth to cover completely the crystal face. Thus, perfect layer-by-layer growth takes place.

When the surface area which is in contact with the supersaturated vapor is large, a large amount of nuclei are formed on the crystal surface on one and the same level. During the growth of the first layer nuclei, a certain size Λ can be reached at which second-layer nuclei can form on top. The average time elapsed from the nucleation of the first-layer nucleus to the appearance of the second-layer nucleus is $\tau = \Lambda/v$. The latter should be inversely proportional to the frequency of nucleation on top of the first-layer nucleus $\bar{J}_0 = J_0 l^2$, or in other words, $\Lambda/v \cong 1/\bar{J}_0$. Thus we find that the critical size for second-layer nucleation is $\Lambda_c = (v/J_0)^{1/3}$ [2.99]. Obviously, when the surface coverage by first-layer nuclei is $\Lambda_c^2 N_s \ll 1$, where N_s is the saturation nucleus density, nuclei of the second, third, etc. layers can form before significant coalescence of the first-layer nuclei takes place. The crystal surface will be rough with many layers growing simultaneously. Multilayer growth takes place. The number N of simultaneously growing layers depends on v and J_0. If v is large or J_0 is small, Λ_c will be large and the surface roughness will be small, and vice versa.

In the above physical picture it is assumed that the probabilities of attachment of atoms to a step from both the upper and lower terrace are equal. In other words, it is accepted that the barrier which inhibits the incorporation of the atoms to the step and in turn leads to the kinetic regime discussed above is one and the same from both sides of the step. It was at the beginning of 1966 when *Ehrlich* and *Hudda* discovered that the above is completely incorrect [2.100]. They found with the help of field-ion microscopy (the first method which allowed the visualization of single atoms, invented by *Erwin Müller* in the early 1950s) [2.101], that an atom approaching the step from the upper terrace is repulsed by the step. The additional barrier E_{ES}, known now in the literature as the Ehrlich–Schwoebel barrier, was measured later by *Wang* and *Tsong*, who reported values of the order of 0.15–0.2 eV for Re, Ir, and W [2.102]. Much later *Wang* and *Ehrlich* reported that the steps attract the atoms approaching them from the lower terrace [2.103]. The same authors observed in the case of Ir(111) that the atoms, instead of being repelled from the descending step, were in fact attracted by it. Thus they found another, *push-out*, mechanism of step-down diffusion in which the second-level atom pushes out the edge atom and occupies the position of the latter rather than making a jump [2.104]. The atoms thus sample the potential profiles shown in Fig. 2.14a in the case of step-down jumping and in Fig. 2.14b in the case of the push-out mechanism.

The physics behind these effect are easy to understand if we compare interlayer diffusion with the same phenomenon on terraces. It is clear that an atom jumping down the step from the upper terrace will be less coordinated from the side of the lower terrace. On the contrary, an atom approaching the step from the lower terrace will be additionally attracted from the atoms belonging to the upper atomic plane. In the case of the push-out mechanism the atoms taking part in the process *respect a fundamental rule of chemistry – minimizing the breaking of bonds* [2.105].

Schwoebel immediately grasped the importance of the discovery of Ehrlich and Hudda and published later in the same year a paper dealing with the effect of the step-down diffusion barrier on the bunching of steps during evaporation [2.106, 107]. He went even further to foresee the push-out mechanism long before Ehrlich observed it experimentally [2.106].

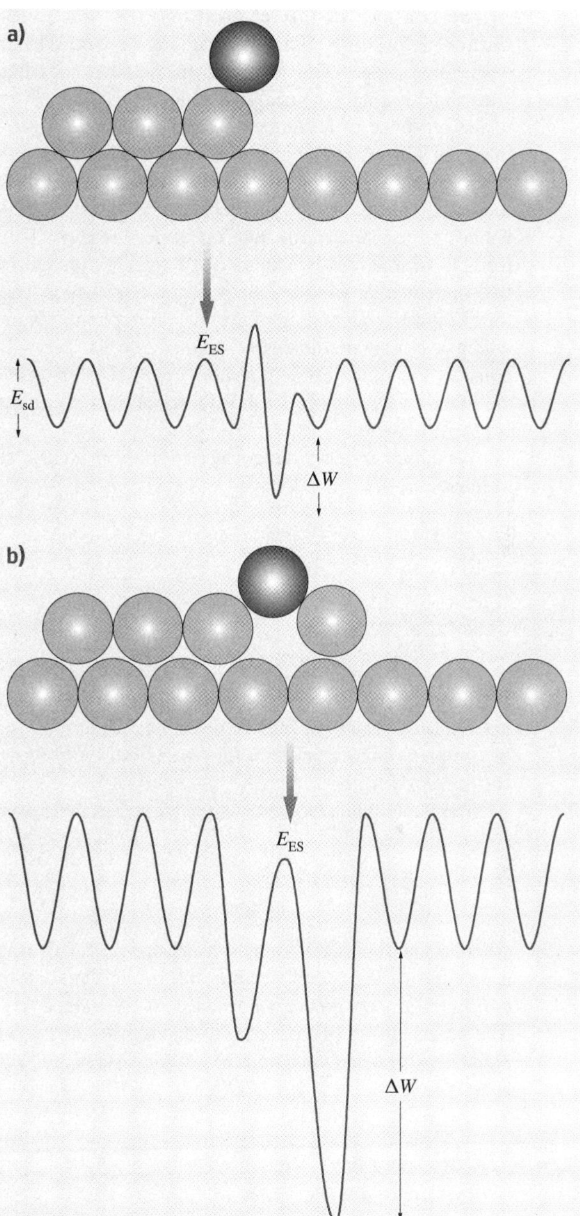

Fig. 2.14a,b Schematic potential diagrams for atoms moving toward ascending and descending steps. **(a)** Traditional view of the Ehrlich–Schwoebel barrier for atoms joining a descending step by a jump and short-range attractive behavior of the ascending step, **(b)** view of the potential sampled by an atom joining a descending step by a push-out mechanism

We consider in this chapter only the traditional Ehrlich–Schwoebel effect of repulsion of atoms from descending steps. The push-out mechanism together with an additional barrier from the lower terrace owing to the presence of surfactant atoms which have decorated the step (the reverse Ehrlich–Schwoebel effect) is considered in [2.108]. The additional ES barrier inhibits the flow of atoms from upper terraces downwards, thus enhancing the nucleation rate on upper terraces. This leads to formation of mounds consisting of concentric two-dimensional islands, one on top of the other, and thus to strong roughening of the surface, a phenomenon which was first predicted by *Villain* [2.109]. We will consider the same problem as above, defining the critical island size Λ for second-layer nucleation accounting for the ES barrier.

We define Λ in the same way as above but writing it in integral form

$$\int_0^{\Lambda} \frac{\bar{J}_0(\rho)}{v(\rho)} \mathrm{d}\rho = 1 , \tag{2.57}$$

where

$$v(\rho) = \frac{\mathrm{d}\rho}{\mathrm{d}t} = \frac{R}{2\pi\rho N_\mathrm{s} N_0} \tag{2.58}$$

is the rate of growth of the first-layer islands in the case of complete condensation before nuclei on their upper surfaces are formed.

The nucleation frequency $\bar{J}_0$ is defined as before as

$$\bar{J}_0 = 2\pi \int_0^{\rho} J_0(r,\rho) r \,\mathrm{d}r , \tag{2.59}$$

where J_0 is the nucleation rate as given by (2.50). It is a function of the island's radius ρ through the adatom concentration on the upper surface of the island N_1. The latter can be determined by solving the diffusion equation (in polar coordinates) in the absence of re-evaporation

$$\frac{\mathrm{d}^2 N_1}{\mathrm{d}r^2} + \frac{1}{r}\frac{\mathrm{d}N_1}{\mathrm{d}r} + \frac{R}{D_\mathrm{s}} = 0 . \tag{2.60}$$

The solution reads

$$N_1 = A - \frac{R}{4D_\mathrm{s}} r^2 , \tag{2.61}$$

where the integration constant should be determined by the boundary condition

$$j = -D_\mathrm{s}\left(\frac{\mathrm{d}N_1(r)}{\mathrm{d}r}\right)_{r=\rho} , \tag{2.62}$$

where $j = j_+ - j_-$ is the net flux of atoms to the descending step which encloses the island, j_+ and j_- being the attachment and detachment fluxes.

Bearing in mind Fig. 2.14 j_+ and j_- read

$$j_+ = a\nu N_\mathrm{st} \exp\left(-\frac{E_\mathrm{sd} + E_\mathrm{ES}}{k_\mathrm{B}T}\right) ,$$

$$j_- = a\nu N_\mathrm{k} \exp\left(-\frac{\Delta W + E_\mathrm{sd} + E_\mathrm{ES}}{k_\mathrm{B}T}\right) ,$$

where N_st is the adatom concentration in the vicinity of the step, ν is the attempt frequency, N_k is the concentration of atoms in a position (presumably kink position) for easy detachment from the step, and $\Delta W = \varphi_{1/2} - E_\mathrm{des}$ is the energy to transfer an atom from a kink position onto the terrace.

The total flux j then reads

$$j = a\nu\left(N_\mathrm{st} - N_1^\mathrm{e}\right) \exp\left(-\frac{E_\mathrm{sd}}{k_\mathrm{B}T}\right)\frac{1}{S} , \tag{2.63}$$

where $S = \exp(E_\mathrm{ES}/k_\mathrm{B}T)$, and $N_1^\mathrm{e} = N_\mathrm{k}\exp(-\Delta W/k_\mathrm{B}T)$ is the equilibrium adatom concentration (see (2.12)).

Combining (2.62) and (2.63) and bearing in mind that $N_\mathrm{st} = A - R\rho^2/4D_\mathrm{s}$ yields [2.110]

$$N_1 = N_1^\mathrm{e} + \frac{R}{4D_\mathrm{s}}\left(\rho^2 + 2\rho a S - r^2\right) . \tag{2.64}$$

As seen in the case of negligible ES barrier ($2aS/\rho \ll 1$), (2.64) turns into

$$N_1 = N_1^\mathrm{e} + \frac{R}{4D_\mathrm{s}}\left(\rho^2 - r^2\right) . \tag{2.65}$$

The adatom concentration on top of the island surface has a profile of a dome with a maximum above the island's center ($r = 0$) and reaches its equilibrium value N_1^e near the island's edge ($r = \rho$). It follows that second-layer nucleation is favored around the middle of the island.

In the other extreme ($2aS/\rho \gg 1$) we neglect the difference $\rho^2 - r^2$ and obtain

$$N_1 \approx \frac{R}{2D_\mathrm{s}} \rho a S .$$

This means that the adatom population on top of an island with repelling boundaries is uniformly distributed all over the surface of the island and a nucleation event can occur with equal probability at any point of it.

We substitute (2.64) into (2.50) and the latter into (2.59) to obtain after integration [2.110]

$$\bar{J}_0 = A\left[\left(\rho^2 + 2\rho a S\right)^{i^*+2} - \left(2\rho a S\right)^{i^*+2}\right] , \tag{2.66}$$

where

$$A = \frac{\pi\alpha^*}{(i^*+2)} D_s N_0^2 \exp\left(\frac{E^*}{k_B T}\right)\left(\frac{R}{4D_s N_0}\right)^{i^*+1} .$$

As seen, a negligible ES barrier ($2aS \ll \rho$) turns (2.66) into

$$\bar{J}_0 = A\rho_1^{2(i^*+2)} . \tag{2.67}$$

The condition for layer-by-layer growth (formation of one nucleus for the time $T = R/N_0$ of deposition of a complete monolayer)

$$N = \int_0^T \bar{J}_0(\rho_1)\,dt = 1 \tag{2.68}$$

gives for the number of the growth pyramids the expression [2.111] (for a review see [2.21])

$$N_s = \frac{1}{4\pi} C^* N_0 \left(\frac{D}{F}\right)^{-\chi} \exp\left(\frac{E_{i^*}}{(i^*+2)k_B T}\right) , \tag{2.69}$$

where C^* is a very weak function of i^* of the order of unity. The above equation is in fact (2.54) with the familiar scaling exponent (2.55).

In the other extreme ($2aS \gg \rho$) we take the last two terms of the expansion of the sum in (2.66) and the latter turns into

$$\bar{J}_0 = B\rho^{i^*+3} , \tag{2.70}$$

where

$$B = \pi\alpha^* D_s N_0^2 \exp\left(\frac{E^*}{k_B T}\right)\left(\frac{RaS}{2D_s N_0}\right)^{i^*+1} .$$

Following the above procedure gives for this case [2.112]

$$N_s = \frac{1}{\pi} C^* N_0 \left(\frac{D}{F}\right)^{-\chi} \exp\left(\frac{2[E_{i^*} + (i^*+1)E_b]}{(i^*+3)k_B T}\right) , \tag{2.71}$$

where C^* is another very weak function of i^* of the order of unity. We again obtained (2.54) but the scaling exponent is given by (2.56).

We can now calculate the critical radii of the islands for second-layer nucleation in both cases of low (subscript "0") and high (subscript "ES") Ehrlich–Schwoebel barrier. Substituting (2.67), (2.70) and (2.58) into (2.57) gives after integration [2.110]

$$\Lambda_0 = aC_0 \left(\frac{D}{F}\right)^{i^*/2(i^*+3)} , \tag{2.72}$$

with

$$C_0 \cong \left(\frac{N_0 e^{-E^*/k_B T}}{\alpha^* N_s}\right)^{1/2(i^*+3)} , \tag{2.73}$$

for the case of negligible ES barrier, and

$$\Lambda_{ES} = aC_{ES} \left(\frac{D}{F}\right)^{i^*/(i^*+5)} S^{-(i^*+1)/(i^*+5)} , \tag{2.74}$$

with

$$C_{ES} \cong \left(\frac{N_0 e^{-E^*/k_B T}}{\alpha^* N_s}\right)^{1/(i^*+5)} , \tag{2.75}$$

for the other limiting case of a significant ES barrier.

Let us compare N_s and Λ in both cases. For this purpose we take typical values for the quantities involved: $N_0 = 1\times10^{15}\,\mathrm{cm^{-2}}$, $R = 1\times10^{13}\,\mathrm{cm^{-2}\,s^{-1}}$, $F = R/N_0 = 1\times10^{-2}\,\mathrm{s^{-1}}$, $E_{sd} = 0.4\,\mathrm{eV}$, $E_{ES} = 0.2\,\mathrm{eV}$, $T = 400\,\mathrm{K}$, $i^* = 1$, and $E^* = 0$. Then, in the case of $E_{ES} = 0$, $N_s \approx 6\times10^{10}\,\mathrm{cm^{-2}}$ and $\Lambda_0 \approx 180\,\text{Å}$. In the other extreme, $N_s \approx 1\times10^{12}\,\mathrm{cm^{-2}}$ and $\Lambda_{ES} \approx 50\,\text{Å}$ is 3 times smaller. We conclude that with a significant ES barrier a larger density of islands is formed which have much smaller critical size for second-layer nucleation. Mounding rather than planar growth is expected.

It is of interest to check the above theory. For this purpose we calculate the number n of atoms on the surface of the base island when its radius has just reached the critical value Λ. We integrate the adatom concentration (2.64) on the island's surface

$$n = 2\pi \int_0^{\Lambda} n_s(r, \Lambda) r\,dr$$

and find

$$n = \frac{\pi F}{8D} N_0^2 \Lambda^4 \left(1 + \frac{4aS}{\Lambda}\right) .$$

We will consider as examples two surfaces of fcc crystals: (100) and (111). The reason is that the (100) surfaces are characterized by a large terrace diffusion barrier and a small step-edge barrier. This is the reason why, during growth, (100) surfaces demonstrate as a rule oscillations of the intensity of the specular beam, which are an indication of layer-by-layer growth. On the contrary, the smoother (111) surfaces are characterized with small intralayer diffusion barriers and large interlayer barriers. The result is a roughening of the crystal surface from the very beginning of deposition and a monotonous decrease of the intensity of the specular beam [2.112].

We consider first the case of Cu(001) [2.113]. The authors have measured the step kinetics of a pyramid consisting of 2-D islands, one on top of the other, and determined the critical radius $\Lambda \approx 3\times10^{-5}$ cm of the uppermost island at which the next layer nucleus is formed ($T = 400$ K, $F = 0.0075\,\mathrm{s}^{-1}$, $E_{sd} = 0.4$ eV, $a = 2.55\times10^{-8}$ cm, $N_0 = 1.53\times10^{15}\,\mathrm{cm}^{-2}$). Comparison with the theory produced the value $E_{ES} = 0.125$ eV. Then, by using the above formula we find for the number of atoms which gives rise to the new monolayer nucleus the value $n = 70$. Note that $aS/\Lambda \approx 0.03$, which confirms the above statement that the kinetics at fcc(001) surfaces is not dominated by the interlayer diffusion and the profile of the adatom concentration looks like a dome.

We consider next the case of Pt(111) [2.114]. Bott, Hohage, and Comsa observed by scanning tunneling microscopy (STM) the appearance of second-layer nuclei at surface coverages of 0.3 (425 K, $N_s = 3.37\times10^{10}\,\mathrm{cm}^{-2}$) and 0.8 (628 K, $N_s = 3.5\times10^{9}\,\mathrm{cm}^{-2}$) ($R = 5\times10^{12}\,\mathrm{cm}^{-2}\,\mathrm{s}^{-1}$). The activation energy for terrace diffusion is well known to be 0.25–0.26 eV [2.115, 116]. Values for E_{ES} varying from 0.12 eV (see [2.117]) to 0.44 eV have been estimated [2.118]. The average number of atoms on the island's surface as computed with the help of the above equation for n turned out to be of the order of 1×10^{-2}, i.e., much less than unity, which is unphysical. In fact n becomes greater than unity when $E_{ES} > 0.5$ eV, which means that the atoms at the island's periphery must overcome a total barrier of about 0.75 eV, which is too large to be believed. In contrast to the previous case, however, $aS/\Lambda \gg 1$, which means that it is interlayer diffusion that dominates the kinetics, and the adatom population on top of the island is spatially uniform.

Whereas the Cu(001) case is physically reasonable, the (111) case looks puzzling. In order to solve the problem of the high ES barrier *Krug* et al. accounted for the probabilistic nature of the main processes involved [2.117]. The authors have taken into account the fact that the atoms arrive randomly on the island's surface with an area $\pi\rho^2$ but not at equal intervals $\Delta t = 1/\pi\rho^2 R$ as is implicitly assumed in the model described above. Second, the time τ that the atoms reside on the island before rolling over and joining the descending edge is also a random quantity. The latter is directly proportional to the island's periphery $2\pi\rho$ and inversely proportional to the rate of step-down diffusion $\omega = a\nu\exp[-(E_{sd}+E_{ES})/k_BT]$, i.e., $\tau \approx 2\pi\rho/\omega = 2\pi\rho aS/D_s$. We introduce further the time $\tau_{tr} = \pi\rho^2/D_s$ required for an atom to visit all sites of the island. The condition $\tau/\tau_{tr} \gg 1$ is equivalent to $2aS/\rho \gg 1$, which is in fact the condition for nucleation kinetics dominated by step-down diffusion (see (2.70)). Assuming $i^* = 1$ (the dimers are stable and immobile) it is concluded that, as soon as two atoms are present simultaneously on the island's surface, their encounter is inevitable. Thus the necessary and sufficient condition for the atoms to meet each other and give rise to a stable cluster is $\tau_{tr} \ll \tau$. Then the probability of nucleation p_{nuc} is equal to the probability p_2 for two adatoms to be present simultaneously on the island. p_2 is determined by the condition that the time of arrival t_2 of the second atom be shorter than the time t_1 of departure of the first atom. Assuming that t_1 and t_2 are randomly distributed around the average values τ and Δt, respectively, one obtains after integration

$$p_{nuc} = \frac{1}{\tau\Delta t}\int_0^\infty \mathrm{d}t_1\,\mathrm{e}^{-t_1/\tau}\int_0^{t_1}\mathrm{d}t_2\,\mathrm{e}^{-t_2/\Delta t} = \frac{\tau}{\tau+\Delta t}\,.$$

Two limiting cases are possible. The case $\tau \gg \Delta t$ and $p_{nuc} \approx 1$ is trivial; it means that the ES barrier is infinitely high and there will always be at least one atom on top of the island. The physically interesting case is when $\Delta t \gg \tau$ and $p_{nuc} = \tau/\Delta t$. Then the nucleation frequency $\bar{J}_0 = \pi\rho^2 R p_{nuc}$ reads

$$\bar{J}_0 \propto \frac{aR^2\rho^5 S}{D_s}\,. \tag{2.76}$$

This equation should be compared with (2.70). With $i^* = 1$ the latter gives

$$\bar{J}_0 \propto \frac{a^2R^2\rho^4 S^2}{D_s}\,. \tag{2.77}$$

Comparing both formulae shows that the mean-field expression (2.77) is $aS/\rho \gg 1$ times larger than the probabilistic one (2.76). The explanation is simple. Equation (2.77) is based on the implicit assumption that on top of the island there is a time-averaged number (smaller than unity but constant) of atoms all the time. As shown above this is indicative of a large ES barrier whose mathematical expression is just $aS/\rho \gg 1$. In fact the island's surface is empty most of the time and is sometimes populated by a single atom, and it very rarely happens that during this time a second atom arrives. Once two atoms are simultaneously present on the island a nucleus is formed with a probability close to unity. That is why the authors coined for this model the term *the lonely adatom model*. The problem of second-layer nucleation has been intensively studied [2.119, 120]. It has been found that the mean-field approach is applicable for critical nuclei consisting of more than three atoms. If this is not the case ($i^* = 1, 2$), the random character of the processes involved becomes significant.

2.6 Mechanism of Clustering in Heteroepitaxy

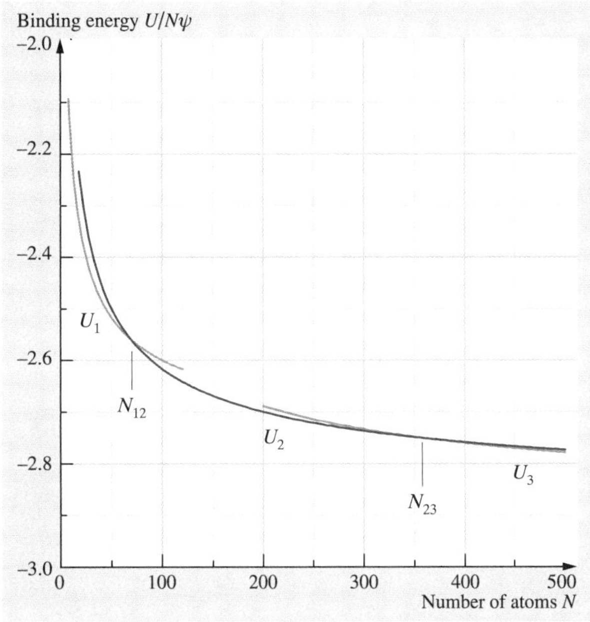

Fig. 2.15 Plot of the binding energy per atom in units of the energy of a single first-neighbor bond ψ of monolayer, bilayer, and trilayer islands with simple cubic lattice as a function of the total number of atoms. The wetting parameter $\phi = 0.1$ (after [2.35]) ▶

We consider first the growth of a heteroepitaxial thin film by the mechanism of Volmer–Weber. As the wetting is incomplete the thermodynamics requires 3-D islanding directly on top of the substrate. We study the stability of islands with different thickness beginning from one monolayer against their volume (or total number of atoms). In other words we study the behavior of the binding energy $-U_i$ in (2.27), which is equal to the *surface* energy term Φ up to a constant $i\varphi_{1/2}$ [2.35].

We study for simplicity a Kossel crystal with (100) substrate orientation. The same result is obtained by using any other lattice and substrate orientation [2.35]. As a first approximation we omit the effect of the lattice misfit. As discussed above the strain energy makes as a rule a minor contribution with the same sign to the difference of the cohesive ψ and adhesive ψ' energies. As another approximation we consider our crystal in a *continuous* way, assuming that the shape remains a complete square irrespective of the number of atoms in it. We calculate first the binding energies of monolayer, bilayer, and trilayer islands with a square shape of the base and consisting of a total of N atoms. Restricting ourselves to nearest-neighbor bonds the energies read

$$\frac{U_1}{N\psi} = -3 + \phi + \frac{2}{\sqrt{N}},$$

$$\frac{U_2}{N\psi} = -3 + \frac{\phi}{2} + \frac{2\sqrt{2}}{\sqrt{N}},$$

$$\frac{U_3}{N\psi} = -3 + \frac{\phi}{3} + \frac{2\sqrt{3}}{\sqrt{N}},$$

where ϕ is the wetting function (2.18).

We plot the above energies as a function of N and find that monolayer-high islands are stable against

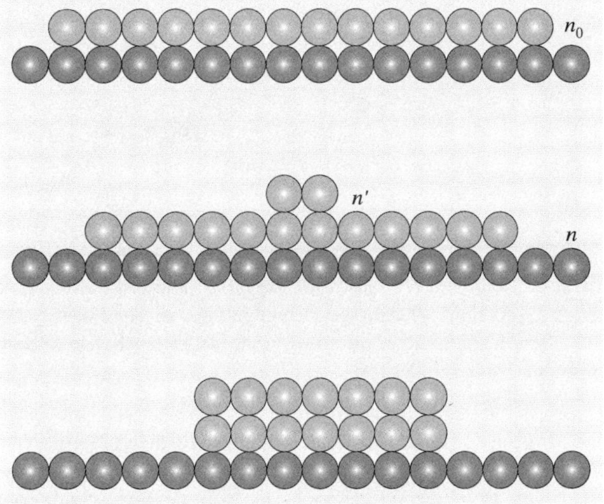

Fig. 2.16 Schematic process for the evaluation of the activation energy of the mono–bilayer transformation. The initial state is a square monolayer island with n_0 atoms in the edge. The intermediate state is a monolayer island with n atoms in the edge plus a second level island with n' atoms in the edge so that $n^2 + n'^2 = n_0^2$. The final state is a complete bilayer island ▶

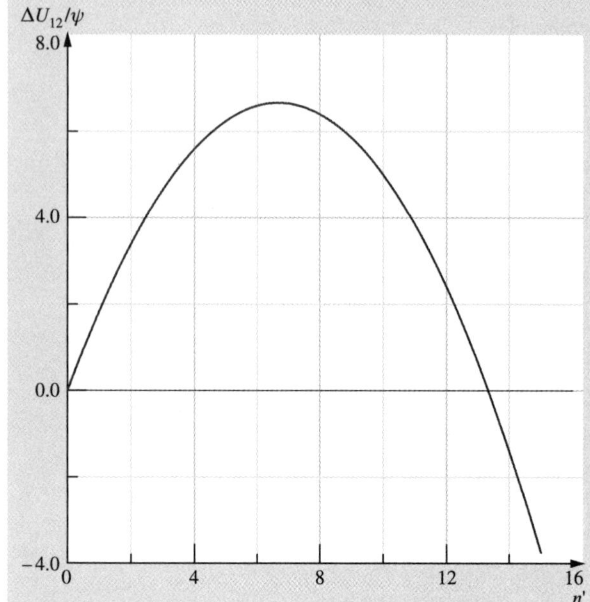

Fig. 2.17 The energy change which accompanies the mono–bilayer transformation in Volmer–Weber growth (after [2.35])

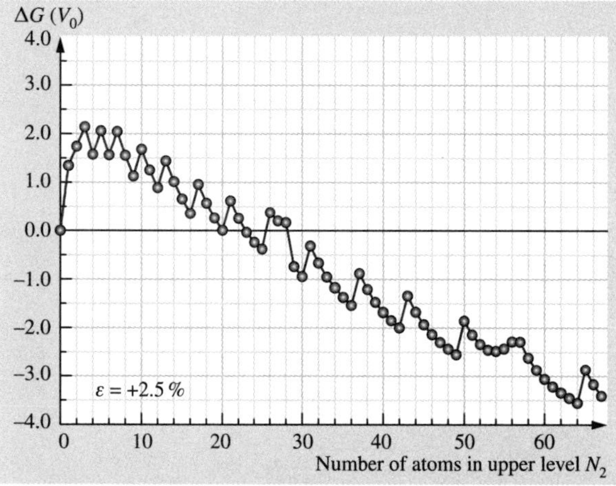

Fig. 2.18 Mono–bilayer transformation curve in Stranski–Krastanov growth representing the energy change in units of bond energy as a function of the number of atoms in the upper level. The lattice misfit is 2.5% (after [2.122])

bilayer islands up to a critical size denoted by N_{12} (Fig. 2.15). The bilayer islands are stable from this size up to a second critical size N_{23}, beyond which trilayer islands become stable, etc. These critical sizes are inversely proportional to the square of the wetting function and go to infinity when $\phi \to 0$. The latter means that, at $\phi = 0$, 3-D islands will not be able to form. Instead, layer-by-layer growth is expected according to the thermodynamics at complete wetting. At finite values of ϕ a mono–bilayer transformation should take place when $N > N_{12}$. A bi–trilayer transformation is expected to occur when $N > N_{23}$, etc. It is very important to note that *monolayer-high islands appear as necessary precursors for 3-D islands* [2.121].

We study further the mechanism of transformation of monolayer to bilayer islands, assuming the following imaginary process illustrated in Fig. 2.16 [2.35]. Atoms detach from the edges of the monolayer islands, which are larger than N_{12} and thus unstable against bilayer islands, diffuse on top of them, aggregate, and give rise to second-layer nuclei. The latter grow further at the expense of the atoms detached from the edges of the lower islands. The process continues up to the moment when the upper island completely covers the lower-level island. The energy change associated with the process of transformation at a particular stage is given by the difference between the energy of the incomplete bilayer island and that of the initial monolayer island

$$\frac{\Delta U_{12}(n')}{\psi} = -n'^2\phi - \frac{n'^2}{n_0} + 2n' , \tag{2.78}$$

where the approximation $n_0 + n = 2n_0$ is used in the beginning of the transformation, n_0, n, and n' being the numbers of atoms in the edge of the initial monolayer island, in the lower edge of the incomplete bilayer island, and in the edge of the second-layer island, respectively (Fig. 2.16).

Equation (2.78) is plotted in Fig. 2.17. As seen, it displays a maximum at some critical size

$$n'^* = \frac{n_0}{1 + n_0\phi} . \tag{2.79}$$

The height of the maximum is given by

$$\Delta U_{12}^* = \frac{n_0}{1 + n_0\phi}\psi = n'^*\psi , \tag{2.80}$$

as should be expected by the classical consideration of the nucleation process (2.36). It follows that the mono–bilayer transformation is a nucleation process.

The same physics functions in the clustering during the Stranski–Krastanov growth of thin films beyond the wetting layer [2.122]. The Stranski–Krastanov growth represents a growth of A on strained A. The strained wetting layer of A is formed on the surface of another crystal B with different lattice parameter. The 3-D islands which form on the wetting layer are fully strained in the middle but relaxed at the side-walls and edges. The atoms near the edges of the base are displaced from the positions they should occupy if the islands were completely strained to fit the wetting layer. As a result the adhesion of the atoms near the edges of the base to the substrate (the wetting layer) is weaker compared with the atoms in the middle of the island's base. Therefore, the average wetting is incomplete, $0 < \phi < 1$, which is the thermodynamic condition for clustering. The detachment of atoms from the edges and the formation of a cluster in the second level beyond some critical size is energetically favorable. The numerically calculated energy accompanying this process is shown in Fig. 2.18 [2.122]. The atoms interact through a pair potential of Morse type whose anharmonicity can be varied by adjusting two constants that govern separately the repulsive and attractive branches, respectively [2.123, 124]. The 3-D crystallites have fcc lattice and (100) surface orientation, thus possessing the shape of a truncated square pyramid. As seen, a critical nucleus consisting of three atoms is formed, beyond which the energy goes down as in an ordinary nucleation process. The misfit dependence of the critical size N_{12}, the nucleus size, and the work for nucleus formation are shown in Fig. 2.19 [2.122]. The nucleation character of the transformation is clearly observed. The energy barrier and the number of atoms in the cluster with highest energy increase steeply with decreasing lattice misfit, which in this case plays the role of the supersaturation. The number N_{12} also goes to infinity, illustrating the critical behavior of the transition from monolayer (2-D) to bilayer (3-D) islands.

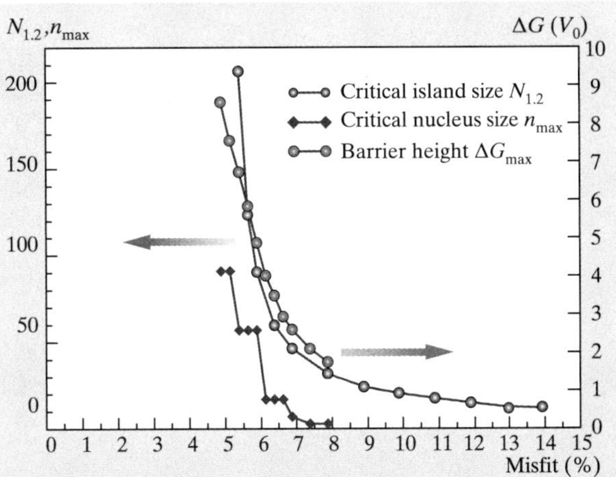

Fig. 2.19 Misfit dependence of the critical size N_{12}, the critical nucleus size (both expressed in number of atoms), and the nucleation barrier (in units of ψ) for compressed overlayers. The initial size of the monolayer island is 20×20 atoms (after [2.122])

It should be pointed out that the mono–bilayer transformation of islands under tensile stress does not display a nucleation behavior, particularly at lower absolute values of the misfit. However, this problem is outside the scope of the present review and will not be discussed.

2.7 Effect of Surfactants on Nucleation

It was found long ago that very often epitaxial films grow in a layer-by-layer mode and show better quality when the vacuum is poor [2.125, 126]. Much later *Steigerwald* et al. found that intentionally adsorbed oxygen on Cu(001) suppresses agglomeration and interdiffusion upon deposition of Fe [2.127]. The significance of these observations was immediately grasped and the very next year *Copel* et al. reported that preadsorption of As drastically alters the mode of growth of Ge on Si(001) and of Si on Ge(001) by suppressing the clustering in the Stranski–Krastanov and Volmer–Weber modes of growth, respectively [2.128]. They suggested an interpretation of their observations in terms of the change of the wetting of the substrate by the overlayer due to the effect of the third element and used the term *surfactant* to stress the thermodynamic nature of the phenomenon. Intensive studies and heated debate concerning the effect of the third elements on the thermodynamics and kinetics of the processes followed. It was shown that the surfactants change not only the thermodynamics but also the kinetics of the processes involved [2.5, 129]. Nevertheless, the term surfactant was widely accepted in the literature. We explore here the effect of surfactants on nucleation in the simpler case of homoepitaxy. Accounting for the unlike substrate requires only the inclusion of a term containing the wetting function (2.19) into the work of nucleus formation.

We calculate first the work for nucleus formation by using the following imaginary process

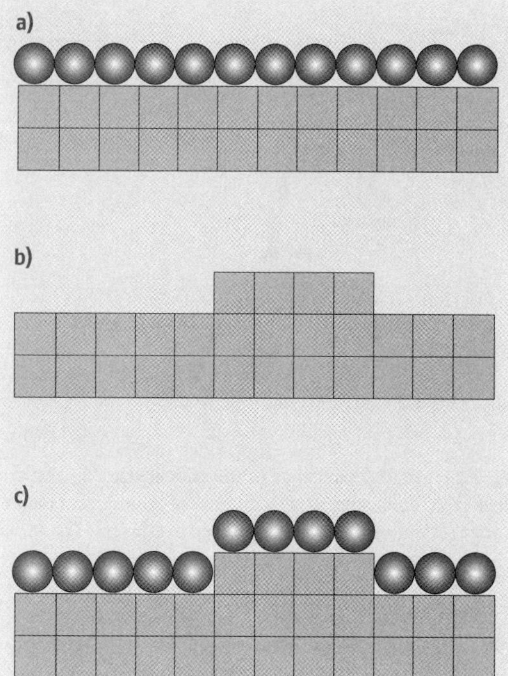

Fig. 2.20a–c Calculation of the Gibbs free energy change for nucleus formation on a surfactant-precovered surface. (**a**) The initial surface covered with a complete monolayer of surfactant atoms denoted by *filled circles*; (**b**) the surfactant layer is evaporated and a cluster consisting of i atoms is created; (**c**) the surfactant layer is condensed back and a cluster consisting of i surfactant atoms is formed on top (after [2.130])

(Fig. 2.20) [2.130]. In order to illustrate the essential physics for simplicity we first make use of the classical nucleation theory. The initial state is a surface of the crystal (C) covered by a complete monolayer of surfactant (S) atoms. We first evaporate reversibly and isothermally all S atoms. Then on the clean surface we produce a cluster consisting of i C atoms. Assuming a square shape with edge length l the work for cluster formation in absence of a surfactant reads

$$\Delta G_0 = -i\Delta\mu + 4l\varkappa_c ,$$

where $\varkappa_c$ is the specific edge energy.

We condense back the S atoms. We gain energy $-4ls\varkappa_c$ due to saturation of the dangling bonds at the cluster periphery by the S atoms, and spend energy $4l\varkappa_s$ to create the new step which surrounds the cluster consisting of S atoms. The work for nucleus formation then reads [2.130]

$$\Delta G_s = \Delta G_0 - 4ls\varkappa_c + 4l\varkappa_s , \tag{2.81}$$

where $\varkappa_s$ is the specific edge energy of the S cluster and the parameter

$$s = 1 - \frac{\omega}{\omega_0}$$

accounts for the saturation of the dangling bonds by S atoms. It is a measure of the *surfactant efficiency*, as the quantities

$$\omega = \tfrac{1}{2}(\psi_{cc} + \psi_{ss}) - \psi_{sc} \tag{2.82}$$

and

$$\omega_0 = \tfrac{1}{2}\psi_{cc}$$

are the energies of the S-saturated and unsaturated dangling bonds, respectively. The subscripts "cc," "ss," and "sc" denote the bond energies C–C, S–S, and S–C, respectively.

Looking at (2.82) it becomes clear that it in fact represents the energetic parameter that determines the enthalpy of mixing of the two species C and S. It must be positive in order to allow the segregation of the surfactant. In the absence of a surfactant $\psi_{ss} = \psi_{sc} = 0$, $\omega = \omega_0$, and $s = 0$. In the other extreme, $\psi_{ss} + \psi_{cc} = 2\psi_{sc}$ and $s = 1$. Thus the parameter s varies from 0 at complete inefficiency to 1 at complete efficiency. (In general the parameter s can be greater than unity, which means $\omega < 0$. However, this means an alloying of the surfactant with the growing crystal, which will have deleterious consequences for the quality of the overlayer and should be avoided.)

It follows from (2.81) that, in the case of surfactant-mediated growth, the Gibbs free energy for nucleus formation contains two more terms that have opposite signs and thus compete with each other. The s-containing term accounts for the decrease of the edge energy of the cluster owing to the saturation of the dangling bonds by the surfactant atoms. The energy $4l\varkappa_s$ of the dangling bonds of the periphery of the cluster, consisting of S atoms, which is unavoidably formed on top of the 2-D nucleus due to the segregation of the surfactant, increases the work of cluster formation.

Finding a solution for a small number of atoms in the critical nucleus in the atomistic extreme is straightforward. We make use of (2.27)

$$\Phi = i\varphi_{1/2} - U_i$$

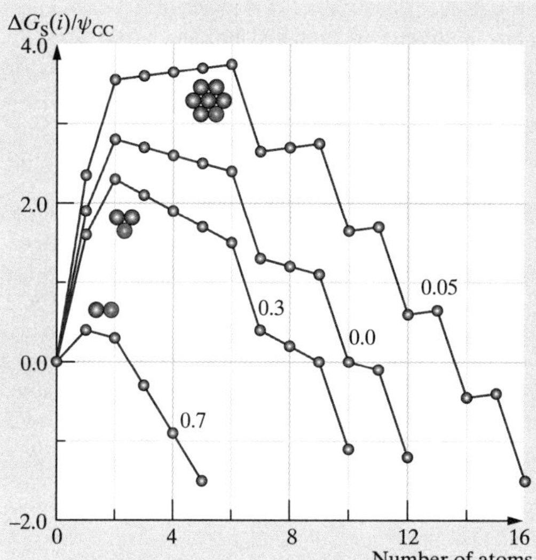

Fig. 2.21 Change of the Gibbs free energy for cluster formation relative to the work needed to disjoin two C atoms versus the number of atoms on the (111) surface of a fcc crystal. The value of the surfactant efficiency s is denoted by figures on each curve. The structure of the nucleus is given by the *filled circles*. The *gray circles* denote the atoms that turn the critical nuclei into smallest stable clusters (after [2.130])

for the edge energy of both clusters instead of using the capillary term for the edge energy $\varkappa$.

The binding energy U_i can be divided into lateral energy E_i and desorption energy E_{des} (assuming additivity of the bond energies)

$$U_i = E_i + iE_{\text{des}} ,$$

and for Φ one obtains

$$\Phi = i\Delta W - E_i ,$$

where $\Delta W = \varphi_{1/2} - E_{\text{des}}$ is the energy to transfer an atom from a kink position onto the terrace.

We then substitute Φ for $4l\varkappa_{\text{c}}$ in (2.81) to obtain

$$\Delta G_{\text{s}}(i) = -i\Delta\mu + i(1-s)\Delta W - (1-s)E_i + \Phi_{\text{s}} , \tag{2.83}$$

where Φ_{s} has the meaning of the edge energy $4l\varkappa_{\text{s}}$ of the surfactant cluster.

Figure 2.21 shows the dependence of $\Delta G_{\text{s}}(i)$ in units of the crystal bond strength, ψ_{cc}, on the cluster size i for the (111) surface of fcc metals ($\varphi_{1/2} = 6\psi_{\text{cc}}$, $E_{\text{des}} = 3\psi_{\text{cc}}$, $\Delta W = 3\psi_{\text{cc}}$), with $\psi_{\text{ss}}/\psi_{\text{cc}} = 0.2$, constant supersaturation $\Delta\mu = 1.1\psi_{\text{cc}}$, and different values of s denoted by figures on each curve. As seen, $\Delta G_{\text{s}}(i)$ represents a broken line (as should be expected for a small number of atoms, cf. Fig. 2.10), displaying a maximum at $i = i^*$. Under clean conditions ($s = 0$) the critical nucleus consists of two atoms. When s is very small ($= 0.05$, the surfactant is almost inefficient), the number of atoms in the critical nucleus equals six due to the contribution of the edge energy of the surfactant cluster $4l\varkappa_{\text{s}}$. The work of formation of the critical nucleus also increases. Increasing s to 0.3 due to decrease of the edge energy of the cluster leads to a decrease of the nucleation work and i^* becomes again equal to two. At some greater value of s ($= 0.7$), $i^* = 1$ and the aggregation becomes irreversible.

We see that the critical nucleus size differs under one and the same conditions (temperature, rate of deposition) in the absence and presence of a surfactant. In general, we should expect a decrease of the nucleus work and, in turn, a steep increase of the nucleation rate. As a result a larger density of smaller 2-D islands will form. The latter can coalesce and cover completely the surface before formation of nuclei of the upper layer. Thus surfactants can induce layer-by-layer growth by enhancing the nucleation rate [2.131, 132].

The rate of nucleation reads (see (2.42))

$$J_{\text{s}} = \omega_{\text{s}}^* \Gamma N_0 \exp\left(-\frac{\Delta G_{\text{s}}(i^*)}{k_{\text{B}}T}\right) , \tag{2.84}$$

where ω_{s}^* is the flux of atoms to the critical nucleus in the presence of a surfactant, and $\Gamma \cong 1$ is the Zeldovich factor. $\Delta G_{\text{s}}(i^*)$ is given by (2.83) with $i = i^*$.

Bearing in mind that $\Delta\mu = k_{\text{B}}T \ln(N_1/N_1^{\text{e}})$, where N_1 and N_1^{e} are the real and the equilibrium adatom concentrations, we can write

$$\Delta\mu = k_{\text{B}}T \ln\left(\frac{N_1}{N_0}\right) - k_{\text{B}}T \ln\left(\frac{N_1^{\text{e}}}{N_0}\right) , \tag{2.85}$$

where N_1^{e} is given by (2.12).

Combining (2.83–2.85) and (2.12) gives

$$J_{\text{s}} = \omega_{\text{s}}^* \Gamma N_0 \left(\frac{N_1}{N_0}\right)^{i^*} \times \exp\left(\frac{i^* s\Delta W + (1-s)E^* - \Phi_{\text{s}}}{k_{\text{B}}T}\right) . \tag{2.86}$$

In the absence of a surfactant, $s = 0$, we obtain the familiar expression (2.50) bearing in mind that $\omega_{\text{s}}^* = \omega^* = \alpha^* D_{\text{s}} N_1$.

Note that the presence of the surfactant is not accounted for only by the s-containing terms in the exponential. It is the flux ω_s^* that strongly depends on the mechanism of transport of crystal atoms to the critical nucleus [2.133, 134]. In the case when the transport of atoms to the critical nucleus takes place under the condition of reversible exchange/deexchange of S and C atoms (the time of de-exchange is much smaller than the time of deposition of complete monolayer and atoms have time to perform many exchange/de-exchange events) the nucleus density is given by [2.134] (see for more details [2.21])

$$N_S = N_{s,0} \exp\left(-\frac{\chi}{i_*}\frac{E_S}{k_B T}\right) , \tag{2.87}$$

where $N_{s,0}$ and χ are given by (2.71) and (2.56), and E_S combines all energy contributions that depend on the presence of the surfactant. Within the framework of the classical nucleation theory the latter is given by

$$\begin{aligned} E_S &= -4ls\varkappa_c + 4l\varkappa_s + E_{ex}^* \\ &\quad - i^*\left[(E_{dex} - E_{ex}) - \left(E_{sd}^0 - E_{sd}\right)\right] , \end{aligned} \tag{2.88}$$

where E_{ex} and E_{dex} are the barriers for exchange and de-exchange far from growing nuclei, E_{ex}^* is the barrier for exchange at the edge of the critical nucleus, and E_{sd}^0 and E_{sd} are the barriers for diffusion on clean surface and on top of the surface of the surfactant monolayer. As seen, the first two terms in E_S are of thermodynamic origin whereas the last two terms are of purely kinetic origin.

It follows that the exponential multiplying $N_{s,0}$ can be smaller or larger than unity depending on the sign of E_S. The latter in turn depends on the interplay of the energies involved. We consider in more detail the case of Sb-mediated growth of Si(111) [2.98, 135]. For this case *Kandel* and *Kaxiras* computed the values $E_{dex} = 1.6$ eV, $E_{ex} = 0.8$ eV, and $E_{sd} = 0.5$ eV [2.136]. The value of $E_{sd}^0 = 0.75$ eV has been calculated from experimental data by *Voigtländer* et al. [2.98]. Thus a value of 0.55 eV was found for the difference $(E_{dex} - E_{ex}) - (E_{sd}^0 - E_{sd})$. We recall that $-4ls\varkappa_c = sE^* - i^*s\Delta W$, where ΔW is of order of the half of the heat of evaporation, which for Si is equal to 4.72 eV [2.137]. It can be shown by inspection that $i^*\Delta W$ is always larger than E^*. Thus, when $i^* = 1$, $E^* = 0$ and $\Delta W \cong 2.3$ eV, and when $i^* = 2$, $E^* = 2.3$ eV and $i^*\Delta W \cong 4.6$ eV, etc. The value of s is close to unity as evaluated from the surface energies of Sb and Si available in the literature. It is thus concluded that it is the decrease of the edge energy of the nuclei $4ls\varkappa_c$ due to the saturation of the dangling bonds with S atoms which plays the major role and determines the sign of E_S [2.21]. The latter explains the larger density of 2-D nuclei in surfactant-mediated growth of Si(111) compared with growth in clean conditions [2.98].

Kandel and *Kaxiras* assumed that the exchange/de-exchange processes influence the kinetics of nucleation by affecting the diffusivity of the atoms and derived an expression for an effective diffusion coefficient including the respective barriers [2.5]

$$D_{eff} \cong D_s^0 \exp\left(-\frac{(E_{dex} - E_{ex}) - \left(E_{sd}^0 - E_{sd}\right)}{k_B T}\right) ,$$

and concluded that the atom diffusivity is inhibited due to $(E_{dex} - E_{ex}) > (E_{sd}^0 - E_{sd})$, which leads to increase of the nucleus density according to the scaling relation (2.54). As discussed above the more rigorous analysis shows that it is the thermodynamic term in (2.88) that controls the effect of the surfactant rather than the kinetic barriers.

2.8 Conclusions and Outlook

As shown above the nuclei of the new phase, particularly on surfaces, represent small clusters whose structure, shape, energy, and even size are still unclear. A large amount of work remains to be done in order to study the stability of small clusters of materials with different chemical bonds and crystal lattices as a function of their structure, shape, and size.

References

2.1 T. Michely, J. Krug: *Islands, Mounds and Atoms: Patterns and Processes in Crystal Growth Far from Equilibrium* (Springer, Berlin Heidelberg 2003)

2.2 R. Kern, G. LeLay, J.J. Metois: Basic mechanisms in the early stages of epitaxy. In: *Current Topics in Materials Science*, Vol. 3, ed. by E. Kaldis (North-Holland, Amsterdam 1979) pp. 131–419

2.3 A. Pimpinelli, J. Villain: *Physics of Crystal Growth* (Cambridge Univ. Press, Cambridge 1998)

2.4 P. Politi, G. Grenet, A. Marty, A. Ponchet, J. Villain: Instabilities in crystal growth by atomic or molecular beams, Phys. Rep. **324**, 271–404 (2000)

2.5 D. Kandel, E. Kaxiras: The surfactant effect in semiconductor thin film growth, Solid State Phys. **54**, 219–257 (2000)

2.6 M. Zinke-Allmang, L.C. Feldman, M.H. Grabow: Clustering on surfaces, Surf. Sci. Rep. **16**, 377–463 (1992)

2.7 A.-L. Barabási, H.E. Stanley: *Fractal Concepts in Surface Growth* (Cambridge Univ. Press, Cambridge 1995)

2.8 R. Kaischew, I.N. Stranski: Über den Mechanismus des Gleichgewichts kleiner Kriställchen II, Z. Phys. Chem. **B26**, 114–116 (1934), in German

2.9 J.K. Nørskov, K.W. Jacobsen, P. Stoltze, L.B. Hansen: Many-atom interactions in metals, Surf. Sci. **283**, 277–282 (1993)

2.10 W. Kossel: Zur Energetik von Oberflächenvorgängen, Nachrichten der Gesellschaft der Wissenschaften Göttingen, Mathematisch-Physikalische Klasse, Band 135 (1927), in German

2.11 I.N. Stranski: Über das Wachsen der Kristalle, Ann. Sofia Univ. **24**, 297–315 (1927), in Bulgarian

2.12 I.N. Stranski: Zur Theorie der Kristallwachstums, Z. Phys. Chem. **136**, 259–277 (1928), in German

2.13 R. Kaischew: On the history of the creation of the molecular-kinetic theory of crystal growth, J. Cryst. Growth **51**, 643–650 (1981)

2.14 A. Dupré: *Théorie Méchanique de la Chaleur* (Gauthier-Villard, Paris 1869) p. 369, in French

2.15 E. Bauer: Phänomenologische Theorie der Kristallabscheidung an Oberflächen I, Z. Krist. **110**, 372–394 (1958), in German

2.16 I.N. Stranski, K. Kuleliev: Beitrag zur isomorphen Fortwaschung von Ionenkristallen aufeinander, Z. Phys. Chem. A **142**, 467–476 (1929), in German

2.17 I.N. Stranski, L. Krastanov: Zur Theorie der orientierten Ausscheidung von Ionenkristallen aufeinander, Monatsh. Chem. **71**, 351–364 (1938), in German

2.18 F.C. Frank, J.H. van der Merwe: One-dimensional dislocations I. Static theory, Proc. R. Soc. Lond. Ser. A **198**, 205–216 (1949)

2.19 F.C. Frank, J.H. van der Merwe: One-dimensional dislocations II. Misfitting monolayers oriented overgrowth, Proc. R. Soc. Lond. Ser. A **198**, 216–225 (1949)

2.20 M. Volmer, A. Weber: Keimbildung in übersättigten Gebilden, Z. Phys. Chem. **119**, 277–301 (1926), in German

2.21 I. Markov: *Crystal Growth for Beginners*, 2nd edn. (World Scientific, New Jersey 2003)

2.22 M.H. Grabow, G.H. Gilmer: Thin film growth modes. Wetting and cluster nucleation, Surf. Sci. **194**, 333–346 (1988)

2.23 B. Voigtländer: Fundamental processes in Si/Si and Ge/Si epitaxy studied by scanning tunneling microscopy during growth, Surf. Sci. Rep. **43**, 127–254 (2001)

2.24 W.K. Burton, N. Cabrera, F.C. Frank: The growth of crystals and the equilibrium structure of their surfaces, Philos. Trans. R. Soc. Lond. Ser. A **243**, 299–358 (1951)

2.25 J.W. Gibbs: *On the Equilibrium of Heterogeneous Substances, Collected Works* (Longmans Green, New York 1928)

2.26 R.L. Dobrushin, R. Kotecky, S. Shlosman: *Wulff Construction: A Global Shape from Local Interactions* (American Mathematical Society, Providence 1993)

2.27 R. Kaischew: Equilibrium shape and work of formation of crystalline nuclei on substrates, Commun. Bulg. Acad. Sci. (Phys.) **1**, 100–133 (1950), in Bulgarian

2.28 I.N. Stranski, R. Kaischew: Gleichgewichtsformen homeopolarer Kristalle, Z. Kristallogr. **78**, 373–383 (1931), in German

2.29 R. Peierls: Clustering in adsorbed films, Phys. Rev. B **18**, 2013–2015 (1978)

2.30 Y.I. Frenkel: *Kinetic Theory of Liquids* (Dover, New York 1955)

2.31 R. Kaischew: On the thermodynamics of crystalline nuclei, Commun. Bulg. Acad. Sci. (Phys.) **2**, 191–202 (1951), in Bulgarian

2.32 I.N. Stranski: Zur Berechnung der spezifischen Oberflächen-, Kanten- und Eckenenergien an kleinen Kristallen, Ann. Sofia Univ. **30**, 367–375 (1936), in German

2.33 N. Cabrera, R.V. Coleman: Theory of crystal growth from the vapor. In: *The Art and Science of Growing Crystals*, ed. by J.J. Gilman (Wiley, New York 1963) pp. 3–28

2.34 H.-C. Jeong, E.D. Williams: Steps on surfaces: Experiment and theory, Surf. Sci. Rep. **34**, 171–294 (1999)

2.35 S. Stoyanov, I. Markov: On the 2D–3D transition in epitaxial thin film growth, Surf. Sci. **116**, 313–337 (1982)

2.36 S. Toschev, M. Paunov, R. Kaischew: On the question of formation of three-dimensional and two-dimensional nuclei in crystallization on substrates, Commun. Dept. Chem. Bulg. Acad. Sci. **1**, 119–129 (1968), in Bulgarian

2.37 I. Markov, R. Kaischew: Influence of the supersaturation on the mode of crystallization on crystalline substrates, Thin Solid Films **32**, 163–167 (1976)

2.38 I. Markov, R. Kaischew: Influence of the supersaturation on the mode of thin film growth, Krist. Tech. **11**, 685–697 (1976)

2.39 A. Crottini, D. Cvetko, L. Floreano, R. Gotter, A. Morgante, F. Tommasini: Step height oscillations during layer-by-layer growth of Pb on Ge(001), Phys. Rev. Lett. **79**, 1527–1530 (1997)

2.40 V.V. Voronkov: Movement of elementary step by formation of one-dimensional nuclei, Sov. Phys. Crystallogr. **15**, 13–19 (1970), in Russian

2.41 F.C. Frank: Nucleation-controlled growth one a one-dimensional growth of finite length, J. Cryst. Growth **22**, 233–1236 (1974)

2.42 J. Zhang, G.H. Nancollas: Kink densities along a crystal surface step at ow temperatures and under nonequilibrium conditions, J. Cryst. Growth **106**, 181–190 (1990)

2.43 S. Stoyanov: Formation of bilayer steps during growth and evaporation of Si(001) vicinal surfaces, Europhys. Lett. **11**, 361–366 (1990)

2.44 I. Markov: Kinetics of MBE growth of Si(001)1 × 1, Surf. Sci. **279**, L207–L212 (1992)

2.45 P. Vekilov: Kinetics and mechanisms of protein crystallization at the molecular level, Methods Mol. Biol. **300**, 15–52 (2005)

2.46 R. Becker, W. Döring: Kinetische Behandlung der Keimbildung in übersättigten Dämpfen, Ann. Phys. **24**, 719–752 (1935), in German

2.47 J.W. Christian: *The Theory of Transformations in Metals and Alloys*, 3rd edn. (Pergamon, New York 2002), Parts I and II

2.48 D. Kashchiev: *Nucleation* (Butterwords, Oxford 2000)

2.49 S.W. Benson: *The Foundations of Chemical Kinetics* (McGraw-Hill, New York 1960)

2.50 M. Volmer: *Kinetik der Phasenbildung* (Theodor Steinkopf, Dresden 1939), in German

2.51 J. Lothe, G.M. Pound: Reconsideration of nucleation theory, J. Chem. Phys. **36**, 2080–2085 (1962)

2.52 G.M. Pound, M.T. Simnad, L. Yang: Heterogeneous nucleation of crystals from vapor, J. Chem. Phys. **22**, 1215–1219 (1954)

2.53 J.G. Dash: Clustering and percolation transitions in helium and other thin films, Phys. Rev. B **15**, 3136–3146 (1977)

2.54 J.M. Liang, L.J. Chen, I. Markov, G.U. Singco, L.T. Shi, C. Farrell, K.N. Tu: Crystallization of amorphous $CoSi_2$ thin films I. Kinetics of nucleation and growth, Mater. Chem. Phys. **38**, 250–257 (1994)

2.55 J.P. Hirth, G.M. Pound: *Condensation and Evaporation, Progress in Materials Science* (MacMillan, New York 1963)

2.56 E. Korutcheva, A.M. Turiel, I. Markov: Coherent Stranski–Krastanov growth in 1+1 dimensions with anharmonic interactions: An equilibrium study, Phys. Rev. B **61**, 16890–16901 (2000)

2.57 J.E. Prieto, I. Markov: Thermodynamic driving force of formation of coherent three-dimensional islands in Stranski–Krastanov growth, Phys. Rev. B **66**, 073408 (2002)

2.58 D. Walton: Nucleation of vapor deposits, J. Chem. Phys. **37**, 2182–2188 (1962)

2.59 S. Stoyanov: Nucleation theory for high and low supersaturations. In: *Current Topics in Materials Science*, Vol. 3, ed. by E. Kaldis (North-Holland, Amsterdam 1979) pp. 421–462

2.60 S. Stoyanov: On the atomistic theory of nucleation rate, Thin Solid Films **18**, 91–98 (1973)

2.61 S. Toschev, I. Markov: An experimental study of nonsteady state nucleation, Ber. Bunsenges. Phys. Chem. **73**, 184–188 (1969)

2.62 A. Milchev, S. Stoyanov: Classical and atomistic models of electrolytic nucleation: comparison with experimental data, J. Electroanal. Chem. **72**, 33–43 (1976)

2.63 A. Milchev, J. Malinowski: Phase formation – Stability and nucleation kinetics of small clusters, Surf. Sci. **156**, 36–43 (1985)

2.64 D. Kashchiev: On the relation between nucleation, nucleus size and nucleation rate, J. Chem. Phys. **76**, 5098–5102 (1982)

2.65 B. Müller, L. Nedelmann, B. Fischer, H. Brune, K. Kern: Initial stages of Cu epitaxy on Ni(100): postnucleation and a well-defined transition in critical island size, Phys. Rev. B **54**, 17858–17865 (1996)

2.66 J.A. Venables: *Introduction to Surface and Thin Film Processes* (Cambridge Univ. Press, Cambridge 2000)

2.67 J.L. Robins, T.N. Rhodin: Nucleation of metal clusters on ionic surfaces, Surf. Sci. **2**, 320–345 (1964)

2.68 R. Kaischew, B. Mutaftschiev: Über die elektrolytische Keimbildung des Quecksilbers, Electrochim. Acta **10**, 643–650 (1965), in German

2.69 B. Lewis, D. Campbell: Nucleation and initial growth behavior of thin film growth, J. Vac. Sci. Technol. **4**, 209–218 (1967)

2.70 M.J. Stowell: The dependence of saturation nucleus density on deposition rate and substrate temperature in the case of complete condensation, Philos. Mag. **21**, 125–136 (1970)

2.71 I. Markov: The influence of surface diffusion processes on the kinetics of heterogeneous nucleation, Thin Solid Films **8**, 281–292 (1971)

2.72 R.A. Sigsbee: Vapor to condensed-phase heterogeneous nucleation. In: *Nucleation*, ed. by A.C. Zettlemoyer (Marcel Dekker, New York 1969) pp. 151–224

2.73 I. Markov, A. Boynov, S. Toschev: Screening action and growth kinetics of electrodeposited mercury droplets, Electrochim. Acta **18**, 377–384 (1973)

2.74 A.N. Kolmogorov: Statistical theory of crystallization of metals, Izv. Akad. Nauk USSR (Otd. Phys. Math. Nauk) **3**, 355–359 (1937), in Russian

2.75 M. Avrami: Kinetics of phase change. I. General theory, J. Chem. Phys. **7**, 1103–1112 (1939)

2.76 M. Avrami: Kinetics of phase change. II. Transformation-time relations for random distribution of nuclei, J. Chem. Phys. **8**, 212–224 (1940)

2.77 M. Avrami: Kinetics of phase change III. Granulation, phase change and microstructure of phase change, J. Chem. Phys. **9**, 177–184 (1941)

2.78 W. Johnson, R. Mehl: Reaction kinetics in processes of nucleation and growth, Trans. Am. Inst. Min. Metal. Eng. **135**, 416–458 (1939)

2.79 I. Markov, D. Kashchiev: The role of active centers in the kinetics of new phase formation, J. Cryst. Growth **13/14**, 131–134 (1972)

2.80 I. Markov, D. Kashchiev: Nucleation on active centres I. General theory, J. Cryst. Growth **16**, 170–176 (1972)

2.81 G. Zinsmeister: A contribution to Frenkel's theory of condensation, Vacuum **16**, 529–535 (1966)

2.82 G. Zinsmeister: Theory of thin film condensation, Part b: Solution of the simplified condensation equations, Thin Solid Films **2**, 497–507 (1968)

2.83 G. Zinsmeister: Theory of thin film condensation, Part c: Aggregate size distribution in islands films, Thin Solid Films **4**, 363–386 (1969)

2.84 G. Zinsmeister: Theory of thin film condensation, Part d: Influence of variable collision factor, Thin Solid Films **7**, 51–75 (1971)

2.85 D.R. Frankl, J.A. Venables: Nucleation on substrates from the vapor phase, Adv. Phys. **19**, 409–456 (1970)

2.86 J.A. Venables: Rate equations approaches to thin film nucleation and growth, Philos. Mag. **27**, 697–738 (1973)

2.87 S. Stoyanov, D. Kashchiev: Thin film nucleation and growth theories: A confrontation with experiment. In: *Current Topics in Materials Science*, Vol. 7, ed. by E. Kaldis (North-Holland, Amsterdam 1981), pp. 69–141

2.88 G.S. Bales, D.C. Chrzan: Dynamics of irreversible island growth during submonolayer epitaxy, Phys. Rev. B **50**, 6057–6067 (1994)

2.89 G.S. Bales, A. Zangwill: Self-consistent rate theory of submonolayer homoepitaxy with attachment/detachment kinetics, Phys. Rev. B **55**, R1973–R1976 (1997)

2.90 J.G. Amar, F. Family, P.M. Lam: Dynamic scaling of the island-size distribution and percolation in a model of submonolayer molecular beam epitaxy, Phys. Rev. B **50**, 8781–8797 (1994)

2.91 J.G. Amar, F. Family: Critical cluster size: Island morphology and size distribution in submonolayer epitaxial growth, Phys. Rev. Lett. **74**, 2066–2069 (1995)

2.92 H. Brune, G.S. Bales, J. Jacobsen, C. Boragno, K. Kern: Measuring surface diffusion from nucleation island densities, Phys. Rev. B **60**, 5991–6006 (1999)

2.93 J.A. Venables, G.D.T. Spiller, M. Handbücken: Nucleation and growth of thin films, Rep. Prog. Phys. **47**, 399–460 (1984)

2.94 D. Kandel: Initial stages of thin film growth in the presence of island-edge barriers, Phys. Rev. Lett. **78**, 499–502 (1997)

2.95 C. Ratsch, P. Šmilauer, A. Zangwill, D.D. Vvedensky: Submonolayer epitaxy without a critical nucleus, Surf. Sci. **329**, L599–L604 (1995)

2.96 S. Iwanari, K. Takayanagi: Surfactant epitaxy of Si on Si(111) surface mediated by a Sn layer I. Reflection electron microscope observation of the growth with and without a Sn layer mediate the step flow, J. Cryst. Growth **119**, 229–240 (1992)

2.97 I.-S. Hwang, T.-C. Chang, T.T. Tsong: Exchange-barrier effect on nucleation and growth of surfactant mediated epitaxy, Phys. Rev. Lett. **80**, 4229–4232 (1998)

2.98 B. Voigtländer, A. Zinner, T. Weber, H.P. Bonzel: Modification of growth kinetics in surfactant mediated epitaxy, Phys. Rev. B **51**, 7583–7591 (1995)

2.99 A.A. Chernov: *Modern Crystallography III*, Springer Series in Solid State Sciences, Vol. 36 (Springer, Berlin 1984)

2.100 G. Ehrlich, F.G. Hudda: Atomic view of surface self-diffusion: tungsten on tungsten, J. Chem. Phys. **44**, 1039–1049 (1966)

2.101 E. Müller: Das Feldionenmikroskop, Z. Phys. **131**, 136–142 (1951), in German

2.102 S.C. Wang, T.T. Tsong: Measurements of the barrier height on the reflective W(110) plane boundaries in surface diffusion of single atoms, Surf. Sci. **121**, 85–97 (1982)

2.103 S.C. Wang, G. Ehrlich: Atom condensation at lattice steps and clusters, Phys. Rev. Lett. **71**, 4174–4177 (1993)

2.104 S.C. Wang, G. Ehrlich: Atom incorporation at surface clusters: an atomic view, Phys. Rev. Lett. **67**, 2509–2512 (1991)

2.105 P. Feibelman: Surface diffusion by concerted substitution, Comments Condens. Matter. Phys. **16**, 191–203 (1993)

2.106 R. Schwoebel, E.J. Shipsey: Step motion on crystal surfaces, J. Appl. Phys. **37**, 3682–3686 (1966)

2.107 R. Schwoebel: Step motion on crystal surfaces II, J. Appl. Phys. **40**, 614–618 (1966)

2.108 I. Markov: Kinetics of surfactant mediated epitaxial growth, Phys. Rev. B **50**, 11271 (1994)

2.109 J. Villain: Continuum models of crystal growth from atomic beams with and without desorption, J. Phys. I France **1**, 19–42 (1991)

2.110 J. Tersoff, A.W. Denier van der Gon, R.M. Tromp: Critical island size for layer-by-layer growth, Phys. Rev. Lett. **72**, 266–269 (1994)

2.111 S. Stoyanov: Layer growth of epitaxial films and superlattices, Surf. Sci. **199**, 226–242 (1988)

2.112 I. Markov: Surface energetics from the transition from step-flow growth to two-dimensional nucleation in metal homoepitaxy, Phys. Rev. B **56**, 12544–12552 (1997)

2.113 R. Gerlach, T. Maroutian, L. Douillard, D. Martinotti, H.-J. Ernst: A novel method to determine the Ehrlich–Schwoebel barrier, Surf. Sci. **480**, 97–102 (2001)

2.114 M. Bott, T. Hohage, G. Comsa: The homoepitaxial growth of Pt on Pt(111) studied by STM, Surf. Sci. **272**, 161–166 (1992)

2.115 P. Feibelmann, J.S. Nelson, G.L. Kellogg: Energetics of Pt adsorption on Pt(111), Phys. Rev. B **49**, 10548–10556 (1994)

2.116 M. Bott, T. Hohage, M. Morgenstern, T. Michely, G. Comsa: New approach for determination of diffusion parameters of adatoms, Phys. Rev. Lett. **76**, 1304–1307 (1996)

2.117 J. Krug, P. Politi, T. Michely: Island nucleation in the presence of step-edge barriers: Theory and applications, Phys. Rev. B **61**, 14037–14046 (2000)

2.118 I. Markov: Method for evaluation of the Ehrlich–Schwoebel barrier to interlayer transport in metal homoepitaxy, Phys. Rev. B **54**, 17930–17937 (1996)

2.119 J. Rottler, P. Maass: Second layer nucleation in thin film growth, Phys. Rev. Lett. **83**, 3490–3493 (1999)

2.120 S. Heinrichs, J. Rottler, P. Maass: Nucleation on top of islands in epitaxial growth, Phys. Rev. B **62**, 8338–8359 (2000)

2.121 C. Priester, M. Lannoo: Origin of self-assembled quantum dots in highly mismatched heteroepitaxy, Phys. Rev. Lett. **75**, 93–96 (1995)

2.122 J.E. Prieto, I. Markov: Quantum dots nucleation in strained-layer epitaxy: Minimum energy pathway in the stress-driven two-dimensional to three-dimensional transformation, Phys. Rev. B **72**, 205412 (2005)

2.123 I. Markov, A. Trayanov: Epitaxial interfaces with realistic interatomic forces, J. Phys. C **21**, 2475–2493 (1988)

2.124 I. Markov: Static multikink solutions in a discrete Frenkel–Kontorova model with anharmonic interactions, Phys. Rev. B **48**, 14016–14019 (1993)

2.125 J.W. Matthews, E. Grünbaum: The need for contaminants in the epitaxial growth of gold on rock salt, Appl. Phys. Lett. **5**, 106–108 (1964)

2.126 E. Grünbaum: Epitaxial growth of single-crystal films, Vacuum **24**, 153–159 (1973)

2.127 D.A. Steigerwald, I. Jacob, W.F. Egelhoff Jr.: Structural study of the epitaxial growth of fcc-Fe films, sandwiches and superlattices on Cu(100), Surf. Sci. **202**, 472–492 (1988)

2.128 M. Copel, M.C. Reuter, E. Kaxiras, R.M. Tromp: Surfactants in epitaxial growth, Phys. Rev. Lett. **63**, 632–635 (1989)

2.129 I. Markov: Surfactants in semiconductor heteroepitaxy: thermodynamics and/or kinetics?. In: *NATO ASI Series: Collective Diffusion on Surfaces: Correlation Effects and Adatom Interactions*, ed. by M. Tringides, Z. Chvoy: (Kluwer, Dordrecht 2001) pp. 259–271

2.130 I. Markov: Kinetics of nucleation in surfactant-mediated epitaxy, Phys. Rev. B **53**, 4148–4155 (1996)

2.131 G. Rosenfeld, R. Servaty, C. Teichert, B. Poelsema, G. Comsa: Layer-by-layer growth of Ag on Ag(111) induced by enhanced nucleation: A model study for surfactant-mediated growth, Phys. Rev. Lett. **71**, 895–898 (1993)

2.132 H.A. van der Vegt, J. Vrijmoeth, R.J. Behm, E. Vlieg: Sb-enhanced nucleation in homoepitaxial growth of Ag(111), Phys. Rev. B **57**, 4127–4131 (1998)

2.133 I. Markov: Scaling behavior of the critical terrace width for step-flow growth, Phys. Rev. B **59**, 1689–1692 (1999)

2.134 I. Markov: Nucleation and step-flow growth in surfactant mediated homoepitaxy with exchange/de-exchange kinetics, Surf. Sci. **429**, 102–116 (1999)

2.135 M. Horn-von Hoegen, J. Falta, R. Tromp: Surfactants in Si(111) homoepitaxy, Appl. Phys. Lett. **66**, 487–489 (1995)

2.136 D. Kandel, E. Kaxiras: Surfactant mediated crystal growth of semiconductors, Phys. Rev. Lett. **75**, 2742–2745 (1995)

2.137 R. Hultgren, P.D. Desai, D.T. Hawkins, M. Gleiser, K.K. Kelley, D.D. Wagman: *Selected Values of the Thermodynamic Properties of the Elements* (American Society for Metals, Metals Park 1973)

3. Morphology of Crystals Grown from Solutions

Morphology

Francesco Abbona, Dino Aquilano

Growth from solutions is widely used both in research laboratories and in many industrial fields. The control of crystal habit is a key point in solution growth as crystals may exhibit very different shapes according to the experimental conditions. In this chapter a concise review is given on this topic. First, the equilibrium shape is rather deeply developed due to its primary importance to understand crystal morphology, then the growth shape is treated and the main factors affecting the crystal habit are briefly illustrated and discussed. A rich literature completes the chapter.

Interest in the crystal habit of minerals dates back a long time in the history of mankind. A detailed history on this topics and crystallization in general is given by *Scheel* [3.1]; here only a short account of crystal morphology is presented. Crystal habit, which attracted the interest of great scientists such as Kepler, Descartes, Hooke, and Huygens, is relevant from the scientific point of view, since it marks the beginning of crystallography as a science. Its birth can be dated to 1669 when the Danish scientist Niels Steensen, studying in Florence the quartz and hematite crystals from Elba island, suggested the first law of crystallography (constancy of the dihedral angle) and the mechanism of face growth (layer by layer). A century later this law was confirmed by Romé de l'Isle. At the end of the 18th century the study of calcite crystals led the French abbé René Just Haüy to enunciate the first theory on crystal structure and to discover the second law (rational indices). It is worth noticing that these early scholars met with great difficulty in studying crystal habit since, contrary to botany and zoology where each species has its own definite morphology, the crystal habit of minerals is strongly variable within the same species. In the first part of the 19th century the study of crystal habit led to the development of the concept of symmetry and the derivation of the 32 crystal classes. Bravais, by introducing the idea of the crystal lattice, was the first to try to relate crystal habit to internal structure (the Bravais law, saying that the crystal faces are lattice planes of high point density). At the end of the 19th century research on internal symmetry ended with the derivation of the 230 space groups. In this century research on crystallization, mainly from solution but also from melt, went on and interlaced with progress in other disciplines (chemistry, physics, thermodynamics, etc.). We should recall the important contributions by Gibbs (1878), Curie (1885), and Wulff (1901) on the equilibrium form of crystals, which was tackled later from an atomistic point of view by *Stranski* [3.2] and *Stranski* and *Kaischew* [3.3, 4].

The relationship between morphology and internal structure (the Bravais law) was treated by *Niggli* [3.5] and developed by *Donnay* and *Harker* [3.6], who considered the space group instead of the Bravais lattice type as a factor conditioning the crystal morphology. From about 1950 onwards, interest in crystal growth increased due to the role of crystals in all kinds of industry and the discovery of relevant properties of new crystalline compounds. Besides the technological progress, a milestone was the publication in 1951 of the first theory on growth mechanisms of flat crystal faces by *Burton*, *Cabrera*, and *Frank* (BCF) [3.7].

Also, the crystal habit was receiving growing attention due to theoretical interest and industrial needs. The Donnay–Harker principle is exclusively crystallographic. A chemical approach was adopted by Hartman and Perdok; looking at crystal structure as a network of periodic bond chains (PBC) they published in 1955 a method that is still fundamental to studies of theoretical crystal morphology [3.8–10]. The method, at first qualitative, was made quantitative through the calculation of the broken bond energy and, since about 1980, has been integrated with the statistical mechanical theory of Ising models which led to the integrated Hartman–Perdok roughening transition theory [3.11], later applied to modulated crystals [3.12]. These methods do not take into account the external habit-controlling factors, namely the effects of fluid composition and supersaturation, which are explicitly considered in the interfacial structure (IS) analysis [3.13]. An improvement in predicting morphology was represented by the application of ab initio calculations to the intermolecular interactions between tailor-made additives and crystal surface [3.14].

Computer facilities have promoted tremendous advances in all kinds of calculation necessary in the different sectors of crystal growth, enabling progress in theoretical approaches and sophisticated simulations which are now routine practice. A relevant instrumental advance was achieved when atomic force microscopy (AFM) was applied to study the features of crystal faces, giving new impulse to a topic that had always been the center of thorough research [3.15–18].

This chapter is devoted to the morphology of crystals grown from solution. In the first part, the theoretical equilibrium and growth shapes of crystals are treated from the thermodynamic and atomistic points of view. In the second part the factors affecting crystal habit will be considered with some specific examples. High-temperature solution growth, mass, and protein crystallization are excluded to limit the scope of the chapter.

3.1 Equilibrium Shape

When equilibrium is reached between a crystalline phase and its surroundings, the statistical amount of growth units exchanged between the two phases is the same and does not change with time. This implies that the crystallized volume remains constant, but nothing is specified about many important questions, such as:

1. The surface of the crystals, i. e., how large its extension is and which $\{hkl\}$ forms enter the equilibrium shape (ES).
2. The difference, if any, between the stable ES of a crystal immersed in either a finite or infinite mother phase and the unstable shape obtained when the activation energy for nucleation is reached.
3. How does the ES change when some adhesion is set up between the crystal and a solid substrate?
4. How can solvent and impurity concentrations affect the ES?

To address these questions, a few elementary concepts must be fixed to structure our language and a simple but effective crystal model adopted in the following.

3.1.1 The Atomistic Approach: The Kossel Crystal and the Kink Site

Let us consider a perfect monoatomic, isotropic, and infinite crystal. The work needed to separate an atom occupying a *mean lattice site* from all its n neighbors is $\varphi^{\text{sep}} = \sum_i^n \psi_i$, where ψ_i is the energy binding one atom to its ith neighbor. We will see later on that this peculiar site really exists and is termed a *kink*. The potential energy (per atom) of the crystal will be $\varepsilon_{\text{p}}^{\text{c}\infty} = -(1/2)\varphi^{\text{sep}}$. The simplest model, valid for homopolar crystals, is due to *Kossel* [3.19]. Atoms are replaced by elementary cubes bounded by pair interactions, $\psi_1, \psi_2, \ldots, \psi_n$: the separation work between the first, second, and nth neighbors, with the pair potential decreasing with distance, $\psi_1 > \psi_2 > \ldots > \psi_n$ (Fig. 3.1a). In the first-neighbors approximation, the separation work for an atom lying in the crystal bulk is $\varphi^{\text{sep}} = 6\psi_1$. Thus, $\varepsilon_{\text{p}}^{\text{c}\infty} = -3\psi_1$. On the other hand, $\varepsilon_{\text{p}}^{\text{c}\infty}$ represents the variation of the potential energy that an atom undergoes when going from the vapor to a *mean lattice site*, which coincides with a well-defined surface site, as suggested by *Kossel* [3.19] and *Stranski* [3.2]. Once an atom has entered this special site, the potential energy variation of the considered system is equal to $-3\psi_1$ and so the separation work for

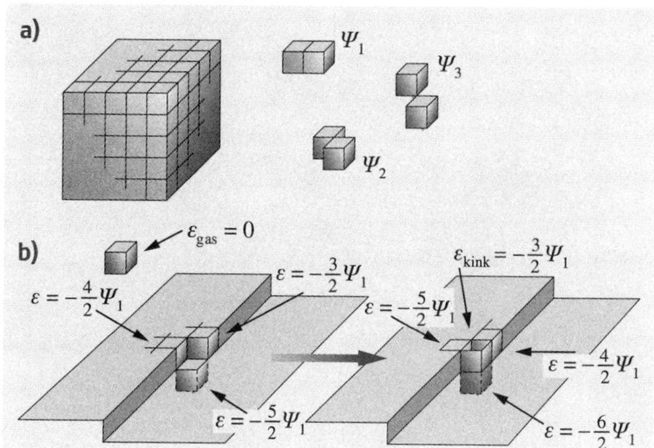

Fig. 3.1 **(a)** Kossel crystal; separation work between first (ψ_1), second (ψ_2), and third (ψ_3) neighbors. **(b)** When an atom enters a kink, there is a transition in the potential energy, the difference between final and initial stage being $-3\psi_1$ (first neighbors)

an atom occupying this site is $\varphi_{\text{c}\infty} = 3\psi_1$ (Fig. 3.1b). A *kink* is the name adopted worldwide for this site, for practical reasons. Different historical names have been given: *repetitive step* [3.2, Z. Phys. Chem.] and *half-crystal position* [3.2, Annu. Univ. Sofia], both related to the physics of the site. In fact, deposition or evaporation of a growth unit onto/from a kink reproduces another kink, thus generating an equal probability for the two processes [3.20]. Moreover, the chemical potential (μ) of a unit in a kink is equal to that of the vapor. Hence, *kinks are crystal sites in a true* (and not averaged) *thermodynamic equilibrium*, as will be shown below.

3.1.2 Surface Sites and Character of the Faces

Flat (F) faces. A crystal surface, in equilibrium with its own vapor and far from absolute zero temperature, is populated by steps, adsorbed atoms, and holes. In the Kossel model all sites concerning the adsorption and the outermost lattice level are represented (Fig. 3.2). The percentage of corner and edge sites is negligible for an infinite crystal face, and hence we will confine our attention to the *adsorption and incorporation sites*. Crystal units can adsorb either on the surface terraces (ad_s) or on the steps (ad_l), with the same situation occurring for the incorporation sites (in_s, in_l).

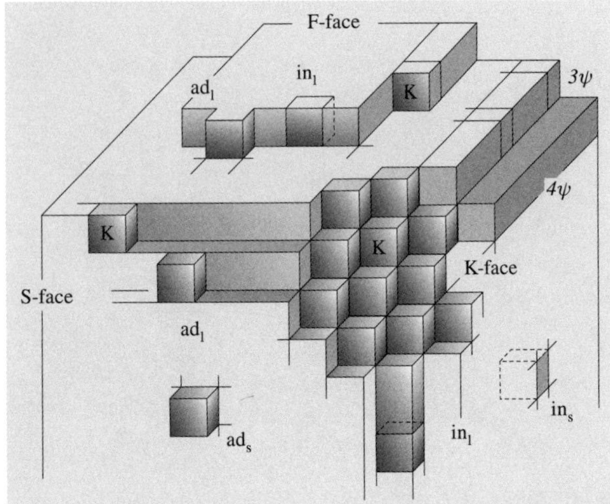

Fig. 3.2 The different types of faces of a Kossel crystal: {100}-F, {111}-K, and {110}-S faces. Adsorption (ad_s, ad_l) and incorporation (in_s, in_l) sites are shown on surfaces and steps. The uniqueness of the K (kink) site is also shown

The binding energies of *ad-sites* and *in-sites are complementary to one another*

$$\varphi_{\text{ad}_\text{s}} + \varphi_{\text{in}_\text{s}} = \varphi_{\text{ad}_\text{l}} + \varphi_{\text{in}_\text{l}} = 2\varphi_{\text{kink}} \rightarrow \varphi_{\text{ad}} + \varphi_{\text{in}} = 2\varphi_{\text{kink}} , \tag{3.1}$$

which is generally valid since it depends neither on the type of face, nor on the crystal model, nor on the kind of lattice forces [3.21, p. 56]. The interaction of the unit in the kink with the crystal (φ_{kink}) consists of two parts. The first represents its *attachment energy* (φ_{att}) with all the *crystal substrate*, and coincides with that of an ad-unit, which implies

$$\varphi_{\text{att}} = \varphi_{\text{ad}} . \tag{3.2a}$$

The second is its *slice energy* (φ_{slice}), i. e., the interaction with the half of the outermost crystal slice, $\varphi_{\text{slice}} = (\omega/2)$, where ω is the interaction of the unit with all of its slice. Thus

$$\varphi_{\text{in}} = \varphi_{\text{att}} + \omega , \tag{3.2b}$$

and, from relation (3.1)

$$\varphi_{\text{kink}} = \varphi_{\text{att}} + \varphi_{\text{slice}} . \tag{3.2c}$$

Relation (3.2c) states that φ_{att} and φ_{slice} *of a growth unit are complementary to one another*. In fact, since φ_{kink} is constant for a given crystal, the higher the lateral interaction of one unit, the lower its interaction with the subjacent crystal. This criterion is of the utmost importance for understanding the growth morphology of crystals. Moreover, the binding of a growth unit must fulfil the qualitative inequality: $\varphi_{\text{ad}} < \varphi_{\text{kink}} < \varphi_{\text{in}}$. The quantitative treatment was elegantly addressed by *Kaischew* [3.3, 4], who calculated the coverage degree (θ_i) and other related quantities for every i-site of the surface drawn in Fig. 3.2

$$\theta_i = \{1 + \exp[(\varphi_{\text{kink}} - \varphi_i)/(k_\text{B}T)]\}^{-1} , \tag{3.3}$$

where k_B is the Boltzmann constant. For a (001) Kossel surface and within the first-neighbors approximation, having assumed for the binding energy the standard value $\psi_1 = 4k_\text{B}T$ (valid for Au crystals not far from the melting point), the set of results shown in Table 3.1 was obtained.

From Table 3.1 it follows that:

1. Kinks are the only sites in thermodynamic equilibrium, being half filled and half empty at the same time.
2. Ad-units form a very dilute layer (row) which moves randomly on the surface (step edge) and hence cannot belong to the crystal.
3. In-units belong to the crystal, from which they may escape, generating a temporary hole, with a very low exchange frequency with respect to the other sites.

Looking at the face as a whole, the face profile can neither advance nor move backwards: hence, the face is in *macroscopic equilibrium*. Fluctuations around the equilibrium cannot change its flatness since the lifetime of the growth units in the ad-sites is very short and the vacancies generated among the in-sites are filled again in

Table 3.1 Coverage degree (3.3) and exchange frequency of growth units in the main surface sites of the (001) face of a Kossel crystal, assuming $\psi_1 = 4k_\text{B}T$ (after [3.21]). The exchange frequency is the reciprocal of the mean time between two successive evaporation (or condensation) events on the same i-site (i. e. s^{-1} indicates the number of exchanges per unit time in a given site)

Type of surface site	Separation work	Coverage degree θ_i	Exchange frequency (s^{-1})
$\text{ad}_{\text{surface}}$	ψ_1	0.0003	3.06×10^7
ad_{ledge}	$2\psi_1$	0.0180	3.02×10^7
kink	$3\psi_1$	1/2	1.54×10^7
in_{ledge}	$4\psi_1$	0.9820	5.55×10^6
$\text{in}_{\text{surface}}$	$5\psi_1$	0.9997	1.03×10^4

even shorter time. So, this kind of *equilibrium face* has been named an F-*type* (flat) *face*.

Kinked (K) and Stepped (S) Faces. The uniqueness of F-faces is even more evident when considering the behavior of the {111} form of a Kossel crystal, near the equilibrium. Only kinks can be found on this surface and hence only one type of binding exists ($3\psi_1$) among growth units, within the first neighbors. Since in this case no units exhibit bonds in their slice, $\omega = 0$, which implies: $\varphi_{\text{ad}} = \varphi_{\text{kink}} = \varphi_{\text{in}}$. With every ad-unit transforming into an in-unit, the surface profile is not constrained and hence fluctuates, with the mother phase, around the equilibrium. This interface is diffuse and the corresponding faces are termed K (kinked) *faces*.

The behavior of the {110} form may be thought of as midway between that of F- and K-faces, since only ledge-type sites exist, apart from the kinks. Any fluctuation near the equilibrium can lead either to the evaporation of an entire [100] step or to the growth of a new one. In the first case, it is sufficient that a unit leaves an in-ledge site to promote step evaporation, while in the second case the formation of an ad-ledge site automatically generates two kinks, allowing the filling of a new step. Both processes are not correlated, even for contiguous steps, since there are no lateral bonds ($\omega = 0$) in the outermost (110) slice; thus, steps can form (or disappear) independently of each other and may bunch, giving rise to an undulating profile around the zone axis. Parallel steps being the feature of this kind of surface, the corresponding faces are termed S-*type* (stepped) *faces*.

3.1.3 The Equilibrium Crystal – Mother Phase: The Atomistic Point of View

Here we will deal with the equilibrium between a crystal and its vapor; however, *our conclusions can be basically applied to solutions and melts as well.* Let us consider a Kossel crystal built by n^3 units (each having mass m and vibration frequency ν). Since the work to separate two first neighbors is ψ, the mean evaporation energy of the n-sized crystal is easily calculated

$$\langle \Delta H \rangle_{cn} = 3\psi[1-(1/n)] = \varphi_{cn} \,. \tag{3.4a}$$

Then, for an infinite-sized crystal,

$$\langle \Delta H \rangle_{c\infty} = 3\psi = \varphi_{c\infty} = \text{const} \,. \tag{3.4b}$$

This means that the *units belonging to the crystal surface* reduce the value of the mean evaporation energy and so they *cannot be neglected when dealing with finite crystals.*

An Infinite Crystal and Its Mother Phase

As shown in Appendix 3.A, the equilibrium pressure (p_{eq}^{∞}) between a monoatomic vapor and its infinite crystalline phase decreases with its evaporation work $\varphi_{c\infty} = (\varepsilon_{\text{V}} - \varepsilon_{c\infty})$, according to

$$p_{\infty}^{\text{eq}} = [(2\pi m)^{3/2}(k_{\text{B}}T)^{-1/2}\nu^3]\exp(-\varphi_{c\infty}/(k_{\text{B}}T)) \,, \tag{3.5a}$$

ε_{V} and $\varepsilon_{c\infty}$ being the potential energy of a unit in the vapor and in the infinite crystal, respectively. The term $p\text{d}V$ can be neglected in $\langle \Delta H \rangle_{c\infty}$ with respect to the term ($\text{d}U$). Assuming, as a reference level, $\varepsilon_{\text{V}} = 0$, it is easy to show that $\langle \Delta H \rangle_{c\infty} = \varphi_{c\infty} = -\varepsilon_{c\infty}$.

The Finite Crystal – The Link to the Thermodynamic Supersaturation

When dealing with finite crystals (3.5a) transforms simply by changing $\varepsilon_{c\infty}$ with ε_{cn}, which is the potential energy of a unit in the finite crystal. It ensues that $\varphi_{cn} = (\varepsilon_{\text{V}} - \varepsilon_{cn})$. The frequency ($\nu$) does not vary from large to small crystal size, so

$$p_n^{\text{eq}} = (2\pi m)^{3/2}(k_{\text{B}}T)^{-1/2}\nu^3 \exp(-\varphi_{cn}/(k_{\text{B}}T)) \,. \tag{3.5b}$$

From (3.5a) and (3.5b) the following fundamental relation is obtained:

$$p_n^{\text{eq}} = p_{\infty}^{\text{eq}} \exp[(\varphi_{c\infty} - \varphi_{cn})/(k_{\text{B}}T)] \,. \tag{3.5c}$$

Since $\varphi_{\infty} > \varphi_n$, (3.5c) shows that *the equilibrium pressure for finite crystals is higher than that for infinite ones.* This can also be written

$$\varphi_{c\infty} - \varphi_{cn} = k_{\text{B}}T \ln\left(p_n^{\text{eq}}/p_{\infty}^{\text{eq}}\right) = k_{\text{B}}T \ln\beta \,, \tag{3.6}$$

where $\beta = p_n^{\text{eq}}/p_{\infty}^{\text{eq}} = (p_{\infty}^{\text{eq}} + \Delta p)/p_{\infty}^{\text{eq}} = 1+\sigma$ is the *supersaturation ratio* of the vapor with respect to the finite crystal. The (percentage) distance from equilibrium is $\sigma = (\Delta p/p_{\infty}^{\text{eq}})$, the exceeding pressure being $\Delta p = p_n^{\text{eq}} - p_{\infty}^{\text{eq}}$.

Equilibrium can also be viewed in terms of chemical potentials. Using the Helmholtz free energy, the chemical potentials, per unit, of the infinite and finite crystal read: $\mu_{c\infty} = -\varphi_{c\infty} - Ts_{c\infty}$ and $\mu_{cn} = -\varphi_{cn} - Ts_{cn}$. The vibrational entropies per unit, $s_{c\infty}$ and s_{cn}, are very close. Thus $\varphi_{c\infty} - \varphi_{cn} = \mu_{cn} - \mu_{c\infty} = \Delta\mu$ (Fig. 3.3). Hence, the following master equation for the equilibrium is obtained:

$$\Delta\mu = k_{\text{B}}T \ln\beta \,, \tag{3.7}$$

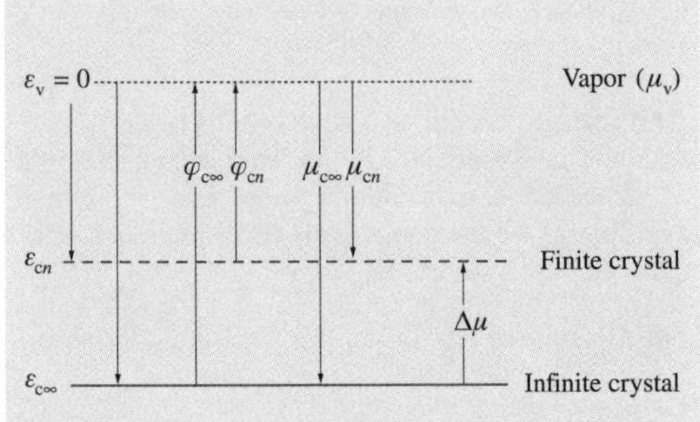

Fig. 3.3 Potential energy ε, evaporation work φ, and chemical potential μ of a growth unit in the vacuum, in a *mean site* of both finite and infinite crystal. $\Delta\mu = \mu_{cn} - \mu_{c\infty}$ is the thermodynamic supersaturation

where $\Delta\mu$ is the *thermodynamic supersaturation*. In heterogeneous systems a unit spontaneously goes from the higher chemical potential (μ') to the lower one (μ''). During the transition a chemical work ($\mu'' - \mu') = -\Delta\mu$ is gained, per growth unit.

The equilibrium between a finite crystal and its surroundings is analogous to the equilibrium of a spherical liquid drop of radius r (finite condensed phase 2) immersed in its own vapor (infinite dispersed phase 1). The phenomenological treatment is detailed in [3.21], where the two different equilibria are compared in the same way as we dealt with the atomistic treatment. Hence, one obtains the Thomson–Gibbs formula for droplets

$$\Delta\mu = k_B T \ln(p/p_{eq}) = \Omega_2 p_\gamma = 2\Omega_2(\gamma/r)\,, \quad (3.8)$$

where:

1. p_{eq} is the pressure of the vapor in equilibrium with a flat liquid surface
2. γ and Ω_2 are the surface tension at the drop–vapor interface and the molecular volume of the drop, respectively
3. The capillarity pressure p_γ at the drop interface defined by Laplace's relation ($p_\gamma = 2\gamma/r$) equilibrates the difference between the internal pressure of the drop (p_r) and the actual vapor pressure (p): $p_\gamma = (p_r - p)$.

The ratio (p/p_{eq}) is nothing else than β. When working with ideal or nonideal solutions, β is expressed by the concentrations (c/c_{eq}) or by the activities (a/a_{eq}), respectively. When a crystal is considered instead of a liquid drop, the system is no longer isotropic and then the radius r represents only the *size* of the crystal, as we will see later on. Nevertheless, the Thomson–Gibbs formula continues to be valid and expresses the relation among the deviation $\Delta\mu$ of the solution from saturation, the tension γ_{cs} of the crystal–solution interface, and the size of the crystals in equilibrium with the solution.

3.1.4 The Equilibrium Shape of a Crystal on a Solid Substrate

This topics has been deeply treated by *Kern* [3.22], who considered simultaneously both mechanical (capillary) and chemical (thermodynamic) equilibrium to obtain the ES of a crystal nucleating on a substrate from a dispersed phase. In preceding treatments, the Curie–Wulff condition and the Wulff theorem [3.23] only took into account the minimum of the crystal surface energy, the crystal volume remaining constant. According to [3.22], when n_A units of a phase A (each having volume Ω) condense under a driving force $\Delta\mu$ on a solid substrate B (heterogeneous nucleation) to form a three-dimensional (3-D) crystal (Fig. 3.4), the corresponding variation of the free Gibbs energy reads

$$\Delta G^{3\text{-D}}_{\text{hetero}} = -n_A \times \Delta\mu + \left(\gamma_i^A - \beta_{adh}\right) S_{AB} + \sum_j \gamma_j^A S_j^A\,, \quad (3.9)$$

where the second and the third term represent the work needed to generate the new crystal–substrate interface of area S_{AB} and the free crystal surfaces (of surface tension γ_j^A and area S_j^A), respectively.

The term $(\gamma_i^A - \beta_{adh})S_{AB}$ comes from the balance between the surface work lost ($-\gamma_B \times S_{AB}$) and gained ($\gamma_{AB} \times S_{AB}$) during nucleation. It is obtained from Dupré's formula: $\gamma_{AB} = \gamma_B + \gamma_i^A - \beta_{adh}$, where γ_{AB} is the crystal/substrate tension, γ_B is the surface tension of the substrate, γ_i^A is the surface tension of

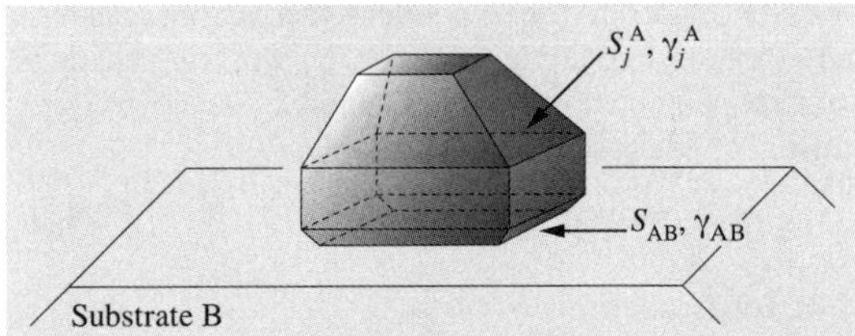

Fig. 3.4 Surface parameters involved in the balance of the free Gibbs energy variation when n_A units of a phase A condense on a solid substrate B to form a 3-D crystal (heterogeneous nucleation)

the i-face of the A crystal (when considered not in contact with the substrate), and β_{adh} stands for the specific crystal/substrate adhesion energy. At the (unstable) equilibrium of the nucleation any variation of $\Delta G^{\text{3-D}}_{\text{hetero}}$ must vanish. Then, under the reasonable assumption that also the specific surface tensions do not vary for infinitesimal changes of the crystal size,

$$\mathrm{d}\left(\Delta G^{\text{3-D}}_{\text{hetero}}\right) = -\mathrm{d}n_{\text{A}} \times \Delta\mu + \left(\gamma_i^{\text{A}} - \beta_{\text{adh}}\right)\mathrm{d}S_{\text{AB}} + \sum_j \gamma_j^{\text{A}}\,\mathrm{d}S_j^{\text{A}} = 0\,. \tag{3.10}$$

The fluctuation $\mathrm{d}n_{\text{A}}$ is related to those of the face areas ($\mathrm{d}S_j^{\text{A}}$ and $\mathrm{d}S_{\text{AB}}$) and to their distances (h_j and h_{s}) with respect to the crystal center. Then, (3.10) may be written in terms of $\mathrm{d}S_j^{\text{A}}$ and $\mathrm{d}S_{\text{AB}}$. Its solution is a continuous proportion between the energies of the faces and their h_j and h_s values

$$\frac{\gamma_1^{\text{A}}}{h_1} = \frac{\gamma_2^{\text{A}}}{h_2} = \cdots = \frac{\gamma_j^{\text{A}}}{h_j} = \frac{\gamma_i^{\text{A}} - \beta_{\text{adh}}}{h_{\text{s}}} = \text{const} = \frac{\Delta\mu}{2\Omega}\,. \tag{3.11}$$

This is the *unified Thomson–Gibbs–Wulff (TGW) equation*, which provides the ES of a crystal nucleated on a solid substrate:

1. The ES is a polyhedron limited by faces whose distances from the center are as shorter as lower their γ values.
2. The distance of the face in contact with the substrate will depend not only on the γ value of the lattice plane parallel to it, but also on its adhesion energy.
3. The faces entering the ES will be only those limiting the *most inner* polyhedron, its size being determined once $\Delta\mu$ and one out of the γ values are known.

The analogy between the crystal ES and that of a liquid drop on solid substrates is striking. It is useful to recall Young's relation for the mechanical equilibrium of a liquid drop on a substrate (Fig. 3.5)

$$\gamma_{\text{sl}} = \gamma_{\text{lv}} \cos\alpha + \gamma_{\text{sv}}\,, \tag{3.12a}$$

where α is the contact angle and γ_{sl}, γ_{lv}, and γ_{sv} are the surface energies of the substrate–liquid, liquid–vapor, and substrate–vapor interfaces, respectively. Besides, from Dupré's relation one obtains

$$\gamma_{\text{sl}} = \gamma_{\text{sv}} + \gamma_{\text{lv}} - \beta_{\text{adh}}\,. \tag{3.12b}$$

Since $-1 \le \cos\alpha \le 1$, the range of the adhesion energy (wetting) must fulfil the condition

$$2\gamma_{\text{lv}} \ge \beta_{\text{adh}} \ge 0\,. \tag{3.12c}$$

Adhesion values affect the sign of the numerator in the term $\left(\gamma_i^{\text{A}} - \beta_{\text{adh}}\right)/h_{\text{s}}$ (3.11).

The ES of the crystal is a nontruncated polyhedron when the crystal/substrate adhesion is null, as occurs for homogeneous nucleation. However, as the adhesion increases, the truncation increases as well, reaching its maximum when $\beta_{\text{adh}} = \gamma_i^{\text{A}}$. If the wetting

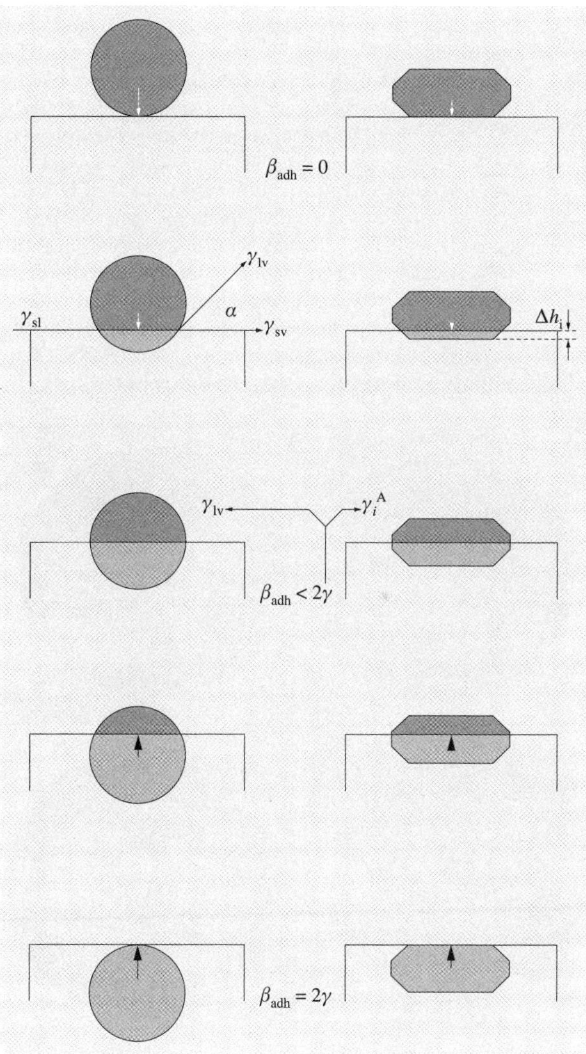

Fig. 3.5 Analogy between the equilibrium shape of a liquid drop on a solid substrate and that of a crystal, both heterogeneously nucleated. The adhesion energy β_{adh} rules both the contact angle of the drop with the substrate and the *crystal truncation*

Part A | 3.1

further increases the truncation decreases, along with the thickness of the crystal cup. When β_{adh} reaches its extreme value, $2\gamma_i^A$, the crystal thickness reduces to a *monomolecular* layer.

The Equilibrium Shape of a Finite Crystal in Its Finite Mother Phase

Microscopic crystals can form in fluid inclusions captured in a solid, as occurs in minerals [3.25], especially from solution growth under not low supersaturation and flow. If the system fluctuates around its equilibrium temperature, the crystal faces can exchange matter among them and with their surroundings: then crystals will reach their ES, after a given time. *Bienfait* and *Kern* [3.24], starting from an inspired guess by *Klija* and *Lemmlein* [3.26], first observed the ES of NH_4Cl, NaCl, and KI crystals grown in small spherical inclusions (10–100 μm) filled by aqueous solution (Fig. 3.6). The crystals contained in each inclusion (initially dendrites) evolve towards a single convex polyhedron and the time to attain the ES is reasonable only for microscopic crystals and for droplet diameter of a few millimeters. The ES so obtained did not correspond to the maximum of the free energy (unstable equilibrium) but to its minimum, and then to a stable equilibrium. Finally, it was shown that both unstable and stable ESs are homothetic but with different sizes.

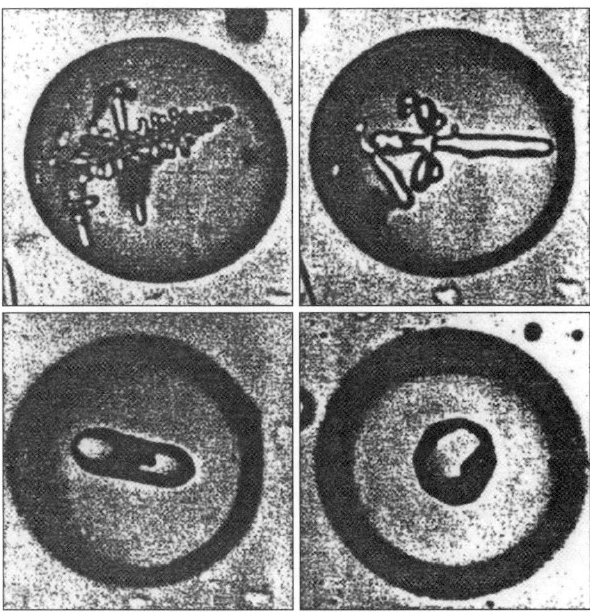

Fig. 3.6 The evolution towards equilibrium of NH_4Cl dendrites formed in an aqueous solution droplet (closed system) (after [3.24]). The total surface energy is minimized in passing from the dendritic mass to a single convex polyhedron at constant volume and T (equilibrium shape). Droplet size: 100 μm

3.1.5 The Stranski–Kaischew Criterion to Calculate the Equilibrium Shape

Without Foreign Adsorption

In the preceding sections, the surface tensions of the $\{hkl\}$ forms have been considered to be independent of crystal size. This is true when the crystal exceeds microscopic dimensions, but is no longer valid for those sizes which are very interesting both in the early stages of nucleation and in the wide field of nanosciences. In these cases, it should be reasonable to drop the use of the surface tension values, which are macroscopic quantities, to predict the equilibrium shape of micro- and nanocrystals. To face this problem, it is useful to recall the brilliant path proposed by *Stranski* and *Kaischew* [3.21, p. 170]. Their method, named the *criterion of the mean separation works*, is based on the idea that the mean chemical potential $\langle\mu\rangle_{c,m} = (1/m)\sum_{j=1}^{m}\mu_{j,c}$ averaged over all m units building the outermost layer of a finite facet, must be constant over all the facets, once the phase equilibrium is achieved. The chemical potential of a unit in a kink (Appendix 3.A) is

$$\mu_{c\infty} = -\varphi_{kink} - k_B T \ln \Omega_c + \mu^0 , \tag{3.13a}$$

and, by analogy, in a j-site of the surface

$$\mu_{j,c} = -\varphi_{j,c} - k_B T \ln \Omega_j + \mu^0 . \tag{3.13b}$$

The mean vibrational volumes being the same for every crystal sites, one can write for a generic site and especially at low temperature

$$\mu_{j,c} \approx -\varphi_{j,c} + \text{const} . \tag{3.14}$$

At equilibrium between a small crystal and its vapor: $\mu^{gas} = \langle\mu\rangle_{c,m}$. Subtracting the equality which represents the equilibrium between an infinite crystal and its saturated vapor ($\mu^{gas}_{saturated} = \mu_{c\infty}$) and applying relation (3.14), one can finally obtain

$$\Delta\mu = \mu^{gas} - \mu^{gas}_{saturated} = \langle\mu\rangle_{c,m} - \mu_{c\infty} \approx \varphi_{kink} - \langle\varphi\rangle_{c,m} .$$

That represents the *Thomson–Gibbs formula, valid for every face of small-sized crystals*

$$\varphi_{kink} - \langle\varphi\rangle_{c,m} \approx \Delta\mu = k_B T \ln \beta , \tag{3.15}$$

which allows one to determine the β value at which a unit (lying on a given face) can belong to the ES. Using (3.15), the ES can be determined without using the γ values of the different faces.

Let n_{01} and n_{11} be the number (not known a priori) of units in the most external $\langle 01 \rangle$ and $\langle 11 \rangle$ rows of a 2-D Kossel crystal (Fig. 3.7). Within the second neighbors, the mean separation works for these rows are

$$\begin{aligned}\langle\varphi\rangle_{01} &= (1/n_{01})[2\psi_1(n_{01}-1)+\psi_1+2\psi_2 n_{01}] \\ &= 2\psi_1+2\psi_2-(\psi_1/n_{01})\,, \qquad (3.16a)\\ \langle\varphi\rangle_{11} &= (1/n_{11})[2\psi_2(n_{11}-1)+\psi_2+2\psi_1 n_{11}] \\ &= 2\psi_1+2\psi_2-(\psi_2/n_{11})\,. \qquad (3.16b)\end{aligned}$$

The separation work from the kink is $\varphi_{\text{kink}} = 2\psi_1 + 2\psi_2$ and hence from (3.15) it ensues that

$$\begin{aligned}\Delta\mu &= \varphi_{\text{kink}} - \langle\varphi\rangle_{01} = \varphi_{\text{kink}} - \langle\varphi\rangle_{11} \\ &= (\psi_1/n_{01}) = (\psi_2/n_{11})\,, \qquad (3.16c)\end{aligned}$$

which represents both the phase equilibrium and the ES of the 2-D crystal. In fact the ratio between the lengths of the most external rows is obtained as

$$(n_{01}/n_{11}) = (\psi_1/\psi_2)\,. \qquad (3.17)$$

Equation (3.17) is nothing other than Wulff's condition $(h_{01}/h_{11}) = (\gamma_{01}/\gamma_{11})$ applied to this small crystal (3.11) [3.21, p. 172].

The *criterion of the mean separation work* can also answer a question fundamental to both equilibrium and growth morphology: how can we predict whether a unit is stable or not in a given lattice site? Let us consider, as an example, the unit lying at corner X of the 2-D Kossel crystal (Fig. 3.7). Its separation work, within the second neighbors, reads $\varphi_X = 2\psi_1 + \psi_2$. Stability will occur only if the separation work of the unit X is higher than the mean separation work of its own row, i. e., $\varphi_X \geq \langle\varphi\rangle_{01}$ and hence, from (3.16c), $\varphi_X \geq \varphi_{\text{kink}} - \Delta\mu$. It ensues that $2\psi_1 + \psi_2 \geq 2\psi_1 + 2\psi_2 - \Delta\mu$. Finally, one obtains $\Delta\mu = k_B T \ln\beta \geq \psi_2$, which transforms to

$$\beta \geq \beta^* = \exp(\psi_2/k_B T)\,. \qquad (3.18)$$

This means that, when β is lower than the critical β^* value, the unit must escape from the site X, thus generating an ES which is no longer a square, owing to the beginning of the $\langle 11 \rangle$ row. In other words, the absolute size (n_{01}, n_{11}) of the crystal homothetically decreases with increasing β (ψ_1 and ψ_2 being constant), as ensues from (3.16c). Since $\psi_1 > \psi_2$, $n_{01} > n_{11}$ and the ES will assume an octagonal shape dominated by the four equivalent $\langle 01 \rangle$ sides, the octagon reducing to the square

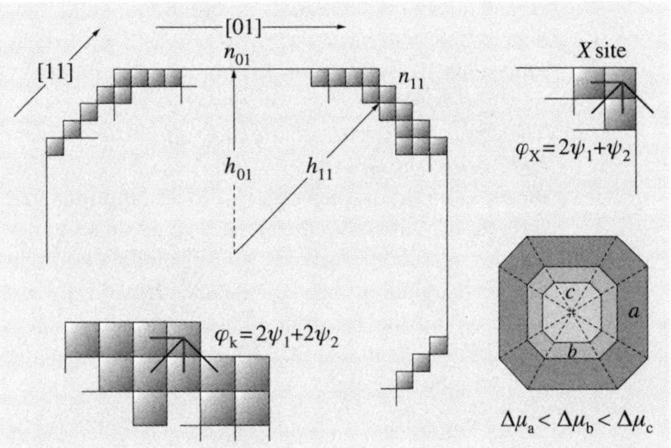

Fig. 3.7 To derive the equilibrium shape of a 2-D Kossel crystal by the criterion of the mean separation work, only the 1st, 2nd, ..., n-th-neighbors interactions are needed. The figure illustrates the scheme for the second-nearest neighbors approximation, the kink energy (φ_{kink}), the stability criterion for a unit X occupying a corner site and, finally, the 2-D equilibrium shape and size for $(\psi_1/\psi_2) = 1.5$ and for increasing supersaturation ($\Delta\mu$) values

when the number of units along the $\langle 11 \rangle$ sides is reduced to $n_{11} = 1$. As $\Delta\mu = (\psi_2/n_{11})$, this occurs when $\Delta\mu = \psi_2$, which exactly reproduces what we have just found in (3.18).

With Foreign Adsorption

In growth from solution a second component (the solvent) intervenes in the interfacial processes, since its molecules interact strongly with the crystallizing solute. Here we are interested in studying how the ES of a crystal is affected by the presence of a foreign component. Two approaches exist in order to give a full answer to this problem:

1. The *thermodynamic* approach, which allows one to forecast the variation $d\gamma$ of the surface tension γ of a face due to the variation $d\mu_i$ of the chemical potential of component i of the system, when it is adsorbed. To calculate $d\gamma$ for a flat face one has to apply Gibbs' theorem [3.22, p. 171]

$$d\gamma = -s^{(s)}\,dT - \sum_i \Gamma_i\,d\mu_i\,, \qquad (3.19)$$

where $s^{(s)}$ is the specific surface entropy and $\Gamma_i = -(\partial\gamma/\partial\mu_i)_{T,s,\mu\neq\mu_i}$ corresponds to the excess of the surface concentration of component i. Solving (3.19) is not simple, even at constant T, since one has to know the functional dependence of Γ_i on

μ_i and hence on the activity a_i of component i. This means that one has to know Γ_i, which ultimately represents the adsorption isotherm of component i on a given face.

2. The approach grounded on the *atomistic view of equilibrium* proposed by *Stranski* [3.27, 28]. This model is based on the simplifying assumptions that foreign ad-units have the same size as those building the adsorbing surface (Kossel model) and that only first-neighbor interactions are formed between ad-units and the substrate. Three types of adsorption site are defined (Fig. 3.8), each of them having its own binding energy.

From (3.19) it ensues that adsorption generally lowers the surface tension of the substrate ($\Delta\gamma < 0$), so γ increases when an adsorption layer is reversibly desorbed. Let us denote the desorption work by $w = -\Delta\gamma \times a$, representing the increase per ad-site of the surface tension of the substrate (where a is the mean area occupied by an ad-unit) [3.29–31]. Thermodynamics allows to evaluate w, according to the type of adsorption isotherm [3.21, p. 175]

$$w = -k_B T \ln(1-\theta) - (\omega/2)\theta^2 \quad \text{(Frumkin–Fowler type)}\,, \tag{3.20a}$$

$$w = -k_B T \ln(1-\theta) \quad \text{(Langmuir type)}\,, \tag{3.20b}$$

valid when ω, the lateral interaction of the ad-unit with the surrounding, vanishes and

$$w = -k_B T \times \theta \quad \text{(Henry type)}\,, \tag{3.20c}$$

when the coverage degree in ad-units is low ($\theta \ll 1$). In the last case one can compare the θ values of the different sites remembering that, at given bulk concentration of foreign units, the coverage degree for an isolated ad-unit behaves as $\theta \propto \exp(\varphi_{ads}/(k_B T))$. Here, φ_{ads} is the binding energy of the ad-unit with the substrate. From (3.20c) one can write

$$\frac{w_i}{w_j} = \frac{\theta_i}{\theta_j} = \exp\frac{(\varphi^i_{ads} - \varphi^j_{ads})}{k_B T}\,, \tag{3.21}$$

which shows that the difference in the desorption works is very sensitive to the φ_{ads} value. This can be verified by applying (3.21) to the three sites in Fig. 3.8a of a cubic Kossel crystal and remembering that, in this case, φ_{ads} is equal to ψ_{ads}, $2\psi_{ads}$, and $3\psi_{ads}$, where $\psi_{ads} = k_B T$, $2\times k_B T$, $3\times k_B T, \ldots$ is the energy of one adsorption bond. An important consequence of this reasoning is that *the chemical potential of an infinite crystal* (and hence its solubility) *is not changed by the adsorption of impurities on its surfaces*, as is proved by the balance detailed in Fig. 3.8b, which represents the initial and final stages of the desorption of a foreign unit from a kink site.

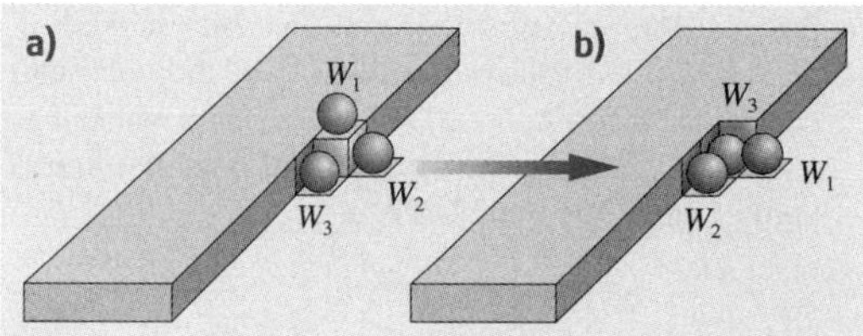

Fig. 3.8 (**a**) The three types of adsorption sites on a Kossel crystal (only 1st neighbors interaction). Each ad-site has its binding energy: $w_1 < w_2 < w_3$. (**b**) Energy balance representing the initial a) and the final b) stage of the desorption of a foreign unit from a kink-site. The binding energy does not vary on the adsorbance (after [3.21])

Let us now evaluate how the ES of a finite crystal changes, by applying the criterion of the mean separation works to the mentioned Stranski adsorption model. The stability of a unit in the corner site X when adsorption occurs (Fig. 3.9a) can be compared with that obtained without adsorption (3.18). The separation work of a unit in X is $\varphi^{ads}_X = 2\psi_1 + \psi_2 + 2w_1 - w_2$, where w_1 and w_2 are the desorption works for the two ad-sites, respectively.

The stability criterion requires $\varphi^{ads}_X \geq \langle\varphi\rangle_{01}$ and hence, from (3.16c), $\varphi^{ads}_X \geq \varphi_{kink} - \Delta\mu$.

Since $\varphi_{kink} = 2\psi_1 + 2\psi_2$, stability occurs only when $\Delta\mu \geq \psi_2 - (2w_1 - w_2)$. This implies

$$\beta^*_{ads} \geq \exp\{[\psi_2 - (2w_1 - w_2)]/(k_B T)\}\,. \tag{3.22}$$

Comparing (3.22) with (3.18) it turns out that the stability of the corner unit occurs at lower β value ($\beta^*_{ads} < \beta^*$) if $w_2 < 2w_1$. This means that, if the impurity fulfils the inequality $w_2 < 2w_1$, the ES is a pure square crystal at a β value lower than that predicted in pure growth medium. The $\langle 11\rangle$ edges begin to appear when the corner units can escape from the crystal (instability of the X-site), i. e., if $\beta < \beta^*_{ads}$. On the contrary, if $w_2 > 2w_1$ the impurity adsorption does not favor the stability of the corner unit and an octagonal ES forms at a β value lower than that found in pure growth medium. Figure 3.9b illustrates how the smoothing of a 2-D K-face can be obtained with foreign adsorption [3.21, pp. 178–189]. The energy difference between the final and initial stages is that which we obtained for the X-site, so the conclusions are obviously those fulfilling (3.22). Figure 3.9c concerns the stability of an ad-unit (site A) on the $\langle 10 \rangle$ edges in the presence of foreign adsorption. The separation work of a unit at A is $\varphi_A^{ads} = \psi_1 + 2\psi_2 + 2(w_2 - w_1)$. The stability criterion for this site requires

$$\beta^*_{ads} \geq \exp\{[\psi_1 - 2(w_2 - w_1)]/(k_B T)\} , \quad (3.23)$$

while, in analogy with (3.18), the stability criterion without impurities reads

$$\beta^* \geq \exp\left(\frac{\psi_1}{k_B T}\right) . \quad (3.24)$$

Thus, the foreign adsorption favors the stability of the growth units at site A if $\beta^*_{ads} < \beta^*$ and hence if $w_2 > w_1$. If this occurs, $\langle 10 \rangle$ edges transform from flat to rough owing to the random accumulation of ad-units.

Transferring these results from 2-D to 3-D crystals, the conditions expressed by (3.22) and (3.23), respectively, rule the transition of character K$\rightarrow$F and F$\rightarrow$K due to foreign adsorption.

The changes in the ES when adsorption occurs can now be calculated, according to the Stranski–Kaischew principle of the *mean separation work*. This means that, when an entire $\langle 10 \rangle$ or $\langle 11 \rangle$ row is removed from a 2-D crystal in the presence of adsorbed impurities, the mean separation works must fulfil the condition $\langle\varphi\rangle_{01}^{ads} = \langle\varphi\rangle_{11}^{ads}$, in analogy with (3.16a) and (3.16b). From calculation it ensues that

$$\left(\frac{n_{01}}{n_{11}}\right)_{ads} = \frac{\psi_1 - 2(w_2 - w_1)}{\psi_2 - (2w_1 - w_2)} , \quad (3.25)$$

which can be compared with the analogous expression (3.17) obtained without foreign adsorption

$$\left(\frac{n_{01}}{n_{11}}\right)_{ads} : \left(\frac{n_{01}}{n_{11}}\right) = \frac{\psi_1 - 2(w_2 - w_1)}{\psi_2 - (2w_1 - w_2)} : \frac{\psi_1}{\psi_2}$$
$$= \frac{\psi_1\psi_2 - \psi_2 \times 2(w_2 - w_1)}{\psi_1\psi_2 - \psi_1 \times (2w_1 - w_2)} . \quad (3.26)$$

Hence the importance of the $\langle 10 \rangle$ edges in the ES increases to the detriment of the $\langle 11 \rangle$ edges, if the condition $2(w_2 - w_1)/(2w_1 - w_2) < \psi_1/\psi_2$ is fulfilled. A simpler solution is obtained within the first-neighbors approximation ($\psi_2 = 0$, $\psi_1 = \psi$). Remembering that, without foreign adsorption, the ES is a pure square, in the presence of impurities some changes should occur. In this case, expression (3.25) reduces to $(n_{01}/n_{11})_{ads}^{1st} = (\psi - 2(w_2 - w_1))/(w_2 - 2w_1)$.

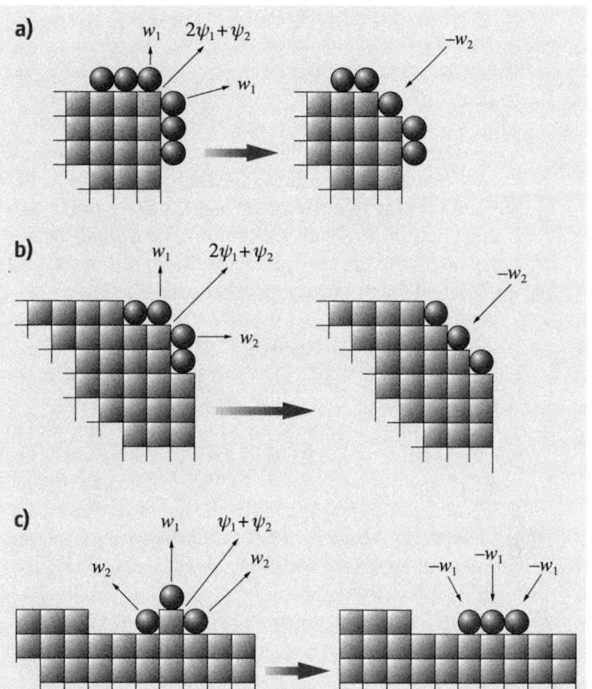

Fig. 3.9a–c The criterion of the mean separation works applied to the Stranski adsorption model in the second-neighbors approximation. (**a**) The first balance corresponds to the stability of the X site (*corner*) in the presence of foreign adsorption. (**b**) The second balance shows that the $\langle 11 \rangle$ row becomes smooth with foreign adsorption if $w_2 > 2w_1$. (**c**) The third balance describe the energies involved in calculating the stability of an ad-unit (site A) on the $\langle 10 \rangle$ edges in the presence of foreign adsorption. The figure has been inspired by [3.21]

The $\langle 11 \rangle$ row will exist if $n_{11} > 0$. Taking into account that necessarily $n_{10} > 0$, one must have simultaneously that $\psi > 2\psi(w_2 - w_1)$ and $w_2 > 2w_1$. The first inequality is verified by (3.23) since the ES of a finite crystal needs a supersaturated mother phase ($\beta^*_{ads} > 1$), so the only way for the $\langle 11 \rangle$ row to exist is for the second inequality also to be true, as found above. Summing up, the method of the *mean separation work* is a powerful tool to predict both qualitatively and quantitatively the ES of crystals, with and without foreign adsorption, without an a priori knowledge of the surface tension of their faces.

3.2 The Theoretical Growth Shape

When working with solution growth one usually has to deal with crystals having complex structures and/or low symmetry. In this case neither the Kossel model nor simple lattices, such as those related to the packing of rigid spheres, can be used to predict the most probable surface profiles. On the other hand, these profiles are needed both to evaluate the ES of crystals and for understanding the kinetics of a face. To do this, structural and energetic approaches have been developed.

3.2.1 The Structural Approach

The first works on theoretical growth morphology were grounded on structural considerations only and led to the formulation of the Bravais–Friedel–Donnay–Harker (BFDH) law [3.5, 6, 32]; see [3.33] for a recent review. According to this law, the larger the lattice distance d_{hkl}, the larger the morphological importance (MI) of the corresponding $\{hkl\}$ form

$$d_{h_1k_1l_1} > d_{h_2k_2l_2} \rightarrow \mathrm{MI}(h_1k_1l_1) > \mathrm{MI}(h_2k_2l_2)\,, \tag{3.27}$$

MI(hkl) being the relative size of a $\{hkl\}$ form with respect to the whole morphology. The inequality may also be viewed as the relative measure of the growth rate of a given form

$$R_{hkl} \propto (1/d_{hkl}) \tag{3.28}$$

once the *effective* d_{hkl} distances, due to the systematic extinction rules, are taken into account. Thus, the BFDH theoretical growth shape of a crystal can be obtained simply by drawing a closed convex polyhedron limited by $\{hkl\}$ faces whose distances from an arbitrary center are proportional to the reciprocal of the corresponding d_{hkl} values [3.5, 6]. The BFDH rule was improved [3.32], considering that many crystal structures show pseudosymmetries (pseudoperiods or subperiods), leading to extra splitting of the d_{hkl} distances, and hence to sublayers of thickness $(1/n)\times d_{hkl}$.

A typical example is that of the NaCl-like structures in which, according to the space group $Fm3m$, the list of d_{hkl} values should be $d_{111} > d_{200} > d_{220}$, etc. Vapor-grown crystals show that the cube is the only growth form, while $\{111\}$ and $\{110\}$ forms can appear when crystals grow from aqueous solutions (both pure and in the presence of specific additives) [3.34]. This was explained [3.32] considering that the face-centered structural 3-D cell can also be thought of as a pseudo unit cell (i. e., a neutral octopole) which, being primitive, leads to the cube as the theoretical growth shape.

The $R_{hkl} \propto (1/d_{hkl})$ structural rule works rather well since it implies an *energetic concept*. In fact, looking at the advancement of a crystal face as a layer-by-layer deposition, the energy released (per growth unit) when a d_{hkl} layer deposits on a fresh face is lower than that released by a sublayer since the interaction of the growth units slows down with their distance from the underlying face. Thus, the rule $R_{hkl} \propto (1/d_{hkl})$ is reasonable under the hypothesis that the face rate is proportional to the energy released when a growth unit attaches to it: $R_{hkl} \propto$ *probability of attachment*. Nevertheless, this is a crude approximation, because neither the lateral interactions of the growth units nor the fact that only the flat faces can grow by lateral mechanism (i. e., layer by layer, as shown in Sect. 3.2.2) are considered.

3.2.2 Crystal Structure and Bond Energy: The Hartman–Perdok Theory

To go beyond these limitations, Hartman and Perdok (HP) looked at crystals as a 3-D arrays of bond chains building straight edges parallel to important $[uvw]$ lattice rows. Thus, *units* of the growth medium (GU) bind among themselves (through bonds in the first coordination sphere), forming more complex units that build, in turn, the crystal and reflect its chemical composition. These *building units* (BU) repeat according to the crystal periodicity, thus giving rise to *periodic bond chains* (PBCs). An example of a PBC is the set of equivalent PBCs running along the edges of the cleavage rhombohedron of calcite; these PBCs can be represented by the sequence shown in Fig. 3.10, where Ca^{2+} and CO_3^{2-} ions are the GUs assumed to exist in solution, the group $CaCO_3$ is the crystal BU, and the vector $\frac{1}{3}[\bar{4}41]$ is the period of the $[\bar{4}41]$ PBC. This PBC is stable, since the

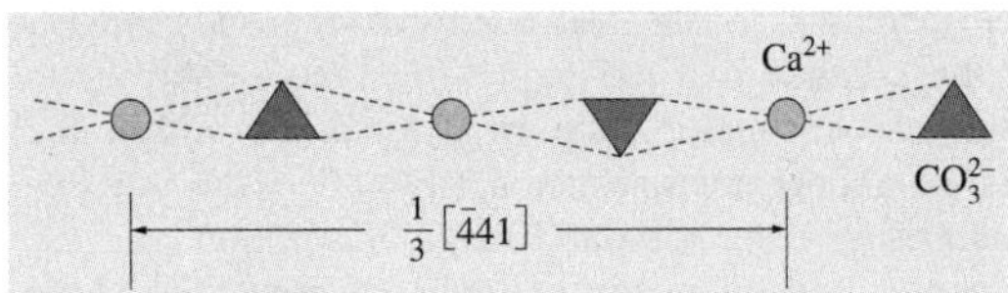

Fig. 3.10 Schematic drawing of the PBC running along the $\langle\bar{4}41\rangle$ edges of calcite crystal. (o) calcium, (Δ) carbonate ions. The PBC is stoichiometric; the repeat period is shown. The dipole moments, perpendicular to the chain axis, cancel each other

resultant dipole moment cancels out perpendicular to its development axis.

When applying the HP method to analyze a crystal structure, one must look, first of all, at the effective d_{hkl} spacing. Then, one has to search for the number of different PBCs that can be found within a slice of thickness d_{hkl}. Three kinds of faces can be distinguished, according to the number n of PBCs running within the d_{hkl} slice ($n \geq 2$, 1 or 0). Looking at the most interesting case ($n \geq 2$), the PBCs contained in this kind of slice have to cross each other, so allowing one to define:

1. An area A_{hkl} of the cell resulting from the intersection of the PBCs in the d_{hkl} slice
2. The *slice energy* (E_{sl}), which is half of the energy released when an infinite d_{hkl} slice is formed; its value is obtained by calculating the interaction energy (per BU) between the content of the A_{hkl} area and the half of the surrounding slice
3. The *attachment energy* (E_{att}), i. e., the interaction energy (per BU) between the content of the area A_{hkl} and the semi-infinite crystal underlying it.

The BUs within the area A_{hkl} are strongly laterally bonded, since they form (at least) two bonds with the end of the two semi-infinite chains (Fig. 3.11). This implies that a BU forming on this kind of faces is likely to be incorporated at the end of the chains, thus contributing to the advancement of the face in (at least) two directions, parallel to the face itself. Hence, the characteristic of these faces will maintain their flat profile, since they advance laterally until their outermost slice is filled. In analogy with what we obtained within the frame of the Kossel crystal model, these are F-faces. Moreover, their E_{sl} is a relevant quantity with respect to their E_{att}, due to the prevailing lateral interactions within the slice. From Fig. 3.11 it ensues that the energy

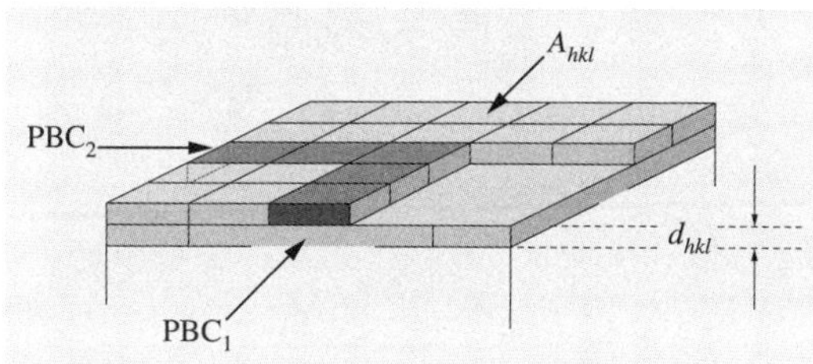

Fig. 3.11 Two PBCs within the slice d_{hkl} intersect in an elementary cell of area A_{hkl}, which occupies a kink site. The interaction of its content with half of the d_{hkl} slice gives the slice energy (E_{sl}); the interaction with all the crystal substrate gives its attachment energy (E_{att})

released (per BU) when the A_{hkl} content definitely belongs to the crystal is the crystallization energy (E_{cr}), which is a constant for a given crystal and hence for all crystal faces [3.35, p. 379]

$$E_{cr} = E_{att}^{hkl} + E_{sl}^{hkl} . \tag{3.29}$$

This relation is of greatest importance to predict the growth shape of crystals, as can be understood when looking at the *kinetic meaning* of E_{att}^{hkl}. In fact, the central HP hypothesis is that, the higher the E_{att}^{hkl} value, the higher the probability that a BU will remain fixed to the (hkl) face, and thus of belonging to the crystal. It ensues that the E_{att}^{hkl} value becomes a relative measure of the normal growth rate of the $\{hkl\}$ form [3.36]

$$R_{hkl} \propto E_{att}^{hkl} . \tag{3.30}$$

From (3.29) and (3.30) it follows that, as E_{sl} increases, both the attachment energy and the advancement rate of the face decrease. Examples par excellence can be found in layered crystal species such as the normal paraffins (C_nH_{2n+2}) and micas. Both cases are characterized by similar packing; in fact, in paraffin crystals, long-chain molecules are strongly laterally bonded within d_{00l} slices, while the interaction between successive slices is very weak; in micas, T–O–T sheets are built by strong covalent and ionic bonds whilst the interaction between them is ruled mainly by weak ionic forces. The best example is calcite, in which the E_{sl} of the $\{10\bar{1}4\}$ rhombohedron reaches 92% of the crystallization energy value and E_{att} reduces to account for the remaining 8%. This striking anisotropy explains, from one hand, the well-known cleavage properties of calcite and, on the other hand, the slowest growing of the $\{10\bar{1}4\}$ form, within a large β range and in the absence of impurities in the mother solution.

It is worth outlining the *similarity between the relation* (3.2c) *ruling the kink energy and the relation* (3.29) *defining the crystallization energy*

$$\varphi_{kink} = \varphi_{att} + \varphi_{slice} \rightarrow E_{cr} = E_{att}^{hkl} + E_{sl}^{hkl} . \tag{3.31}$$

Both relationships can be expressed in energy/BU: the first relation concerns a single GU (atom, ion or molecule), while the second one is extended to a unit cell compatible with the d_{hkl} thickness allowed by the systematic extinction rules. This means that *HP theory permits one to predict the growth morphology of any complex crystal, through a brilliant extension of the kink properties to the unit cell of the outermost crystal layers.*

The example shown in Fig. 3.12a concerns the PBC analysis applied to the lithium carbonate structure (space group $C2/c$). [001] PBCs are found along with another kind of PBC, running along the equivalent set of ⟨110⟩ directions. From this it ensues that the {110} form has F-character, since two kinds of PBCs run within the allowed slice of d_{110} thickness. On the contrary, both {100} and {010} are S-forms, as no bond can be found between successive [001] PBCs within the slices of allowed thickness d_{200} and d_{020}, respectively. Figure 3.12b shows that only the {110} prism exists in the [001] zone of a Li_2CO_3 crystal grown from pure aqueous solution, thus proving that the prediction obtained through the HP method is valid.

The choice of the BU is strategic for predicting both growth and equilibrium shapes. With reference to the preceding example, four different BUs can be found in Li_2CO_3 crystal, due to the distorted fourfold Li^+ coordination. Each of these BUs determines a different profile of the crystal faces and, consequently, different γ and E_{att} values. Hence, one has to search for all possible surface configurations and then calculate their corresponding γ and E_{att} values in order to choose those fulfilling the minimum-energy requirement.

Concerning *methods to find PBCs and face characters*, one has to carry out many procedures, ranging from the original visual method to computer methods, which began to be applied about 30 years ago and reached their highest level of sophistication in elementary graph theory [3.11, 37–40], in which crystallizing GUs are considered as points and bonds between them as lines. A different computer method to find the surface profile with minimum energy was developed by *Dowty* [3.41], who searched for the *plane* parallel to a given *(hkl)* face cutting the minimum number of bonds per unit area, irrespective of the face character. This method has proved interesting as a *preliminary step* for calculating both equilibrium and growth crystal shapes. In the last 50 years, a lot of papers have been produced in which the theoretical growth morphology has been predicted for a wide variety of crystals exhibiting different types of bonds. The reader is invited to consult authoritative reviews on this subject [3.35, 42] and to proceed with caution in accepting predicted morphologies because there is a certain tendency to confuse, in practice, equilibrium and growth morphology.

3.2.3 The Effect of Foreign Adsorption on the Theoretical Growth Shape

In the original HP method, E_{att} is evaluated without considering either the temperature effect or the influence of the growth medium. Neglecting temperature does not imply a crude approximation on the predicted

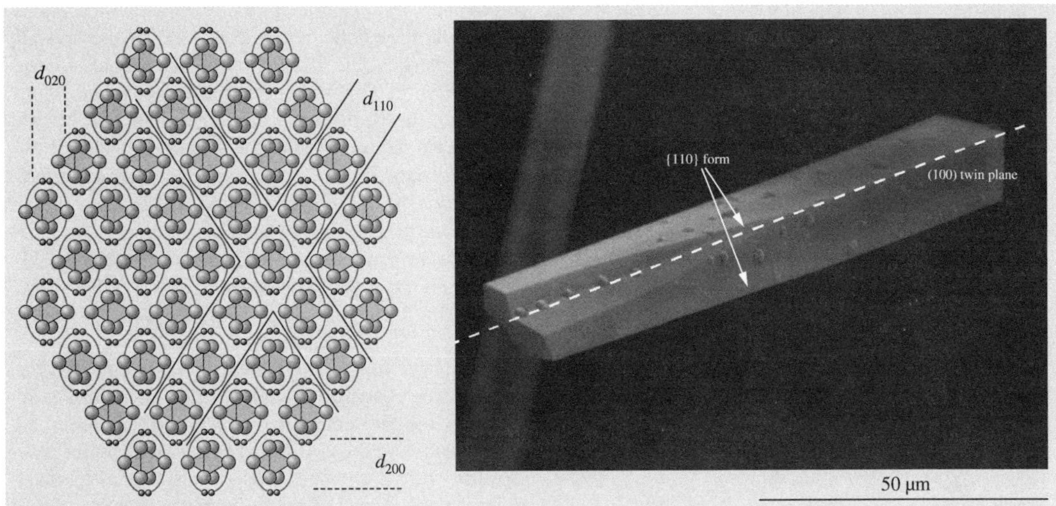

Fig. 3.12a,b Comparison between the experimental morphology, in the [001] zone, of lithium carbonate crystal and the theoretical one (HP method). **(a)** [001] PBCs are seen up–down with bonds among them, within the d_{110} slices. The {110} prism is an F-form; the {100} and {010} are S-forms (no bonds within the slices of d_{200} and d_{020}, respectively). **(b)** Scanning electron microscopy (SEM) image of Li_2CO_3 twinned crystal grown from aqueous solution, showing the dominance of the {110} prism. The 100 twin plane is indicated

equilibrium and growth shape, when dealing with low-temperature solution growth. In fact both γ and E_{att} values are not particularly affected by the entropic term, in this case. On the contrary, a condensed phase around the crystal (the solvent) and/or specific added impurities can deeply modify the behavior of the crystal faces.

The Role of the Solvent

This topic has been carefully examined, first with the aim of predicting *qualitatively* how the crystal–solution interface is modified by the solvent, and then which the slow-growing faces are likely to be. To do this, the roughness of the interface has been quantified in terms of the so-called *α-factor* [3.43, 44] which defines the enthalpy changes taking place when a flat interface roughens. This factor, originally conceived for crystal–melt interface, has been modified for solution growth and is commonly expressed in two ways

$$\alpha = \xi_{hkl}\frac{\Delta H_s}{RT} \quad \text{or} \quad \alpha = \xi_{hkl}\left[\frac{\Delta H_f}{RT_m} - \ln X_s(T)\right] , \tag{3.32a}$$

where ΔH_s represents the heat of solution at saturation, ΔH_f the heat of fusion, X_s the solubility, T_m the temperature of fusion, and ξ_{hkl} is a factor describing the anisotropy of the surface under consideration [3.44]. ξ_{hkl} is evaluated by means of HP analysis, since it is strictly related to the slice energy of the $\{hkl\}$ form

$$\xi_{hkl} = \frac{E_{hkl}^{slice}}{E_{cr}} . \tag{3.32b}$$

Three different situations occur, according to the α value:

1. When $\alpha \leq 3$, the interface is rough and the face behaves as a K- or S-face.
2. If $3 \leq \alpha \leq 5$, the interface is smoother (F-face) and the creation of steps on the surface becomes a limiting factor at low β-values (birth and spread of 2-D nuclei).
3. When $\alpha > 5$ the growth at low β is only possible with the aid of screw dislocations since the barrier for 2-D nucleation is too high.

Equations (3.32a) and (3.32b) clearly show that different $\{hkl\}$ forms should have different *α-factor* values, not only owing to the ξ_{hkl} anisotropy, but also because of the solubility and of the heat of solution. Thus, the crystal morphology will also be dependent on the growth solvent.

As mentioned above, the evaluation of the α-factor is useful for predicting if a crystal form can survive against competition with other forms, but nothing can be deduced on the relative growth rates of the surviving forms. To overcome this drawback, solvent interaction with crystal surfaces was considered quantitatively by *Berkovitch-Yellin* [3.14, 45] who calculated the electrostatic maps of certain faces and identified the most likely faces for adsorption.

A clear example of the role played by the solvent is that concerning the theoretical equilibrium and growth forms of sucrose. We will not consider here its polar $\{hkl\}$ forms to avoid the complications due to the coupling of adsorption and polarity; rather we will confine our attention to the nonpolar $\{h0l\}$ forms. HP analysis shows that the theoretical growth morphology of sucrose agrees with the experimental one, obtained from pure aqueous solution, with the only exception of the $\{101\}$ form [3.46]. In fact, the [010] PBCs are not connected by strong bonds (H-bonds in this case) within the d_{101} slice (Fig. 3.13a) and then $\{101\}$ should behave as an S-form. Nevertheless, the S-character does not agree with its high occurrence frequency ($\approx 35\%$), rather unusual for a stepped form. The way to get out of this discrepancy is composed of two paths.

First, *one has to carry out a quantitative HP analysis* considering the strength of the PBCs running within the d_{101} slice. The energies released when a molecule deposits on the top of different molecular chains (i. e., the end chain energy, ECE) have been calculated. It results that ECE[010] $= -0.525 \times 10^{-12}$ and ECE[$10\bar{1}$] $= -0.077 \times 10^{-12}$ erg/molecule. These interactions being attractive, two PBCs really exist in the d_{101} slice and $\{101\}$ is a F-form, contrarily to what was concluded through qualitative application of the HP method. However, its F-character is weak, due to the strong anisotropy between the two PBCs and its E_{att} value being too high with respect to those of the other $\{h01\}$ forms, so that the $\{101\}$ form cannot belong to the growth shape of the crystal (Table 3.2 and Fig. 3.13b).

Secondly, one has to *consider the specificity of water adsorption* on the $\{101\}$ surfaces. In fact, even if H-bonds do not exist within a d_{101} slice at the crystal–solution interface, water adsorption can occurs between two consecutive [010] chains by means of two strongly adsorbed water molecules over a $\|[010]\|$ period. Then, a new [$11\bar{1}$] PBC forms and the PBC [$10\bar{1}$] results stronger on the outermost crystal layer than in the crystal bulk. Consequently, the F-character of the face is greatly enhanced. This kind of water adsorption is specific to this form, as all other faces in the $\{h0l\}$ zone are built by [010] PBCs strongly bonded among them, without allowing free sites for bonding of interchain wa-

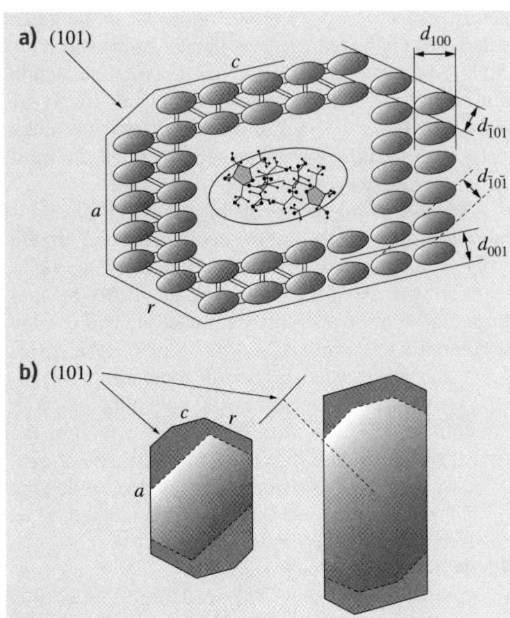

Fig. 3.13 (**a**) Projection along the [010] PBC of the sucrose structure. Each ellipse, containing two sucrose molecules, fixes the boundaries of a PBC. No bond can be found between two consecutive PBCs within a slice d_{101}, whilst bonds occur within the other $\{h0l\}$ forms. (**b**) Theoretical equilibrium and growth shapes of sucrose in the [010] zone, calculated without (*full line*) and with water adsorption (*dotted line*) (after [3.46])

ter molecules. New equilibrium and growth shapes are obtained (Fig. 3.13b), remembering that γ_{hkl} can be obtained from E_{att}^{hkl} through the relationship holding for molecular crystals [3.36], where the second-neighbors interactions are weak

$$E_{\mathrm{att}}^{hkl} \cong \gamma_{hkl} \times 2A_{\mathrm{2D}}^{hkl}/z\,, \tag{3.33}$$

where A_{2D}^{hkl} is the area of the unit cell related to the d_{hkl} slice and z are the molecules within it. Growth isotherms showed that the $\{101\}$ form can grow by a spiral mechanism, thus proving its F-character, and that water desorption is the rate-determining process of its kinetics. Moreover, and for the first time, the *idea of* E_{att} *was also successfully extended to the spiral steps* running on the $\{101\}$ surfaces, thus proving that the attachment energy at the spiral steps determines the growth shape of spirals, especially at low supersaturation values [3.47].

Table 3.2 Calculated surface (γ_{h0l}: erg cm^{-2}) and attachment energies (E_{att}^{h0l}: 10^{-12} erg/molecule) for the $\{h0l\}$ zone of sucrose crystal, in the crystal–vacuum system at $T = 0$ K

Form	$\{100\}$	$\{101\}$	$\{001\}$	$\{10\bar{1}\}$
γ_{h0l}	143	201	206	198
E_{att}^{h0l}	0.96	2.50	1.92	2.01

PBC analysis has been used, in recent times, as a preliminary step for predicting growth morphology in the presence of the solvent. A general and powerful kinetic model was elaborated by the *Bennema* school [3.13] in which growth mechanisms of the faces are considered, that is, spiral growth at low β-values and 2-D polynucleation at high β-values. Furthermore, to analyze the influence of the fluid phase on the crystal morphology, an interfacial analysis has been developed within the framework of inhomogeneous cell models [3.48, 49]. However, this model suffers from some limitations, since it is assumed that in solution growth the solute incorporation into the steps is governed by direct diffusion of molecules into the kinks. Experiments show that this is not always the case. Good examples are those of the growth isotherms obtained by the Boistelle group in Marseille for the normal paraffins. Surface diffusion is the rate-determining step for the $\{110\}$ form of octacosane ($C_{28}H_{58}$) crystals [3.50], while coupled volume and surface diffusion effects dominate the growth rate of $\{001\}$ form of hexatriacontane ($C_{36}H_{74}$) crystals growing from heptane solution [3.51, 52]. Another interesting case is that represented by the complementary $\{110\}$ and $\{\bar{1}\bar{1}0\}$ F-forms of sucrose crystals, which are ruled by volume and surface diffusion, respectively, when growing from pure aqueous solution between 30 and 40 °C [3.53, 54].

The modifications induced on E_{att} by the solvent have been evaluated theoretically by considering the relationship (3.33) holding for molecular crystals and obtaining a new expression for E_{att}, where the maximum number (n_{s}) of solvent molecules interacting with the surface unit cell (S_{hkl}) and their interaction energy (E_{hkl}^{i}) is taken into account [3.55]

$$E_{\mathrm{att}}^{hkl}\ (\text{solvent modified}) = E_{\mathrm{att}}^{hkl} - \left[n_{\mathrm{s}} E_{hkl}^{\mathrm{i}} - N_{\mathrm{A}} S_{hkl} \gamma_{\mathrm{s}}\right] \times Z^{-1}\,, \tag{3.34}$$

Z and N_{A} being the number of molecules in the unit cell and the Avogadro number, respectively. Expression (3.34) can be also easily adapted to the adsorption of an additive, treated as a medium, once the adhesion energy of the solvent has been adequately replaced by that of the additive. This solvent-effect approach was successfully applied to the growth of the α-polymorph of glycine from aqueous solution, since a bi-univocal correspondence was found among the theoretical and experimentally observed F-faces. Moreover, this model explained as well the replacement of the most important {110} form of γ-aminobutyric acid in vacuo by the {120} form in water, and the flattening of the {001} form when a cationic or H-bonding additive is used [3.56].

The most complex and up-to-date method for predicting the growth morphology from solution was proposed by the Bennema school. Two interesting examples will be illustrated here.

In a first paper [3.57] it was shown that both 3-D and surface morphologies of tetragonal lysozyme crystals could be explained by a connected net analysis based on three different bond types corresponding to those used in the Monte Carlo growth simulation. Besides, the E_{att} of the different forms were estimated, along with their step energies. Furthermore, the significant β dependence of the relative growth rates of the {110} and the {101} forms, experimentally observed, was coherently explained on the basis of the multiple connected net analysis. More recently [3.58] a comparison was made between the E_{att} method and Monte Carlo simulations applied to all faces occurring in the growth morphology, with both approaches based on the connected net analysis. This was done considering that the E_{att} method cannot, intrinsically, take into account T, β, and growth mechanisms, while the simulation [3.59] can not only do this, but can also predict growth with or without the presence of a screw dislocation. The comparison was applied to solution growth of the monoclinic polymorph of paracetamol, using four different force fields [3.58]. It resulted that:

- The force field has only a small effect on the morphology obtained by the E_{att} method.
- The morphology so predicted does not resemble the experimentally ones for any of the β regimes.

Monte Carlo simulation gave different results, even under the limiting assumption that surface diffusion could be neglected: the {110}, $\{20\bar{1}\}$, and {100} forms have for all crystal graphs approximately the same growth curves, while all graphs show very different growth behaviors for the other two forms, {001} and {011}, owing to the differences of their step energies. Moreover, the {100} form is the theoretically fastest-growing form, according to the experimental observations. Finally, from the overall comparison between Monte Carlo simulated and experimental morphologies, it emerged that the simulated results were poorer in form, even if the β effect was accounted for. This discrepancy is due to the $\{20\bar{1}\}$ faces, which grow too slowly in the simulations and then assume too large an importance when compared with the other forms.

The Effect of Impurity Adsorption on the Theoretical Growth Shape

A large body of research on this topic is that carried out on NaCl-like crystals when specific ions are added to pure aqueous mother solutions. In the NaCl structure there are strong PBCs, in the equivalent $\langle 100\rangle$ directions, determining the F-character of the cube-{100} form. Other zigzag $\cdots Na^+{-}Cl^-{-}Na^+{-}Cl^-\cdots$ chains run along the $\langle 110\rangle$ directions, within slices of thickness d_{111}, but they are polar chains and thus cannot be considered PBCs. Consequently, the {111} octahedron has K-character and is electrically unstable, being built by faces consisting of alternating planes containing either Na^+ or Cl^- ions. Stability can be achieved by removing 3/4 of the ions of the outermost layer and 1/4 of the subjacent one: thus, the new unit cell of the crystal, a cubic $4\times[Na^+Cl^-]$ octopole, has no dipole moment. Nonetheless, the *reconstructed octahedron maintains its K-character* and cannot appear either at equilibrium or in the theoretical growth shape of the crystal, owing to the too high values of γ_{111}^{NaCl} and $E_{att}^{111(NaCl)}$ with respect to γ_{100}^{NaCl} and $E_{att}^{100(NaCl)}$, respectively [3.60, 61]. Evidence of this behavior was shown by annealing {111} faces of NaCl crystals, near equilibrium with their vapor, and proving that they were structurally similar to those predicted by the reconstructed model [3.62].

A widely different situation emerges when NaCl-like crystals grow in solution in the presence of minor amounts of species such as Cd^{2+}, Mn^{2+}, Pb^{2+}, urea ($CO(NH_2)_2$), Mg^{2+}, and CO_3^{2-}. The most complete contribution on this subject is that of the *Kern* school [3.63–67]. Apart from the influence of β on the growth shape, Cd^{2+} is the most effective impurity, as even a small concentration gives rise to the habit change $\{100\}\rightarrow\{100\}+\{111\}$, as observed by optical microscopy. This change is not due to the random adsorption of Cd^{2+} ions on the {111} surfaces; it was attributed first to 2-D epitaxial layers of $CdCl_2$, which can form matching the {111} surface lattice even if the mother solution is unsaturated with respect to the 3-D crystal phase of $CdCl_2$ [3.66]. Later on, and after measurements of adsorption isotherms, another interpretation was proposed [3.67]: the adsorption 2-D

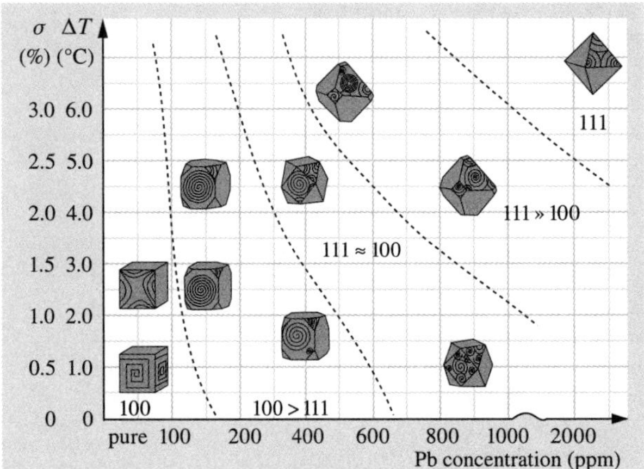

Fig. 3.14 Morphodrome showing the change $\{100\} \rightarrow \{100\}+\{111\}$ of crystal habit of KCl crystals with supersaturation excess ($\sigma = \beta - 1$) and impurity (Pb^{2+} ion) concentration. Surface patterns are also drawn (after [3.71])

epitaxial layer assumes the structure of the mixed salt $CdCl_2 \cdot 2NaCl \cdot 3H_2O$ once the isotherm has reached its saturation value. This hypothesis was supported by the finding, at supersaturation, of 3-D crystallites of the mixed salt epitaxially grown on $\{111\}$ surfaces. However, the existence of the 2-D epitaxial adsorption layer was not proved experimentally. More modern and recent research, based on optical observations, AFM measurements, and in situ surface x-ray diffraction [3.68–70] led to the conclusion that the polar $\{111\}$ surface should be stabilized by a mixed monolayer of Cd^{2+} (occupancy 0.25) and water (occupancy 0.75) in direct contact with the top Cl^- layers of $\{111\}$ NaCl underneath.

Summarizing, the evidence that emerges from this long-debated topic is that: when the surface of the growing crystal undergoes some intrinsic structural instability, such as surface polarity, and the growth medium contains some *suitable* impurities, *more or less ordered and layered structures form at the interface*, lending stability to the surface structure.

Since supersaturation plays a fundamental role in habit change, a sound and practical method was proposed [3.34] to represent the crystal habit as a function of both supersaturation (β) and impurity concentration. This drawing was called a *morphodrome*, as illustrated in Fig. 3.14, showing the changes of crystal habits of KCl crystals with β and impurity (Pb^{2+} ion) concentration [3.71].

A behavior similar to that of NaCl-like crystals is shown by calcite ($CaCO_3$) crystals growing in the presence of Li^+ ions, which generate the $\{10\bar{1}4\} \rightarrow \{10\bar{1}4\}+\{0001\}$ morphological change [3.72]. Also in this case the formation of a 2-D epitaxial layer of the monoclinic Li_2CO_3 seems to be the most reasonable way of stabilizing the $\{0001\}$ form. In fact, AFM observations prove that lithium promotes the generation of quasiperiodic layer growth on the $\{0001\}$ surfaces (K→F character transition), while structural calculation indicates that Li^+ ions coming from the mother phase can perfectly take the place of calcium ions missing in the outermost reconstructed calcite layers. The credibility of this epitaxial model is enhanced by ab initio calculation [3.73] showing that the relaxed CO_3^{2-} ions, belonging to the d_{002} slice of Li_2CO_3 at the calcite/Li_2CO_3 interface, entail the best coupling with the relaxed position of the outermost CO_3^{2-} ions of the reconstructed calcite crystal.

Certainly, the search for an epitaxial model for solution growth is most intriguing when one can both predict and interpret the effect of an impurity on the crystal habit, especially when the solvent may favor the formation of a structured crystal–solution interface. Nevertheless, there are other ways of assessing the impurity effects. One of these is to consider the modifications introduced by the impurity on the *energetics* of the elementary cell of the crystal in its outermost layer. This is the case for *disruptive tailor-made additives* [3.74], which are generally smaller than the host system but with a high degree of molecular similarity (e.g., benzamide/benzoic acid), which can adsorb on specific surface sites and thus influence the attachment energy value associated with the adsorption of subsequent growth layers. On the other hand, the *blocker type of molecular additive*, which is structurally similar but usually larger than the host material, has an end group which differs significantly and hence can be accepted at specific sites on some crystal faces. Thus, the end group (the blocker) prevents incoming molecules getting into their rightful positions at the surface. In fact, in the naphthalene–biphenyl host–additive system the E_{att} values of the different $\{hkl\}$ forms are selectively modified by the blocker additive. Steric repulsion resulting from the atoms of the blocker residing close to, or actually in, the same physical space in the crystal as atoms of the adjacent host molecules, both lowers the corresponding E_{att} value and prevents host molecules from adsorbing, due to the blocking of surface sites [3.75]. Summarizing, the effects of solvent and impurities were globally considered by calculating, through an ad hoc program, modified attachment energy terms, leading to simulated modified morphologies [3.76].

3.3 Factors Influencing the Crystal Habit

It is convenient to state some definitions. By *morphology* we mean the set of $\{hkl\}$ crystal forms occurring in a crystal independent of the surface areas, which is taken into consideration in the *crystal shape*. *Crystal habit* has to do with the dominant external appearance and is related to growth conditions. In the following only crystal habit is considered.

Crystals of the same phase can exhibit a great variety of crystal habits. This was one of the major difficulties in the beginning of crystal study and partly still remains, notwithstanding the enormous theoretical and experimental progress. This subject has both scientific and applied relevance. In many industrial sectors, crystal habit change is necessary to prevent crystal caking, filter crystal precipitates, obtain more convenient crystal products (in terms of shape, size, size repartition, purity, quality, etc.), simplify storage and package, etc. Empiricism played an important role in industrial crystallization in the past, but has been progressively supported and replaced by knowledge of crystal growth mechanisms and phenomenological rules.

Experiments show that crystal faces generally grow layer by layer, as already noticed by Niels Steensen. They move at different rates, and the fast-growing ones are destined to disappear. Therefore the habit of a crystal is determined by the faces having the slowest growth rates. Crystal habit may change either through the relative development of already existing $\{hkl\}$ forms or the appearance of new $\{h'k'l'\}$ forms.

The procedures to study the crystal habit change are well established: experimental crystal habits, grown from different solvents, are compared to the theoretical one, which may be obtained by calculations with different available methods (BFDH, PBC-attachment energy-connected nets, IS analysis) or by growing the crystal from the vapor phase, in which the fluid–solid and fluid–fluid interactions are negligible. Indeed, a complete study should involve the growth kinetics of each face, in order to determine its growth mechanism and the roles of the specific solvent and/or impurity.

As the crystal–solution interface is the critical site for face growth and crystal habit, all available approaches are applied to the study of this surface. A list is given in Chap. 5. The factors influencing the crystal habit are numerous and have different effects, which explains the great habit variability. They are usually classified into two main categories:

1. Internal factors: the crystal structure, on which the surface structures (i. e., the profiles) of the faces depend, and crystal defects
2. External factors, which act from the *outside*: supersaturation, the nature of the solvent, solution composition, impurities, physical conditions (temperature, solution flow, electric and magnetic fields, microgravity, ultrasound, etc.).

There are also mixed factors, such as the free energy of crystal surfaces and edges, which depend on both crystal surface structure (an internal factor) and the growth environment (an external factor). The most important ones are considered separately in the following, even if it is necessary to look at the crystal growth as a whole, complex process, in which a change in one parameter (temperature, solubility, solvent, supersaturation) influences all the others, so that they together affect crystal growth and habit. Let us consider a polymorphic system made of two phases: A and B. Changing, for example, the solvent at constant temperature and concentration, both surface free energy (γ) and supersaturation (β) are changed. If these variations are small, changes concern only the crystal habit of one polymorph (e.g., A). If the variations are large, the nucleation frequencies of the two polymorphs can be so affected that a change in crystal phase occurs and the B polymorph may nucleate. The same considerations apply to the temperature change, which promotes variations in solubility, surface tension, and supersaturation, especially in highly soluble compounds.

3.4 Surface Structure

Each crystal face has a specific surface structure, which controls its growth mechanism. As the crystal habit is limited by the faces having the slowest growth rates, i. e., the F-faces, in the following only the F-faces will be considered. The surface of an F-face is not perfectly flat and smooth, but is covered by steps and other features (hillocks), which condition the growth rate of the face and its development. Indeed, layer growth is possible when the edge energy of a 2-D nucleus is positive [3.2, 27]. The growing steps may be inclined with respect to the surface, forming an acute and obtuse angle which advance at two different velocities, as observed on the {001} face of monoclinic paraffins [3.77]. Surface features (dislocation activity, step bunching) and parameters (step speed, hillock slope) are sensitive to supersaturation and impurities and behave in different ways at low and high supersaturation, with linear and nonlinear dependence [3.78]. Connected to these factors is the morphological instability of steps and surfaces, which is enhanced or prevented by shear flow, depending on the flow direction [3.78–83]. Great theoretical, experimental, and technical contributions to the study of surface phenomena are due to Russian [3.78–81], Dutch [3.82–85], and Japanese [3.15, 16, 86] groups as well as to other researchers [3.87–92] and to those quoted in all these papers. Surface phenomena and morphology have been recently reviewed [3.15–18]. AFM has enormously enlarged this research field, as it allows the observation of the surface features of the growing faces both ex situ and in situ at a molecular level. This new technique is providing a growing number of new data and observations, and at the same time renewing and stimulating interest in surface phenomena and processes, especially step kinetics, impurity effects, edge fluctuations, and stability.

3.4.1 The α-Factor and the Roughening Transition

In Sect. 3.2.2 the concept of α-factor was introduced as a measure of the roughness of a surface and its probable growth mechanism. Knowledge of α is mostly useful; however, it may be not sufficient, as noticed for several alkanes, which show the same α value in different solvents, yet have different growth mechanisms [3.93], and also for the {010} and {001} faces of succinic acid grown from water and isopropyl alcohol (IPA). Each face has the same α-value in both solvents; nevertheless the growth rates are appreciably lower in IPA than in water, owing to different efficiency of hydrogen bonding with IPA and water molecules [3.45]. With increasing temperature (3.32a) the α-factor decreases and may reach values lower than 3.2. In that case the surface loses its flatness, becomes rough, and grows by a continuous mechanism. This transition occurs at a definite roughening temperature which is characteristic for each face. For example, the {110} faces of paraffin $C_{23}H_{48}$ growing from hexane has a roughening temperature of $T^R = 10.20 \pm 0.5$. Below T^R the faces are straight, whereas above it they become rounded even if supersaturation is very low [3.11].

3.4.2 Kinetic Roughening

Beside thermal roughening, a surface may undergo kinetic roughening, which occurs below the roughening temperature when the supersaturation exceeds a critical value. In this case the sticking fraction on the surface is so high and the critical two-dimensional nucleus so small that the surface becomes rough and grows through a continuous mechanism. This behavior was observed on the {100} faces of NaCl in aqueous solutions [3.94] and in naphthalene crystals, which become fully rounded when σ attains 1.47% in toluene solvent. The same does not occur with hexane, due to structural dissimilarity of hexane molecules with respect to naphthalene [3.11]. Four criteria used to identify the beginning of kinetic roughening have been studied by Monte Carlo simulations on a Kossel (100) surface, leading to different values of the critical driving force [3.95].

3.4.3 Polar Crystals

In polar crystals a d_{hkl} slice may present a dipole moment. In that case a correction term, E_{corr}, should be added to the expression for E_{att} to maintain the value of E_{cr} constant [3.35, 36] (Sect. 3.2.2, (3.29)), i. e.,

$$E_{cr} = E_{att} + E_{slice} + E_{corr} , \tag{3.35}$$

with $E_{corr} = 2\pi\mu^2/V$, where V is the volume of the primitive cell and μ the dipolar moment of the slice (per formula unit) [3.96]. The surfaces of the two opposite faces (hkl) and $(\bar{h}\bar{k}\bar{l})$, being structurally complementary, interact in a selective way with the solvent and impurity molecules. The final result is a different development of these faces, which may lead to the occurrence of only one form, as observed in the case of the $(011)/(0\bar{1}\bar{1})$ faces of $N(C_2H_5)_4I$ [3.97] and the $\{100\}$, $\{\bar{1}\bar{1}\bar{1}\}$, $\{0\bar{1}\bar{1}\}$, and $\{1\bar{1}0\}$ faces of $ASO_3 \cdot 6H_2O$ ($A = Co^{2+}$, Ni^{2+}, Mg^{2+}). In this case the water molecules are selectively adsorbed on the opposite faces since they have a different surface distribution of sulfite ions and $A(H_2O)_6^{2+}$ groups [3.98]. The structural differences can be so great that the two opposite faces may grow with a different mechanism, as experimentally shown for the $\{110\}$ and $\{1\bar{1}0\}$ faces of sucrose crystals: the former by volume diffusion with $\Delta G_{cr} = 10$ kcal/mol and the latter by surface diffusion with $\Delta G_{cr} = 21$ kcal/mol [3.54].

3.4.4 Looking at Surfaces with AFM

AFM is becoming a routine technique in growth laboratories. Most experiments are carried out in static conditions, some in the dynamic regime. One of the most studied compounds, besides proteins, is calcite. The $\{10\bar{1}4\}$ cleavage form grows via monomolecular steps, which are differently affected by anion and cation impurities [3.99]. AFM has been used to assess the stability of the $\{111\}$ faces of NaCl in pure and impure aqueous solutions and to attempt to solve the problem of surface reconstruction [3.70, 71]. Through AFM investigation of the $\{100\}$ faces of potassium dihydrogen phosphate KDP, the dependence of macrosteps and hillocks on β was measured and new values of the step edge energy, kinetic coefficients, and activation energies for the step motion were calculated, confirming the models of Chernov and van der Eerden and Müller–Krumbhaar [3.100]. In studying the influence of organic dyes on potassium sulfate the link between the surface features at the nanoscale level and the macroscopic habit change was proved [3.101]. To sum up, AFM analysis enables local details of surface structure and their evolution in real time to be captured, yielding a lot of information, but has the drawback that it does not permit large-scale views of the face, so it has to be integrated with other instrumental (optical and x-ray) techniques.

3.5 Crystal Defects

Defects easily and usually occur in crystals. It is not necessary to emphasize the role of screw dislocations in crystal growth. As concerns edge dislocations, they could affect the growth rate since a strain energy is associated with the Burgers vector and then increases the growth rate. Combined research on the effect of dislocations on crystal growth with in situ x-ray topography was done on ammonium dihydrogen phosphate ADP crystal [3.78]. Edge dislocations were proved to be inactive in step generation on the ADP (010) face, whereas screw dislocations were active. When a dislocation line emerges on a given (*hkl*) F-face, the face grows at higher rate than the other equivalent ones and therefore decreases its morphological importance with respect to the others. In crystals with cubic symmetry the habit may change from cubic to tetragonal prism or square tablet. When a screw dislocation crosses an edge, it becomes inactive [3.78]. Contrary to the current opinion that increasing the growth rate leads to a higher defect density, as supported by Monte Carlo simulations [3.102], large crystals with a high degree of structural perfection can be obtained with the method of *rapid growth*, which consists of overheating a supersaturated solution, inserting a seed conveniently shaped, and strongly stirring the solution submitted to a temperature gradient. The method, applied for the first time in the 1990s, allows the preparation in short time of very large crystals of technologically important compounds such as KDP and deuterated potassium dihydrogen phosphate DKDP up to 90 cm long and nearly free of dislocations. The crystal habit, bounded by $\{101\}$ and/or $\{110\}$ faces, may be controlled by creating dislocation structures during the seed regeneration and changing the seed orientation [3.103, 104]. Crystals grown with the traditional method at low temperature are smaller and rich in striations and dislocations, originated by liquid inclusions. Large perfect crystals can also be quickly grown from highly concentrated boiling water solutions. The method has been successfully applied to some compounds, such as KDP, $Pb(NO_3)_2$, and $K_2Cr_2O_7$. Due to the high growth rates and β values, the crystal habit becomes equidimensional [3.105].

3.6 Supersaturation – Growth Kinetics

The effect of supersaturation on growth morphology is well known, but not yet well understood, since when a system becomes supersaturated, other parameters change in turn, especially in solutions of poorly soluble compounds (phosphates, sulfates, etc.). In this case a change in β involves variations in solution composition, chemical species, and related phenomena (ion coordination, diffusion, etc.) [3.106]. First of all, β is important in controlling both the size and shape of the 3-D and 2-D critical nucleus (3.11) and in determining the growth kinetics. The growth rates of S- and K-faces are linear functions of β. For an F-face the dependence is more complicated, being related to the growth mechanism. The dependence law for R_{hkl} versus β may be parabolic, linear (in the case of a spiral mechanism), exponential (for two-dimensional nucleation) or again linear (when the face grows by a continuous mechanism at high β values). Spectacular habit changes are observed with increasing β. At high β values, first hopper crystals, then twins, then dendrites, and finally spherulites may form [3.107]. All possible cases are gathered in the diagram of R_{hkl} versus σ proposed by *Sunagawa* [3.108] (Chap. 5). The basic kinetic laws for growth controlled by surface diffusion, i. e., in the kinetic regime, are summarized here.

3.6.1 Growth Laws

For the spiral mechanism (BCF theory [3.7]) the growth rate of a (hkl) face is given by

$$R_{hkl} = \frac{v_\infty d_{hkl}}{y_0} , \tag{3.36}$$

where v_∞ is the step velocity, d_{hkl} is the interplanar distance, and y_0 is the equidistance between the spiral steps. The step velocity for growth from solution [3.109] is given by

$$v_\infty = \beta_k c_0 D_s n_{s0} f_0 \frac{\sigma}{x_s} \tanh \frac{y_0}{2x_s} , \tag{3.37}$$

where β_k is a retarding factor for the entry of a growth unit (GU) in the kink, $c_0 = \frac{x_s}{x_0}\ln(2x_s/(1.78a))$ valid for $x_s \ll x_0$ (when $x_s \gg x_0$, $c_0 = 1$), x_s is the mean displacement of the GUs on the surface, x_0 is the mean distance between kinks in the steps, D_s is the diffusion constant of GUs in the adsorption layer, n_{s0} is the number of GUs in the adsorption layer per cm^2 at equilibrium, f_0 is the area of one GU on the surface, and $\sigma = (X - X_s)/X_s$ is the relative supersaturation (where X and X_s are the actual and equilibrium molar fraction, respectively). The step equidistance y_0 is given for low supersaturation by

$$y_0 = fr^* = \frac{f\rho a}{k_B T \ln\beta} \cong \frac{f\rho a}{k_B T\sigma} , \tag{3.38}$$

being f a shape factor, r^* the critical radius of the 2-D nucleus, ρ the edge free energy (erg/cm), and a the shortest distance between GUs in the crystal. The relationship is simplified if the supersaturation β is low, in which case $\ln\beta \cong \beta - 1 = \sigma$.

Then (3.36) may be written as

$$R_{hkl} = C\frac{\sigma^2}{\sigma_1}\tanh\frac{\sigma_1}{\sigma} , \tag{3.39}$$

where C and σ_1 are constants:

$$C = \frac{\beta_k c_0 D_s n_{s0}\Omega}{x_s^2} ,$$

$$\sigma_1 = \frac{9.5\rho a}{\varepsilon k_B T x_s} ,$$

where Ω is the volume of growth unit and ε is related to the number of interacting growth spirals.

When $\sigma \ll \sigma_1$ (i.e. $y_0 \gg x_s$)

$$R = C\frac{\sigma^2}{\sigma_1} \quad \text{(parabolic law)} ,$$

When $\sigma \gg \sigma_1$ (i.e. $y_0 \ll x_s$)

$$R = C\sigma \quad \text{(linear law)} .$$

Other relationships were found by *Chernov* for direct integration in the kink [3.110] and by *Gilmer* et al. [3.111] for coupled volume and surface diffusion.

For the two-dimensional nucleation mechanism

$$\text{mononuclear mechanism } R_{hkl} = K_m J_{2D} , \tag{3.40}$$

$$\text{polynuclear mechanism } R_{hkl} = K_p v^{2/3} J_{2D}^{1/3} , \tag{3.41}$$

where K_m and K_p are the kinetic constants for the two types of mechanism, v is the step speed, J_{2D} is the 2-D nucleation frequency, which is given by

$$J_{2D} = K\exp\left(-\frac{f\rho^2 a^2}{(k_B T)^2\ln\beta}\right) , \tag{3.42}$$

where K and f are the kinetic and shape factors, respectively. Equation (3.42), formerly derived for growth from vapor state, is usually applied to growth from solution.

There is also a general empirical law

$$R_{hkl} = K'\sigma^n , \tag{3.43}$$

where K' is a kinetic temperature-dependent constant; the value of n is related to the mechanism.

3.6.2 Some Experimental Results

As $R_F < R_S < R_K$, in the growth form at a given β only F-faces can occur, although not necessarily all of

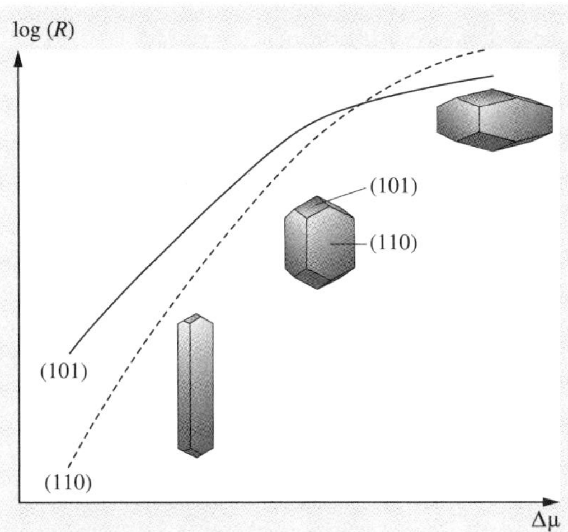

Fig. 3.15 Crossover of the growth rates versus thermodynamic supersaturation for the {110} and {101} faces of tetragonal lysozyme with change in crystal habit (after [3.57])

them. The faces of the same crystal may show a different dependence of R_{hkl} on β and then a crystal habit change can occur. The first static observations were made on NaCl, KCl, CsCl, KI, and many other ionic salts. When a crystal habit is made of F-faces, a higher critical supersaturation is needed for the S- or K-faces to appear [3.63, 112] (Sect. 3.2.2). When an F-face is replaced by another F-face at high supersaturation, the change is progressive. Kinetic measurements of crystal growth reveal that in most cases there is a supersaturation *dead zone* in which no growth occurs, and a critical supersaturation β^* should be reached for growth to start. Changes in relative face size and crystal habit are usually observed with increasing β [3.47, 53]. In n-paraffins the growth rate was found to be dependent also on the platelet thickness [3.93]. When a crossover in the relative growth rates of faces occurs, even radical changes in morphology are observed, as in the case of γ-aminobutyric acid [3.113] and lysozyme crystals (Fig. 3.15) [3.57, 114].

To explain the habit change, a variation in the crystal–solvent interaction is admitted, as for NaCl, CsCl, $M(H_2O)_6 \cdot SiF_6$, and CaF_2 [3.42]. In other cases the effect of supersaturation is attributed to the desolvation kinetics of the solute, as shown for KI, KBr, and KCl [3.115].

Supersaturation also causes changes in surface features, as seen on the {100} faces of $NH_4H_2PO_4$ (ADP), where there are no growth layers below a given β, and elliptical and then parallel layers appear with increasing β [3.116]. Supersaturation affects the activity of dislocation source [3.78]. The intensive studies of surfaces of ADP, KDP, and DKDP reveal that the step speeds on these faces and the hillock slopes may be nonlinear functions of β [3.78, 117]. This nonlinearity has been explained in terms of a complex dislocation step source [3.78], impurity adsorption [3.118], and the generation of kinks at growth steps [3.119]. Not only the step rates but also the kinetic coefficients can depend on β, which enhances the morphological instability [3.87]. It should be noted that, even in the kinetic regime, supersaturation is not constant along the growing face [3.78]. Small fluctuations in β may cause rapid change in the evolution of faces forming small interfacial angles, with the disappearance and reappearance of faces, as observed with potassium dichromate [3.120].

High supersaturation can determine phase transition, as shown in polymorph systems such as calcium carbonate, L-glutamic acid, and L-histidine [3.121]. Interest is also practical: from 2-propanol solutions it is possible through supersaturation to isolate the stable form of stavudine, an antiviral drug used for the treatment of human immunodeficiency virus–acquired immunodeficiency syndrome (HIV-AIDS) [3.122]. However resorting to high supersaturation in order to change the crystal habit is rarely used in industrial crystallization, as there is a risk of unwanted nucleation.

3.7 Solvent

Research into solvent effects on crystal growth and habit is relatively recent. In the years 1940–1967 a few studies were published, as summarized in two surveys papers [3.44, 123]. In the 1960s the first interpretations were geometric–structural [3.124]. The effect of the solvent in determining the mechanism of crystal growth was evaluated through the entropic α-factor and the growth kinetics [3.44, 123]. Since then, interest in this topic has been increasing, also due to the mounting needs of industrial crystallization. Research was then successfully extended to tailor-made additives [3.14, 125] (see also Sect. 3.2.3). Molecular dynamics (MD) simulations are also extensively applied to study the solvent–surface interaction. The solvent itself acts as an impurity; however, due to its importance (e.g., concentration), it is treated separately.

3.7.1 Choice of Solvent

The first essential step for crystallization is the choice of solvent. Knowledge of solubility and of its change with temperature is required. The most common solvent is water, which exerts its influence even in traces in the vapor and liquid phase. When lead molybdate is grown in dry air, it crystallizes as needles, but as platelets in moisturized air [3.127]. Small amounts of water added to organic solvents markedly increase the yield and growth rate of the target polymorph of pharmaceutical compounds [3.128]. Growth in aqueous systems is a complex process. Water molecules are coordinated by cations and anions in solution; in addition, they can be selectively adsorbed at the different sites on the growing faces (Fig. 3.16) [3.109, 126].

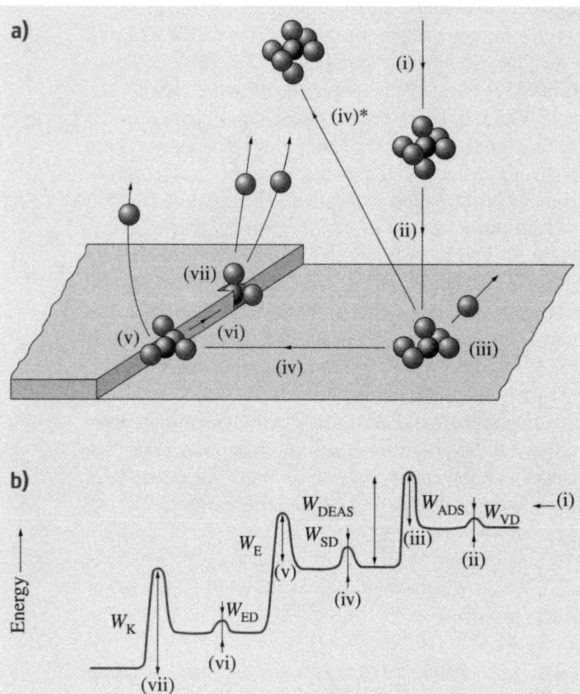

Fig. 3.16a,b The events leading to crystal growth: diffusion in the volume, on the surface, along the step, to the kink with integration; desolvation; and desorption. To each of these processes corresponds an activation energy of different magnitude (after [3.126])

Dehydration and desolvation should then occur, besides diffusion, involving the respective activation energies [3.109]. It should be noted that solvent molecules (and possible impurities) are also involved in surface adsorption processes, with an ensuing competition between the corresponding energies (or relaxation times) of growth units. The bonds engaged may be ionic, hydrogen bond, and van der Waals, with energies of about 15 kcal/mol (for the $H_2O{-}Na^+$ bond), 5 kcal/mol in organic solutions with hydrogen bonds, whereas for the van der Waals bond the strength is one or more orders of magnitude smaller than that of the ionic one. Many other solvents, mostly organic and tailor-made, are nowadays used in the chemical industry. They differ from water in some molecular properties (dielectric constant, dipole moment, size, etc.) and are classified into three main classes: protic (hydrogen donors, e.g., methanol), dipolar aprotic (e.g., acetonitrile), and nonpolar aprotic (e.g., hexane) [3.129].

3.7.2 Change of Solvent

Use is often made of different solvents in crystal growth. Solvent change has several correlated effects (on solubility and supersaturation, interactions with solute and crystal faces, surface free energy, etc.), and modifying the α-factor may change the growth mechanism. If a compound is grown from two different solvents, A and B, there are two values of α, which are related by [3.130]

$$\begin{aligned}\alpha_A &= \alpha_B + 4(\Phi_{B,sf} - \Phi_{A,sf})/(k_B T) \\ &\quad + 2(\Phi_{A,ff} - \Phi_{B,ff})/(k_B T) \\ &\approx \alpha_B + 4(\Phi_{B,sf} - \Phi_{A,sf})/(k_B T)\,. \end{aligned} \tag{3.44}$$

Let $\Phi_{A,ff} \approx \Phi_{B,ff}$. If the solute–solvent interaction energy $\Phi_{B,sf}$ is higher than $\Phi_{A,sf}$, then $\alpha_A > \alpha_B$. For example, the {110} faces of hexamethylene tetramine (HMT) grown in aqueous solution show $\alpha < 2.5$, whereas in ethanolic solutions the value is $\alpha = 3.2$–5.4 [3.130]. The growth mechanisms predicted by these values were confirmed by the growth kinetics [3.131]. Changing the solvent allows one to understand the role of the solvents and the bonds involved, as shown by *m*-nitroaniline. Grown from a nonpolar solvent, the crystal habit matches the theoretical one, based on the sole van der Waals bonds, whereas in polar solvents the electrostatic interactions play a great role [3.132]. By comparing steroid crystals grown from acetone and methanol solutions and performing MD simulations of the solvent–surface interactions, it is found that the contributions of Coulomb and van der Waals bonds are more important than hydrogen bonding in determining the crystals habit [3.133]. Solvents may affect surface roughness [3.129] and the roughening transition of faces, with a change in the transition order too [3.134]. Mixtures of solvents are often used. They can show higher solubility than separate solvents, as happens with L-pyroglutamic acid, which is more soluble in a water–ethanol mixture than in pure water or pure methanol, with a strong change of crystal habit [3.135]. On the other hand, mixtures of water and isopropoxethanol, in which $NaNO_3$ is less soluble, do not affect the shape of the $NaNO_3$ crystals [3.136], whilst adding water to an ethanol–water solution saturated with benzoic acid reduces the solubility of the latter and causes its crystallization with habit change (the drowning-out technique) [3.137]. Solvent can also stabilize polymorphs, as shown with $CaCO_3$ precipitated from water or ethanol solutions [3.138], and promote the formation of twins as well [3.125].

3.7.3 Solvent–Solute

Molecules of solvent and solute always interact, affecting the growth kinetics according to the energy involved. The interaction is more pronounced in supersaturated solutions [3.106]. In aqueous solutions of ionic salts, solute clusters which have the lattice structure of the crystal surface can form [3.139]. In aqueous solutions of hexamethylene tetramine (HMT), each HMT molecule coordinates four water molecules, whereas in alcoholic solutions there is no coordination, with consequences for the growth mechanisms [3.123]. Solute dimers may form, and tetramers, octamers, and even larger -mers were also detected in lysozyme solutions at high supersaturation [3.57]. The strong morphological difference of alizarin crystals grown from alcoholic and other organic solvents is attributed to the partial deprotonation of the alizarin molecules occurring in alcohol solutions, which blocks the needle growth of the side faces, leading to a totally different habit [3.140].

3.7.4 Solvent–Crystal Surface

The approach to the problem of the role played by solvent–crystal interactions is different according to the system being studied and theoretical assumptions. A relevant role is attributed to the α-factor [3.123, 130] with some reservations on its general validity [3.129]. Another way of looking at the solvent–crystal interactions is by considering adsorption of solvent molecules on the surface sites, as shown between the [010] PBCs on {101} faces of sucrose [3.47]. The role of solvents in terms of surface roughening and surface adsorption on crystal morphology is reviewed in [3.44]. A protic solvent can determine both the polymorph and crystal habit, as shown for stavudine, in the molecule of which one N atom is a very strong H-bonding donor and two O atoms are strong H-bonding acceptors. The crystal surfaces interact in different ways with polar and nonpolar solvents [3.122]. The effect of α-butyrolactone as a solvent on cyclotrimethylene trinitramine (RDX) crystals has been explained by invoking a third region between the crystal and solution bulk, i. e., a boundary layer. MD simulations show that the average energy of solvent molecules near the surface is higher than the average energy of the same molecules in the bulk, therefore the potential energy change per unit area can be chosen as a good parameter to use in explaining solvent effects on the crystal habit [3.141].

3.7.5 Mechanisms of Action

Solvent molecules are temporally adsorbed at the various sites of surface: ledges, steps, and kinks, with different lifetimes, which has an effect on growth rates. The formation of surface complexes was suggested for $Hg(CN)_2$ crystals grown from methanol [3.142]. As solutions of n-paraffins in petrol ether behave as a melt, a structural model was proposed, which considers the paraffin crystal as a Kossel crystal and the solvent molecules as in the Ising model, so that they easily adhere by adsorption to steps [3.143]. In polar crystals the complementary forms $\{hkl\}$ and $\{\bar{h}\bar{k}\bar{l}\}$ exhibiting different surface structures selectively interact with the solvent molecules, as in $N(C_2H_5)_4I$ crystals grown from four solvents, which being differently adsorbed onto {011} and $\{0\bar{1}\bar{1}\}$ faces, invert the morphological importance of these faces [3.97]. The morphological difference is relevant for industrial crystallization: the analgesic ibuprofen grown from ethanol shows a pseudohexagonal tabular habit but thin elongated platelets from ethyl acetate, owing to the interactions between ethyl acetate molecules and ibuprofen carboxylic groups emerging at the surface of {100} and {002} faces of the crystal [3.144].

3.8 Impurities

The literature dealing with the effects of impurities on growth kinetics and crystal habit is enormous and increasing. After the publication in 1951 of the BCF theory [3.7] and of *Buckley*'s book on crystal growth [3.145], the advance has been impressive. Fourteen years later a whole book was devoted to the relationships between adsorption, crystal growth, and morphology [3.146]. In Russia this topic has been widely studied, as evident from the series of volumes on crystal growth. The role of impurity is also a specific object of the proceedings of periodic symposia on industrial crystallization. Indeed, a strong impulse to research in this field is provided by mounting industrial demands. Many cases are given and discussed in some review papers [3.83, 147–149] and a summary with a nearly exhaustive literature is found in *Sangwal*'s monograph [3.150]. In recent years, interest has focused on additives and other large molecules used as habit modifiers. A synthetic view is tentatively presented in the following. Admitting a hydrodynamic regime, the role of volume diffusion is neglected as impurities only influence the surface processes.

3.8.1 The Main Factors

Impurities influence both thermodynamic and kinetic factors. Addition of an impurity to a solution may change the surface free energy γ, the edge free energy ρ, and the solubility. Even if an increase of average surface energy is possible, it is usually admitted that ρ decreases with the concentration of the adsorbed impurity C_i. By assuming a Langmuir-type isotherm and equilibrium between the impurities adsorbed on the step and in the bulk, the resulting edge free energy (expressed in ergs, for the sake of simplicity) is given by [3.151]

$$\rho_i = \rho - k_B T \ln C_i \,. \tag{3.45}$$

Similarly, a positive adsorption produces a reduction in the specific free surface energy. It follows that, since ρ_i decreases with impurity concentration C_i, the growth rate of the face should increase.

The effect on solubility depends on C_i. If C_i is low or very low, of the order of ppm, the influence on solubility is negligible. If C_i is higher, of the order of mg/l or g/l, the solubility may change. In the case of solubility increase, the resulting effect is usually an increase in step velocity and the growth rate of faces [3.150]. Solubility and growth rate of ionic crystals may be increased by changing the pH or increasing the ionic strength through addition of soluble salts, e.g., adding NaCl to $CaSO_4 \cdot 2H_2O$ solutions [3.151].

The growth rate depends on the kinetic constant C (3.39) and the step velocity v (3.37). Impurities adsorbed on the surface will cause a decrease of C and hinder the advances of steps by mechanisms that depend on the adsorption site. The retardation factor (3.37) of the steps becomes [3.83]

$$\beta_{st} \cong \beta_{st}^0 \left(1 - \frac{4d}{(\kappa^*)^2}\right) . \tag{3.46}$$

where β_{st}^0 is the retardation factor without adsorption at the step, d is the density of adsorbed impurity (cm^{-2}), and κ^* the critical curvature of the step (cm^{-1}).

As a general result, the growth rate is decreased.

Impurities may therefore have two opposite effects on crystal growth. The final result depends on the supersaturation and impurity concentration, as shown in the classical experiment of $Pb(NO_3)_2$ grown in the presence of methyl blue [3.150]. The increase of the growth rate at low supersaturation and low impurity concentration (the so-called catalytic effect) is attributed to the prevailing thermodynamic factor and a low density of kinks. At higher impurity concentration the resulting effect is a decrease of growth rate due to the dominance of kinetic factors. Similar behaviors were found in other cases [3.150].

3.8.2 Kinetic Models

Various models have been proposed to relate the step velocity to the effect of impurity adsorption. One of the first contributions is due to *Bliznakow* [3.152], who assumes that impurities are adsorbed on some of the active sites on the surface and derives a quantitative relationship between the step velocity and the kink site adsorption

$$v = v_0 - (v_0 - v_m)\theta_{eq} \,, \tag{3.47}$$

in which v_0 is the step velocity without adsorption in the kink; $\theta_{eq} \le 1$ is the coverage degree, in impurity, of the surface, and v_m is the limiting step velocity in impure solution, when all the surface is covered by impurities (i. e., $\theta_{eq} = 1$). If a Langmuir isotherm is valid, then

$$v = v_0 - (v_0 - v_m)\frac{KC_i}{(1 + KC_i)} \,. \tag{3.48}$$

The model was satisfactorily applied to experiments reported by the same author and others [3.146] chiefly on inorganic compounds and confirmed by curves of R versus C_i and the heat of adsorption measured at different temperatures [3.112].

Cabrera and *Vermilyea* (CV) [3.153] proposed a different model, in which the impurities are assumed to be immobile, adsorbed on ledges ahead of steps, where they form a two-dimensional (2-D) lattice. The steps move with average velocity v

$$v = v_0[1 - 2r^*/(d^{-1/2})]^{1/2} , \quad (3.49)$$

where r^* is the critical radius of the 2-D nucleus, and d is the average density of impurities, corresponding to a mean distance $s = d^{-1/2}$. The density of impurities d can also be expressed by the coverage degree of impurity (θ) times the maximum number (n_{max}) of sites available for adsorption per unit area: $d = n_{max}\theta$. From (3.49) it follows that when $s < 2r^*$, the step will stop; when $s > 2r^*$, it squeezes between the two adjacent impurities. At equilibrium between impurities adsorbed on the surface and solution, and if the coverage degree θ is given by a Langmuir isotherm, then

$$\theta = \frac{KC_i}{1 + KC_i} , \quad (3.50)$$

where K is the Langmuir constant. If $v \propto R_{hkl}$ and introducing the relative growth rate η, we have

$$\eta = \frac{v}{v_0} = \frac{R_{hkl}}{R_0} . \quad (3.51)$$

Rearranging (3.49)

$$\frac{1}{(1-\eta^2)^2} = \frac{1}{4(r^*)^2 n_{max}} + \frac{1}{4(r^*)^2 n_{max} K} \frac{1}{C_i} . \quad (3.52)$$

If the CV model is valid, then the previous relation should be satisfied and allows one to obtain n_{max}.

Since r^* is related to the critical supersaturation β^* (3.38), when $KC_i \gg 1$ (i.e. $\theta \to 1$), then

$$\frac{1}{(\ln \sigma^*)^2} = \frac{1}{\left(\ln \sigma^*_{max}\right)^2}\left(1 + \frac{1}{K_1 C_i}\right) . \quad (3.53)$$

When $KC_i \ll 1$, then

$$\ln \sigma^* = \ln \sigma^*_{max}(KC_i)^{1/2} , \quad (3.54)$$

i.e., the critical σ^* necessary for a step to move increases with increasing C_i, which has been experimentally confirmed.

The validity of the CV mechanism has been verified in many experiments, mainly with organic impurities and tailor-made additives, for example, in the growth of $C_{36}H_{74}$ from petroleum ether [3.93] and also in the growth of {101} faces of ADP under low supersaturation [3.118]. It may happen that strongly bound impurities may become incorporated in the crystal.

Sears [3.154], observing that 10^{-5}–10^{-6} molal concentration of FeF_3 was sufficient to poison LiF in aqueous solution, postulated a complete monostep adsorption at the growth steps. He calculated the fractional change of 2-D nucleation rate, which for low step coverage is proportional to the C_i change, and recognized that two opposite effects operate, one tending to increase the growth rate, the other to decrease it.

Albon and *Dunning* [3.155], studying the growth of sucrose in the presence of raffinose, admit adsorption of impurity at kinks, which lowers the step rate. When the impurity distance is less than the diameter d^* of the critical nucleus (measured as an integer number of molecular spacings), the advancing step is blocked. If $p = (1 - C_i)$ is the probability of finding a free site along a step, then the step speed is given by

$$v = v_0\left[d^* - (1 - C_i)d^* + (1 - C_i)\right](1 - C_i)^{d^*} . \quad (3.55)$$

This model gives also a good description of the effects of Cr^{3+} on the growth of the {100} faces of ADP for $C_i > 1.0 \times 10^{-2}\,kg/m^3$ [3.156].

Davey and *Mullin* [3.91] assumed that the rate-determining process is the surface diffusion, which is reduced by impurities absorbed on the ledge, and that the number of absorbed growth units is proportional to the fraction of free surface sites at equilibrium $n_i = n(1 - \theta)$. They derived the relation

$$v/v_0 = 1 - \theta_{eq} . \quad (3.56)$$

By expressing θ_{eq} in terms of Langmuir isotherm, the equation becomes

$$\frac{v_0}{v_0 - v} = 1 + \frac{1}{KC_i} . \quad (3.57)$$

This relation was shown to be valid, for example, for layer velocities in the [001] direction for the {100} faces of ADP in the presence of some ppm of $AlCl_3$, $FeCl_3$, and $CrCl_3$.

Since complete coverage ($\theta_{eq} = 1$) is not a necessary condition to stop crystal growth, *Kubota* and *Mullin* (KM) [3.157, 158] introduced in the Davey–Mullin equation an effectiveness factor α given by

$$\alpha = \frac{r^*}{L} = \frac{\rho a}{k_B T \sigma L} , \quad (3.58)$$

where L is the distance between two adsorption sites and a is the surface area of a growth unit. It follows that, depending on the ratio r^*/L, the factor α may be greater than, equal to, or less than 1. In their model a linear arrangement of adsorption sites is assumed to occur along the step, so that one-dimensional (1-D) coverage is combined with an adsorption isotherm. So we have

$$v/v_0 = 1 - a\theta_{eq} . \tag{3.59}$$

Relating the relative velocity η to C_i by a Langmuir isotherm, we have

$$\frac{1}{1-\eta} = \frac{1}{\alpha K}\frac{1}{C_i} + \frac{1}{\alpha} . \tag{3.60}$$

From this relationship it is possible to evaluate the critical supersaturation necessary to overcome the dead zone

$$\beta^* = \frac{\rho a K C_i}{k_B T L (1 + K C_i)} , \tag{3.61}$$

which can be rearranged as: $1/\beta^* = (C_1/C_i) + C_2$ (where C_1 and C_2 are two constants) [3.158].

The authors provide a number of examples, drawn from the literature, of the validity of their model for growth systems showing different α effectiveness factors [3.157]. The KM model was successively applied to crystal growth of other systems. Very recently, in studying the growth kinetics of the four main F-faces of sucrose in the presence of raffinose, both CV and KM models associated with Langmuir isotherms were tested. The agreement was better with the KM model than with CV, which means that adsorption occurs at kink sites and a spiral growth mechanism operates. The α effectiveness coefficient seems to be a very good parameter to predict the raffinose effects [3.159].

3.8.3 Adsorption Sites

Impurities act through adsorption, as apparent from the above kinetics models, onto the crystal surfaces, at steps and kinks. Adsorption occurs in this order, but only one process is considered determinant in the reaction path. They can be distinguished on the basis of some parameters (adsorption heat, activation energy, lifetime of the adsorbed state). A detailed description of adsorption at the three kinds of growth sites is given in some surveys papers [3.150,160,161]. The overall effect on the crystal habit depends on the effect on growth rates of faces.

Adsorption of impurities at *kinks* reduces the number of kinks available for growth. As growth proceeds through integration at kinks, the step velocity and growth rate are decreased and eventually stopped. Examples and lists of experiments on kink adsorption are given in many papers [3.91,149,150,156,159–161]. Adsorption in kinks, by increasing x_0, the mean distance between kinks, may cause polygonization of growth steps [3.160].

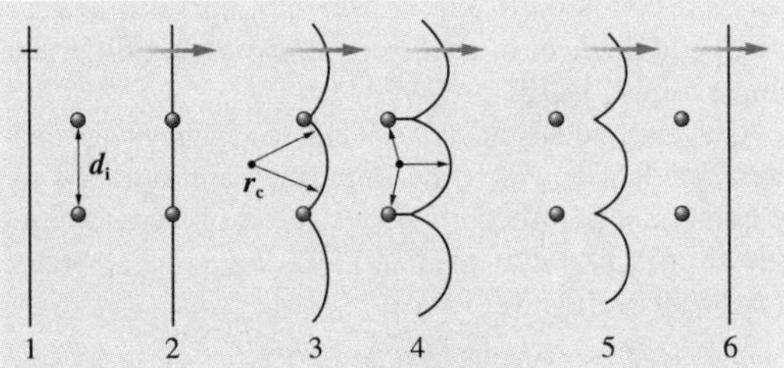

Fig. 3.17 The barrier presented by immobile impurities at the step motion on a ledge. The layer overcomes the impurities at a reduced rate if their distance is higher than the diameter of the critical nucleus

Impurities adsorbed at *steps* represent a steric barrier to diffusion of growth units along the step and to their entry into the kinks (Fig. 3.17). The step can move at a reduced rate only if the distance between two adsorbed impurities along the step is $s > 2r^*$; otherwise they are stopped.

Adsorption at steps is found to have occurred in many experiments [3.155,156,158,160].

Besides the above kinds of adsorption, another possibility arises when the impurity concentration is high and strong lateral interactions occur among the molecules. A 2-D adsorption *layer* may form which has structural similarity with the growing face, as discussed in Sect. 3.2.3. The formation of the adsorption layer is easier with large molecules. The existence of an adsorbed solution layer, strongly structured at the interface, was suggested by Sipyagin and Chernov, who found that the layer may be disturbed by the addition of alcohol, which increases the growth rate [3.93].

3.8.4 Effect of Impurity Concentration and Supersaturation

The impurity effectiveness is very variable. To affect crystal habit, some g/l may be necessary (e.g., Na_2SO_4 for NaCl) or ppm are already sufficient (e.g., $Na_2Fe(CN)_6$ for NaCl). A low or very low impurity concentration (C_i) may increase the growth rate due to its influence on the edge free energy, but higher C_i affecting the kinetic parameters may cause a sharp decrease. This effect is selective, as only some spe-

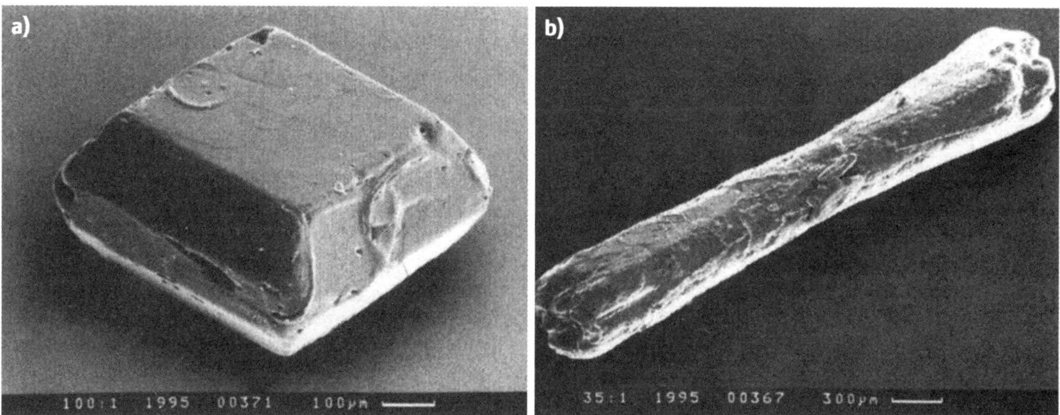

Fig. 3.18 (a) Ammonium sulfate crystals grown from pure solution and (b) in presence of 100 ppm Al^{3+} (after [3.163])

cific faces of a given crystal are concerned, and the crystal habit is modified. This occurs with organic and inorganic impurities (Fig. 3.18). For example, the growth rate of the {101} faces of KDP slowly increases, then decreases with increasing C_i of Fe^{3+} and Cr^{3+}, whereas it is not affected by Co^{2+} and Ni^{2+} [3.162]. Some ppm of Cr^{3+} are able to increase the step bunch spacing on the {100} faces of ADP, and some more to stop the formation of bunches [3.160].

Coupling of supersaturation and impurity concentration can have strong effects. Impurities are usually more efficient at low than at high supersaturation, where competition favors the adsorption kinetics of growth units with respect to that of impurities. For any impurity concentration there is a critical supersaturation σ^* below which growth is stopped (dead zone); the value of σ^* increases with C_i. A distinction is made between regular and irregular growth, according to whether an impurity changes the growth mechanism or

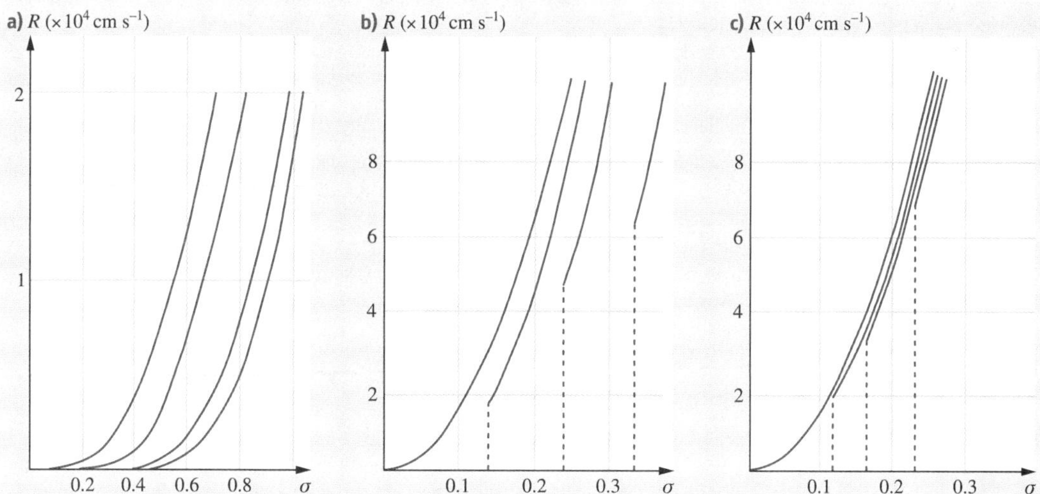

Fig. 3.19a–c Growth curves (R_{hkl} versus σ) for regular (a) and irregular (b,c) growth of different crystals in the presence of small and large impurity molecules, respectively, at increasing impurity concentrations. (a,b) (110) faces of $C_{36}H_{74}$ in the presence of an amine and a copolymer, respectively; (c) different faces of sodium perborate crystals. For a detailed description see [3.93, 150, 161]

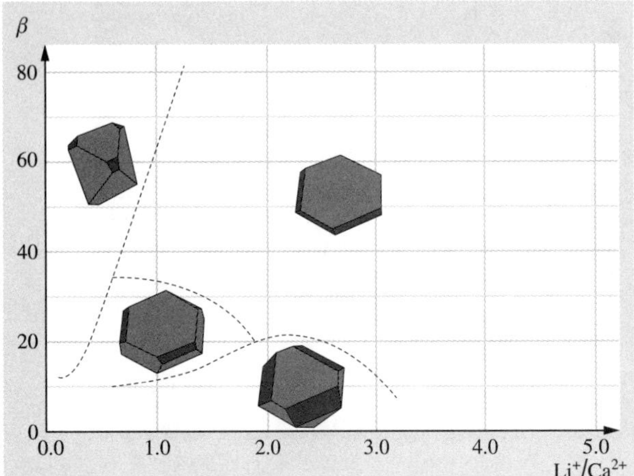

Fig. 3.20 Morphodrome (β versus Li^+/Ca^{2+}) of calcite crystals grown in the presence of Li^+ as impurity (after [3.72])

not, as can be seen from the curves for R_{hkl} versus β (Fig. 3.19).

In regular growth, the curves for impure solutions run parallel to that of pure solution, suggesting an unchanged growth mechanism, whereas in irregular growth the growth rate is stopped at low supersaturation, but starts at higher σ values. In this case impurities adsorbed on crystal surface promote step bunching and originate an irregular crystal morphology [3.93]. Growth rate fluctuations were observed in both static and kinetic regimes [3.78]. Many examples of the effects of supersaturation and impurity concentration are reported in [3.158, 160, 161]. These effects may be represented through morphodromes (Figs. 3.14 and 3.20) (Sect. 3.2.3).

Effects of impurities on 2-D islands lying on the (001) surface of a Kossel crystal have been simulated and a morphodrome calculated in which, due to the roles of edge diffusion and kink poisoning, an initial square crystal changes to a diamond with increasing β and C_i, in qualitative agreement with experimental observations [3.164].

3.8.5 Effect of Impurity Size

Three kinds of impurities are usually distinguished: ions and small molecules, polyelectrolytes, and tailor-made additives, the molecular weights of which are on the order of about 10^2, 10^3, and 10^4, respectively [3.147]. The former may be mobile or immobile, while the two latter types are nearly immobile with great effect on crystal habit and incorporation in the surface layers. A fourth class could be added, the amphiphiles (dyes, surfactants).

Ions

In crystal growth from aqueous solutions ions are often present as impurities. Their interaction with the crystal surface depends on their charge and size and on the structural properties of the crystal faces. For example, the divalent (Co^{2+}, Ni^{2+}, Ba^{2+}) cations generate greater local strains and surface change than the trivalent (Fe^{3+}, Cr^{3+}) ones on the surface of KDP faces due to the incorporation of the former and adsorption of the latter on the surface layer [3.162]. Interactions of divalent (Cd^{2+}, Ni^{2+}) and trivalent (Al^{3+}, Cr^{3+}, Fe^{3+}, La^{3+}, Ce^{3+}) cations with some inorganic crystals (gypsum, KDP, potassium hydrogen phthalate (KAP), etc.) are reported in [3.147]. Ions can be selectively incorporated into the different faces, such as Cr^{3+} and Fe^{3+} more easily included into {100} than into {101} faces, and also into different slopes of a vicinal hillock [3.103]. It is difficult to predict the influence of ionic impurities. For example, a number of divalent cations and anions have no significant effect on the crystal habit of gibbsite, which conversely undergoes dramatic changes in the presence of alkali ions (K^+, Cs^+) [3.165]. The Ba^{2+} and Pb^{2+} cations have opposite effects on the growth of the {111} and {100} faces of $Sr(NO_3)_2$ [3.166]. There is no general criteria to interpret the effect of ions on crystal habit, although a guideline may be the hypothesis of the formation of an ordered adsorption layer at the interface. When an ion is incorporated to form a solid solution, it distributes between the crystal (c) and the solution (s) according to a partition coefficient $K_{eq} = C_c/C_s$. The surface acts as a source or sink for the ion depending on whether K_{eq} is less than or greater than 1. The role of inclusions of isomorphous impurities during crystal growth is widely discussed in a review paper [3.167].

Polyelectrolytes

Polyelectrolytes, which include soluble polymers, proteins, and polysaccharides, are all characterized by an array of polar groups on an open long chain, by which they may be effective at a ppm level. Summarizing the interaction between polyelectrolytes and sparingly soluble salts, two principles were enunciated: the dimensional fit between the functional groups of the impurities and the interionic distance on the surface, and the achievement of stereochemical constraints [3.168].

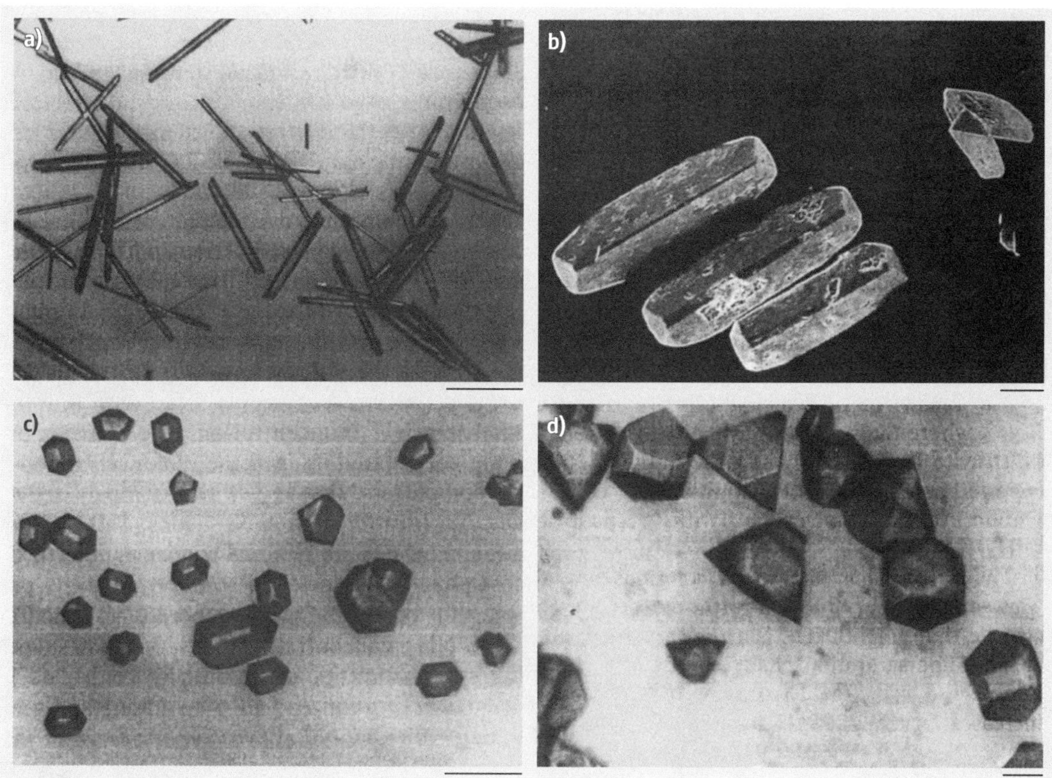

Fig. 3.21a–d Crystals of $Ni(NH_4)PO_4{\cdot}6H_2O$ grown in the presence of increasing ammonia excess. The crystals are elongated along [100] in (**d**) and along [010] in (**a–c**). *Bars*: 40 μm (after [3.170])

Any macromolecule may function selectively as a retardant or promoter, depending on specific conditions. A lot of research has been carried out on the influence of polyelectrolytes and biopolymers on the crystallization of biominerals, such as calcium carbonate and other mineral compounds (gypsum, fluorite, barite) [3.147, 154] and calcium oxalates, components of lithiasis [3.169].

Tailor-Made Additives

A special case of impurities is represented by tailor-made additives, which are used in organic crystallization to change habit. Their main characteristic is the partial resemblance of their molecular structure to that of the crystal molecules (Sect. 3.2.3). A wide list of experiments on tailor-made additives is given in the monograph [3.14]. Normal-paraffins that are used as solvents for the growth of lower and higher homologous can be considered as tailor-made additives [3.93, 129]. Molecules with long and rigid chains are easily adsorbed at the interfaces, whereas molecules with too long chains tend to fold, forming a 2-D heterogeneous nucleus on the surface, which promotes crystal growth [3.48]. To simulate the morphological change due to additives, the built-in approach and the surface docking approach were applied. The former works only if the intermolecular bonding is anisotropic, whereas the latter can be applied to both isotropic and anisotropic bonding [3.171].

3.8.6 Composition of the Solution: pH

In an ionic impurity-free solution, change of habit may occur when there is an excess of either cation or anion or when pH is changed:

1. The former case is presented by calcite, which assumes an elongated habit in the presence of

an excess of Ca^{2+} and becomes tabular when CO_3^{2-} ions are in excess [3.172]. Many other cases are known; we quote silver bromide, which crystallizes as a cube from stoichiometric solution, but as an octahedron in excess bromide [3.173]. $Ni(NH_4)PO_4 \cdot 6H_2O$ crystals grown in excess ammonia change from a needle-like habit to a pseudocubic symmetry due to $Ni(NH_3)_x(H_2O)_{6-x}$ complexes selectively adsorbed at the surfaces (Fig. 3.21) [3.170]. A detailed analysis of the chemical aspects of the impurity effects is given in the paper [3.174].

2. The effect of pH on crystal shape has been studied for a long time, especially for ADP and KDP by Russian researchers [3.175]. The change in pH modifies the concentration ratios of the chemical species in solution, especially of polyprotic acids. The growth rates of the {101} and {100} faces of ADP are differently affected by pH with change in crystal elongation, attributed to the role played by either hydration of NH_4^+, $H_2PO_4^-$ or $NH_4PO_4^{2-}$ ions or the concentration of hydroxonium ions $[H_3O(H_2O)_3]^+$ in solution [3.115]. Addition of Fe^{3+} and Cr^{3+} causes crystal tapering, which is a function of pH. Change in pH does not affect the growth rate of {101} faces of KDP, but when pH differs from the value corresponding to solution stoichiometry, the {100} faces increase in size and crystals become more isometric. This effect is related to the change with pH of surface parameters (step velocity and hillock slope) on the two faces [3.176]. A small pH decrease causes a dramatic increase in growth rate in the [001] direction and a decrease in the [010] direction on L(+)-glutamic acid hydrochloride due to the formation of dimers at low pH, which selectively interact with the two directions [3.177]. The role of pH in crystal growth is fundamental in the presence of amino acids. Glycine exists in aqueous solution as zwitterions ($H_3N^+CH_2COO^-$) in a definite range of pH, in which the critical supersaturation necessary to obtains the {110} form of NaCl attains its minimum value. The effect is explained by admitting adsorption of zwitterions along the [001] directions on the {110} faces [3.65].

3.9 Other Factors

Crystal habit can be affected by (not less important) factors other than those mentioned above. Some of these are briefly considered below, while some (electric field, ultrasound, pressure, microgravity) are neglected.

3.9.1 Temperature

The influence of temperature on crystal growth is multifaceted. Not only does it directly promote growth, as results from the growth rate formulas, but it also affects other factors that act on the growth rates: solubility, surface roughness, solvent/solute-interface interactions, and chemical equilibria. Since these factors work in a selective way on different faces, a change in temperature may cause relevant changes in crystal habit, as shown in experiments since the beginning of the 20th century. Some cases are quoted in [3.115, 126, 178]. Clear anomalies were detected in curves of R versus T for some salts (KCl, $NaClO_3$, and $KClO_3$). Sipyagin, Chernov, and Punin, quoted in [3.93], found that at definite temperatures the adsorbed layer at the crystal surface undergoes structural change, which suggests the occurrence of a third step between volume diffusion and surface diffusion. Significant changes were observed in sucrose crystals, in which the {101} form dominates the other $\{h0l\}$ forms at 30 °C, rendering crystals elongated, whereas at 40 °C it becomes less important and crystals appear isometric. Growth sectors occur at any T and σ (Fig. 3.22) [3.46].

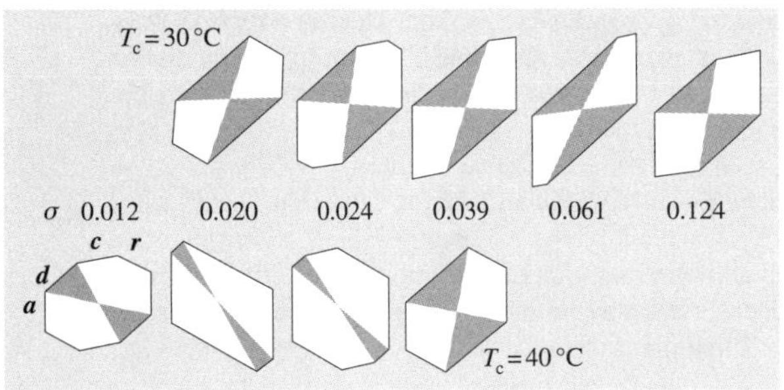

Fig. 3.22 Crystal habit of sucrose seen along [010], grown at 30 and 40 °C and at increasing σ values (after [3.46])

3.9.2 Magnetic Field

Research on the effects of a magnetic field on crystal growth started with the magnetic treatment of hard water in order to prevent scale formation. The number of papers on this topic is increasing due to its practical importance and criticism of results and reproducibility. A magnetic field has been recognized as affecting nucleation and crystal growth rate, polymorphism, and colloidal stability, and is now being applied to crystal growth of proteins and other compounds. In research on many poorly soluble inorganic salts a marked effect was only found for diamagnetic salts of weak acids (carbonates and phosphates): a magnetic field (0.27 T) increased their nucleation and growth rates. No effect on paramagnetic salts was recorded. These results are attributed to faster proton transfer from H-carbonate and H-phosphate ions to water molecules, due to proton spin inversion in the external field [3.179]. Also surface features are sensitive to the effects of magnetic field, as observed in the shape of etch pits and the shift of dislocations on paracetamol crystals [3.180]. Applying a strong magnetic field (10 T) to a lysozyme solution increases its viscosity and birefringence, which suggests molecular ordering with the formation of interconnected network, i. e., gel phase [3.181]. Experiments are also being carried out in gels, where magnetic field orientates $PbBr_2$ nanocrystallites and markedly increases their size [3.182]. Notwithstanding the amount of research in this area, the observed effects are as yet not adequately explained and no definite theory has been presented; indeed this is a new field of research.

3.9.3 Hydrodynamics

Crystal growth can take place under static or dynamic conditions. In the former case, which occurs in solution at rest and in gels, volume diffusion plays a decisive role in the growth kinetics. The latter condition is realized by stirring or shear flow. In this case the growth rate progressively increases with stirring or flow rate up to a limiting value which depends on the supersaturation. Beyond this value, the volume diffusion process does not influence the growth rate and surface diffusion becomes dominant [3.183]. The effects on crystal habit are different. Under static conditions concentration gradients are set up in solution and at the crystal surface, where they promote 2-D nucleation and the formation of hollow crystals and dendrites, depending on the supersaturation. Even in a solution at rest or zero gravity the crystal growth itself engenders convection, which influences morphological stability [3.184]. In a suitable stirred solution there are no concentration gradients except in the boundary layer, which is admittedly covering the crystal surface, and all the crystal faces are in contact with a homogeneous solution. Indeed, supersaturation is not constant along the face even in the kinetic regime [3.78], which may cause inclusions during growth. When a shear flow is applied, there are effects on crystal habit and surface features. The face on which flow is direct grows faster than the opposite face which is fed by a less supersaturated solution. In addition, a shear flow in the direction of step motion favors instability, whereas flow in the opposite direction enhances stability [3.87, 89].

3.10 Evolution of Crystal Habit

When a crystal with a given growth shape is left in contact with its mother solution, reaching saturation, rearrangements occur at the crystal–solution interface in order to achieve the minimum surface energy. A classical example is represented by NH_4Cl dendrites, which when grown in a small drop transform into a single crystal that corresponds to the equilibrium form (Fig. 3.6). Crystal habit evolution can also be observed in the growth from supersaturated solutions of poorly soluble compounds. In Fig. 3.23 the change with time of growth forms of $MgHPO_4 \cdot 3H_2O$, precipitated from solutions of different concentration, is shown. There is a progressive decrease of supersaturation, which goes together with the habit change, which continues even when the solution has become saturated. The general trend, independent of the initial concentrations, is towards the same crystal shape [3.185].

Another kind of evolution takes place when a crystal, grown in a given solvent, is transferred to a supersaturated solution of another solvent. The original faces are slowly replaced by others that are stable in the new solution, as observed for mercuric iodide crystals

Part A | 3.10

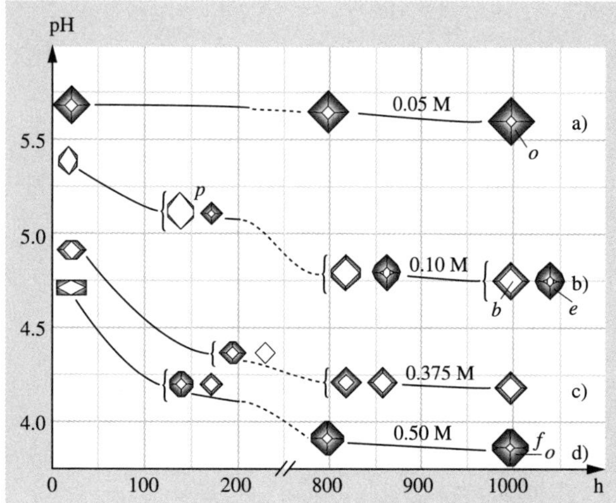

Fig. 3.23 Change of pH and crystal habit versus time of $MgHPO_4 \cdot 3H_2O$ crystals precipitated from equimolar solutions of different initial concentration. Dominant final forms: $o\{111\}$; $b\{010\}$ (after [3.184]) ◀

grown in aqueous solutions and subsequently inserted into methanol solutions [3.142]. A more drastic change occurs when several crystal phases may form from a supersaturated solution, as in the case of Ca-phosphate solutions. The normally stable phase is hydroxyapatite, which is the less soluble phosphate, but, obeying Ostwald's step rule, other more soluble phases nucleate first, even an amorphous phosphate at high pH. These phases are not stable, since the solution is still supersaturated with respect to the less soluble ones. A long process of chemical reactions starts in solution, which involves the nucleation of the more stable crystal phase and the dissolution of the first one [3.186].

3.11 A Short Conclusion

At the end of this concise review it is apparent that the crystal morphology is the result of the complex interaction of several factors – structural, thermodynamic and kinetic –, which make difficult and yet fascinating the study of this field. The great variability of the shape and size of crystals grown from solutions is mainly due to the capability of the crystalline structure of a compound to show different interfaces as a response to the variations introduced by the environment (solvent, impurity,...). The interaction of these interfaces with kinetic factors (such as supersaturation) give rise to the different morphologies and habits for the same crystal phase. On the contrary the shape, size and development of the living organisms are controlled by a genetic program which gives the same pattern to all individuals of a given species. An interesting field of combined research is bio-mineralization, where to the interplay of crystal structure and environment it is necessary to add the biological activity of the living organisms.

3.A Appendix

3.A.1 The Equilibrium Pressure of an Infinite Monoatomic Crystal with Its Own Vapor

From statistical thermodynamics the chemical potential of a molecule in a perfect gas reads

$$\mu^{g} = \varepsilon_{p}^{g} - k_{B}T \ln \Omega_{g} + \mu^{0} . \quad (3.A1)$$

The first term, ε_{p}^{g}, is the potential energy of the gas molecule, while the two last terms represent the entropy of dilution and the thermal entropy, respectively. Ω_g is the mean volume occupied in the gas phase, and $\mu^0 = k_B T \ln[h^3(2\pi m k_B T)^{3/2}]$ depends only on the mass of the molecule and on T. In a perfect monoatomic and infinite Einstein crystal (fulfilling the condition $T \gg (h/k_B)\nu_E$, where ν_E represents the unique vibration frequency) the chemical potential of one atom occupying a mean volume Ω_c is

$$\mu_{c\infty} = \varepsilon_{p}^{c\infty} - k_B T \ln \Omega_c + \mu^0 , \quad (3.A2)$$

where $\varepsilon_{p}^{c\infty}$ is the potential energy of the atom in any lattice site. At crystal–vapor equilibrium the chemical potentials of the two phases are the same, which implies

$$\varepsilon_{p}^{g} - \varepsilon_{p}^{c\infty} = k_B T \ln\left(\frac{\Omega_g}{\Omega_c}\right) .$$

At equilibrium,

$$\Omega_g = k_B T / p_{\infty}^{eq} \quad \text{while} \quad \Omega_c = (k_B T / 2\pi m)^{\frac{3}{2}} \nu_E^{-3} .$$

Being necessarily $\varepsilon_{p}^{g} > \varepsilon_{p}^{c\infty}$, we can define the extraction work of an atom from a mean lattice site (kink) to the vapor $\varphi_{c\infty} = \varepsilon_p^g - \varepsilon_p^c\infty > 0$ and then, from the preceding relation, one obtains the equilibrium pressure of a perfect monoatomic and infinite Einstein crystal

$$p_\infty^{eq} = \left[(2\pi m)^{\frac{3}{2}} (k_B T)^{-\frac{1}{2}} \nu_E^3\right] \exp(-\varphi_{c\infty}/k_B T) . \quad (3.A3)$$

This fundamental result can also be obtained in another way by a kinetic treatment, i. e., equating the frequency of units entering the crystal from the vapor (the Knudsen formula) with that of units leaving it [3.20].

References

3.1 H.J. Scheel: Historical introduction. In: *Handbook of Crystal Growth. 1a. Fundamentals*, ed. by D.T.J. Hurle (North-Holland/Elsevier, Amsterdam 1993) pp. 1–42

3.2 I.N. Stranski: Zur Theorie des Kristallwachstums, Z. Phys. Chem. **136**, 259–278 (1927), in German (see also Annu. Univ. Sofia **24**, 297 (1927))

3.3 I.N. Stranski, R. Kaischew: Gleichgewichtsform und Wachstumsform der Kristalle, Ann. Phys. **415**(4), 330–338 (1935)

3.4 I.N. Stranski, R. Kaischew: Kristallwachstum und Kristallbildung, Phys. Z. **36**, 393–403 (1935), in German

3.5 P. Niggli: *Geometrische Kristallographie des Diskontinuums* (Borntraeger, Leipzig 1919), in German

3.6 J.D.H. Donnay, D. Harker: A new law of crystal morphology extending the law of Bravais, Am. Mineral. **22**, 446–467 (1937)

3.7 W.K. Burton, N. Cabrera, F.C. Frank: The growth of crystals and the equilibrium structure of their surfaces, Philos. Trans. R. Soc. Lond. A **243**, 299–358 (1951)

3.8 P. Hartman, W.G. Perdok: On the relation between structure and morphology of crystals. I, Acta Cryst. **8**, 49–52 (1955)

3.9 P. Hartman, W.G. Perdok: On the relation between structure and morphology of crystals. II, Acta Cryst. **8**, 521–524 (1955)

3.10 P. Hartman, W.G. Perdok: On the relation between structure and morphology of crystals. III, Acta Cryst. **8**, 525–529 (1955)

3.11 P. Bennema: Growth and morphology of crystals: integration of theories of roughening and Hartman–Perdok theory. In: *Handbook of Crystal Growth. 1a. Fundamentals*, ed. by D.T.J. Hurle (North-Holland/Elsevier, Amsterdam 1993) pp. 477–581

3.12 P. Bennema: On the crystallographic and statistical mechanical foundations of the forty-years old Hartman–Perdok theory, J. Cryst. Growth **166**, 17–28 (1996)

3.13 X.Y. Liu, P. Bennema: Prediction of the growth morphology of crystals, J. Cryst. Growth **166**, 117–123 (1996)

3.14 Z. Berkovitch-Yellin, I. Weissbuck, L. Leiserowitz: Ab initio derivation of crystal morphology. In: *Morphology and Growth Unit of Crystals*, ed. by I. Sunagawa (Terra Scientific, Tokyo 1989) pp. 247–278

3.15 I. Sunagawa: Surface microtopography of crystal faces. In: *Morphology of Crystals*, ed. by I. Sunagawa (Terra Scientific, Tokyo 1987) pp. 321–365

3.16 I. Sunagawa: *Crystals, Growth, Morphology and Perfection* (Cambridge Univ. Press, Cambridge 2005)

3.17 W.A. Tiller: *The Science of Crystallization: Microscopic Interfacial Phenomena* (Cambridge Univ. Press, Cambridge 1991)

3.18 K. Sangwal, R. Rodriguez-Clemente: *Surface Morphology of Crystalline Solids* (Trans Tech, Zurich 1991)

3.19 W. Kossel: Zur Theorie des Kristallwachstums, Nachr. Ges. Wiss. Göttingen Math.-Phys. Kl. **1927**, 135–143 (1927), in German

3.20 M. Polanyi, E. Wigner: The interference of characteristic vibrations as the cause of energy fluctuations and chemical change, Z. Phys. Chem. **139**, 439 (1928), quoted in [3.21] pp. 49–51

3.21 B. Mutaftschiev: *The Atomistic Nature of Crystal Growth* (Springer, Berlin 2001)

3.22 R. Kern: The equilibrium form of a crystal. In: *Morphology of Crystals. Part A*, ed. by I. Sunagawa (Terra Scientific, Tokyo 1987) pp. 77–206

3.23 S. Toshev: Equilibrium forms. In: *Crystal Growth: An Introduction*, ed. by P. Hartman (North-Holland, Amsterdam 1973) pp. 328–341

3.24 M. Bienfait, R. Kern: Etablissement de la forme d'équilibre d'un crystal par la méthode de Lemmlein-Klija, Bull. Soc. Fr. Minér. Crist. **87**, 604–613 (1964)

3.25 E. Rodder, P.H. Ribbe (Ed.): *Fluid Inclusions*, Vol. 12 (Mineral. Soc. America, Chelsea 1984)

3.26 G.G. Lemmlein: About the equilibrium form of crystals, Dokl. Akad. Nauk. SSSR **98**, 973–975 (1954), in Russian

3.27 I.N. Stranski: Propriétés des surfaces des cristaux, Bull. Soc. Fr. Minér. Crist. **79**, 359–382 (1956)

3.28 O. Knacke, I.N. Stranski: Kristalltracht und Adsorption, Z. Elektrochem. **60**, 816–822 (1956), in German

3.29 R. Lacmann: Methoden zur Ermittlung der Gleichgewichts- und Wachstumsflächen von homöopolaren Kristallen bei der Adsorption von Fremdstoffen, Z. Krist. **112**, 169–187 (1959), in German

3.30 R. Kaischew, B. Mutaftschiev: Bull. Chem. Dept. Bulg. Acad. Sci. **7**, (1959), quoted in [3.21]

3.31 B. Honigmann: Gleichgewichtsformen von Kristallen und spontane Vergröberungen. In: *Adsorption et Croissance Cristalline* (CNRS, Paris 1965) pp. 141–169

3.32 J.D.H. Donnay, G. Donnay: Structural hints from crystal morphology, 20th Annu. Diffr. Conf. (Pittsburg 1962)
3.33 J. Prywer: Explanation of some peculiarities of crystal morphology deduced from the BFDH law, J. Cryst. Growth **270**, 699–710 (2004)
3.34 R. Kern: Etude du faciès de quelques cristaux ioniques à structure simple. A. Le changement de faciès en milieu pur, Bull. Soc. Fr. Minéral. Crist. **76**, 325–364 (1953)
3.35 P. Hartman: Structure and Morphology. In: *Crystal Growth: An Introduction*, ed. by P. Hartman (North Holland, Amsterdam 1973) pp. 367–402
3.36 P. Hartman, P. Bennema: The attachment energy as a habit controlling factor. I. Theoretical considerations, J. Cryst. Growth **49**, 145–156 (1980)
3.37 F. Harary: *Graph Theory* (Addison Wesley, Reading 1969)
3.38 C.S. Strom: Finding F faces by direct chain generation, Z. Krist. **172**, 11–24 (1985)
3.39 R.F.P. Grimbergen, H. Meekes, P. Bennema, C.S. Strom, L.J.P. Vogels: On the prediction of crystal morphology. I. The Hartman–Perdok theory revisited, Acta Cryst. A **54**, 491–500 (1998)
3.40 S.X.M. Boerrigter, R.F.P. Grimbergen, H. Meekes: *FACELIFT-2.50, a Program for Connected Net Analysis* (Department of Solid State Chemistry University of Nijmegen, Nijmegen 2001)
3.41 E. Dowty: Crystal structure and crystal growth: I. The influence of internal structure on morphology, Am. Mineral. **61**, 448–459 (1976)
3.42 P. Hartman: Modern PBC theory. In: *Morphology of Crystals. Part A*, ed. by I. Sunagawa (Terra Scientific, Tokyo 1987) pp. 269–319
3.43 K.A. Jackson: *Liquid Metals and Solidification* (Am. Soc. Metals, Cleveland 1958) p. 174
3.44 R.J. Davey: The role of the solvent in crystal growth from solution, J. Cryst. Growth **76**, 637–644 (1986)
3.45 Z. Berkovitch-Yellin: Towards the ab initio determination of crystal morphology, J. Am. Chem. Soc. **107**, 8239–8253 (1985)
3.46 D. Aquilano, M. Rubbo, G. Mantovani, G. Sgualdino, G. Vaccari: Equilibrium and growth forms of sucrose crystals in the {*h*0*l*} zone. I. Theoretical treatment of {101}-d form, J. Cryst. Growth **74**, 10–20 (1986)
3.47 D. Aquilano, M. Rubbo, G. Mantovani, G. Sgualdino, G. Vaccari: Equilibrium and growth forms of sucrose crystals in the {*h*0*l*} zone. II. Growth kinetics of the {101}-d form, J. Cryst. Growth **83**, 77–83 (1987)
3.48 X.Y. Liu: Interfacial structure analysis for the prediction of morphology of crystals and implications for the design of tailor-made additives, J. Cryst. Growth **174**, 380–385 (1997)
3.49 X.Y. Liu, E.S. Boek, W.J. Briels, P. Bennema: Analysis of morphology of crystals based on identification of interfacial structure, J. Chem. Phys. **103**(9), 3747–3754 (1995)
3.50 R. Boistelle, A. Doussoulin: Spiral growth mechanisms of the (110) faces of octacosane crystals in solution, J. Cryst. Growth **33**, 335–352 (1976)
3.51 M. Rubbo: Méthodes de mesure et cinétique de croissance et dissolution des faces (001) de l'hexatriacontane en solution. Ph.D. Thesis (Univ. Aix-Marseille III, 1978), in French
3.52 M. Rubbo, R. Boistelle: Dissolution and growth kinetics of the {001} faces of n-hexatriacontane crystals grown from heptane, J. Cryst. Growth **51**, 480–488 (1981)
3.53 D. Aquilano, M. Rubbo, G. Vaccari, G. Mantovani, G. Sgualdino: Growth mechanisms of sucrose from face-by-face kinetics and crystal habit modifications from impurities effect. In: *Industrial Crystallization 84*, ed. by S.J. Jančić, E.J. de Jong (Elsevier, Amsterdam 1984) pp. 91–96
3.54 D. Aquilano, M. Rubbo, G. Mantovani, G. Vaccai, G. Sgualdino: Sucrose crystal growth. Theory, experiments and industrial applications. In: *Crystallization as a Separation Process*, ed. by A.S. Myerson, K. Toyokura (ACS Symp. Ser., Washington 1990) pp. 72–84
3.55 C.H. Lin, N. Gabas, J.P. Canselier, G. Pèpe: Prediction of the growth morphology of aminoacid crystals in solution. I. α-Glycine, J. Cryst. Growth **191**, 791–802 (1998)
3.56 C.H. Lin, N. Gabas, J.P. Canselier: Prediction of the growth morphology of aminoacid crystals in solution. II. γ-Aminobutyric acid, J. Cryst. Growth **191**, 803–810 (1998)
3.57 R.F.P. Grimbergen, E.S. Boek, H. Meekes, P. Bennema: Explanation of the supersaturation dependence of the morphology of lysozyme crystals, J. Cryst. Growth **207**, 112–121 (1999)
3.58 H.M. Cuppen, G.M. Day, P. Verwer, H. Meekes: Sensitivity of morphology prediction to the force field: Paracetamol as an example, Cryst. Growth Des. **4**, 1341–1349 (2004)
3.59 S.X.M. Boerrigter, G.P.H. Josten, J. van der Streek, F.F.A. Hollander, J. Los, H.M. Cuppen, P. Bennema, H. Meekes: MONTY: Monte Carlo crystal growth on any crystal structure in any crystallographic orientation; application to fats, J. Phys. Chem. A **108**, 5894–5902 (2004)
3.60 C.S. Strom, P. Hartman: Comparison between Gaussian and exponential charge distributions in Ewald surface potentials and fields: NaCl, aragonite, phlogopite, Acta Cryst. **A45**, 371–380 (1989)
3.61 D. Aquilano, L. Pastero, M. Bruno, M. Rubbo: {100} and {111} forms of the NaCl crystals coexisting in growth from pure aqueous solution, J. Cryst. Growth **311**, 399–403 (2009)
3.62 D. Knoppik, A. Lösch: Surface structure and degree of coarsening of {111}NaCl surfaces near the thermodynamic equilibrium between crystal and vapour, J. Cryst. Growth **34**, 332–336 (1976)

3.63 R. Kern: Etude du faciès de quelques cristaux ioniques à structure simple. B. Influence des compagnons de cristallisation sur le faciès des cristaux, Bull. Soc. Fr. Minér. Crist. **76**, 391–414 (1953)

3.64 R. Boistelle: Contribution à la connaissance des formes de croissance du chlorure de sodium, Ph.D. Thesis (Nancy, 1966)

3.65 M. Bienfait, R. Boistelle, R. Kern: Le morphodrome de NaCl en solution et l'adsorption d'ions étrangers. In: *Adsorption et Croissance Cristalline* (CNRS, Paris 1965) pp. 577–594

3.66 R. Boistelle, B. Simon: Epitaxies de $CdCl_2{\cdot}2NaCl{\cdot}3H_2O$ sur les faces (100), (110) et (111) des cristaux de chlorure de sodium, J. Cryst. Growth **26**, 140–146 (1974)

3.67 R. Boistelle, M. Mathieu, B. Simon: Adsorption in solution of cadmium ions on {100} and {111} of NaCl. In: *Growth of Crystals*, ed. by A.A. Chernov (Consultants Bureau, New York 1984), vol 12, pp. 99–102

3.68 N. Radenovič, W.J.P. van Enckewort, D. Kaminski, M. Heijna, E. Vlieg: Structure of the {111}NaCl crystal surfaces grown from solution in the presence of $CdCl_2$, Surf. Sci. **599**, 196–206 (2005)

3.69 N. Radenovič, W.J.P. van Enckewort, P. Verwer, E. Vlieg: Growth and characteristics of the {111}NaCl crystal surface grown from solution, Surf. Sci. **523**, 307–315 (2003)

3.70 N. Radenovič: The Role of Impurities on the Morphology of NaCl Crystals. An Atomic Scale View. Ph.D. Thesis (Radboud Univ., Nijmegen 2006)

3.71 L. Li, K. Tsukamoto, I. Sunagawa: Impurity adsorption and habit changes in aqueous solution grown KCl crystals, J. Cryst. Growth **99**, 150–155 (1990)

3.72 L. Pastero, E. Costa, M. Bruno, M. Rubbo, G. Sgualdino, D. Aquilano: Morphology of calcite ($CaCO_3$) crystals growing from aqueous solutions in the presence of Li^+ ions. Surface behavior of the {0001} form, Cryst. Growth Des. **4**, 485–490 (2004)

3.73 M. Bruno, M. Prencipe: Ab-initio quantum-mechanical modeling of the (001), ($\bar{1}$01) and (110) surfaces of zabuyelite (Li_2CO_3), Surf. Sci. **601**, 3012–3019 (2007)

3.74 G. Clydesdale, K.J. Roberts, R. Docherty: Modelling the morphology of molecular crystals in the presence of disruptive tailor-made additives, J. Cryst. Growth **135**, 331–340 (1994)

3.75 G. Clydesdale, K.J. Roberts, K. Lewtas, R. Docherty: Modelling the morphology of molecular crystals in the presence of blocking tailor-made additives, J. Cryst. Growth **141**, 443–450 (1994)

3.76 G. Clydesdale, K.J. Roberts, R. Docherty: HABIT 95 – A program for predicting the morphology of molecular crystals as a function of the growth environment, J. Cryst. Growth **166**, 78–83 (1996)

3.77 D. Aquilano: Complex growth polytypism and periodic polysynthetic twins on octacosane crystals (n-$C_{28}H_{58}$), J. Cryst. Growth **37**, 215–228 (1977)

3.78 A.A. Chernov: Morphology and kinetics of crystal growth from aqueous solutions. In: *Morphology and Growth Unit of Crystals*, ed. by I. Sunagawa (Terra Scientific, Tokyo 1989) pp. 391–417

3.79 A.A. Chernov, T. Nishinaga: Growth shapes and their stability at anisotropic interface kinetics: theoretical aspects for solution growth. In: *Morphology of Crystals. Part A*, ed. by I. Sunagawa (Terra Scientific, Tokyo 1987) pp. 207–267

3.80 S.Y. Potapenko: Moving of steps through impurity fence, J. Cryst. Growth **133**, 147–154 (1993)

3.81 V.V. Voronkov, L.N. Rashkovih: Step kinetics in the presence of mobile adsorbed impurity, J. Cryst. Growth **144**, 107–115 (1994)

3.82 J.P. van der Eerden, H. Müller-Krumbhaar: Step bunching due to impurity adsorption: a new theory. In: *Morphology and Growth Unit of Crystals*, ed. by I. Sunagawa (Terra Scientific, Tokyo 1989) pp. 133–138

3.83 J.P. van der Eerden: Crystal growth mechanisms. In: *Handbook of Crystal Growth. Ia. Fundamentals*, ed. by D.T.J. Hurle (North-Holland, Amsterdam 1993) pp. 307–475

3.84 A.J. Derksen, W.J.P. van Enckevort, M.S. Couto: Behaviour of steps on the (001) face of $K_2Cr_2O_7$ crystals, J. Phys. D: Appl. Phys. **27**, 2580–2591 (1994)

3.85 H.M. Cuppen, H. Meekes, E. van Veenendaal, W.J.P. van Enckevort, P. Bennema, M.F. Reedijk, J. Arsic, E. Vlieg: Kink density and propagation velocity of the [010] step on the Kossel (100) surface, Surf. Sci. **506**, 183–195 (2002)

3.86 K. Tsukamoto: In situ observations of monomolecular steps on crystal growing in aqueous solution, J. Cryst. Growth **61**, 199–209 (1983)

3.87 S.R. Coriell, A.A. Chernov, B.T. Murray, G.B. McFadden: Step bunching: generalized kinetics, J. Cryst. Growth **183**, 669–682 (1998)

3.88 B.T. Murray, S.R. Coriell, A.A. Chernov, G.B. McFadden: The effect of oscillatory shear flow on step bunching, J. Cryst. Growth, **218**, 434–446 (2000)

3.89 S.R. Coriell, G.B. McFadden: Applications of morphological stability theory, J. Cryst. Growth **237–239**, 8–13 (2002)

3.90 R. Ghez, S.S. Iyer: The kinetics of fast steps on crystal surfaces and its application to the molecular beam epitaxy of silicon, IBM J. Res. Dev. **32**, 804–818 (1988)

3.91 R.J. Davey, J.W. Mullin: Growth of the {100} faces of ammonium dihydrogen phosphate crystals in the presence of ionic species, J. Cryst. Growth **26**, 45–51 (1974)

3.92 M. Rubbo: Surface processes and kinetic interaction of growth steps, J. Cryst. Growth **291**, 512–520 (2006)

3.93 B. Simon, R. Boistelle: Crystal growth from low temperature solutions, J. Cryst. Growth **52**, 779–788 (1981)

3.94 M. Bienfait, R. Boistelle, R. Kern: Formes de croissance des halogenures alcalins dans un solvant polaire. In: *Adsorption et Croissance Cristalline* (CNRS, Paris 1965) pp. 515–531, in French

3.95 E. van Veenendaal, P.J.C.M. van Hoof, J. van Suchtelen, W.J.P. van Enckevort, P. Bennema: Kinetic roughening of the Kossel (100) surface, J. Cryst. Growth **198**, 22–26 (1999)

3.96 P. Hartman: The calculation of the electrostatic lattice energy of polar crystals by slice-wise summation, with an application to BeO, Z. Krist. **161**, 259–263 (1982)

3.97 R. Cadoret, J.C. Monier: Influence de l'adsorption des molécules de solvant sur la vitesse normale de croissance des faces opposées appartenant aux formes mérièdres complémentaires $\{hkl\}$ et $\{\bar{h}\bar{k}\bar{l}\}$ d'un cristal non centrosymétrique. In: *Adsorption et Croissance Cristalline* (CNRS, Paris 1965) pp. 559–573

3.98 E. van der Voort, P. Hartman: Morphology of polar $ASO_3 \cdot 6H_2O$ crystals (A = Ni, Co, Mg) and solvent interactions, J. Cryst. Growth **106**, 622–628 (1990)

3.99 A.J. Gratz, P.E. Hillner: Poisoning of calcite growth viewed in the atomic force microscope (AFM), J. Cryst. Growth **129**, 789–793 (1993)

3.100 T.N. Thomas, T.A. Land, T. Martin, W.H. Casey, J.J. DeYoreo: AFM investigation of the step kinetics and hillock morphology of the {100} face of KDP, J. Cryst. Growth **260**, 566–579 (2004)

3.101 M. Moret: Influence of organic dyes on potassium sulphate crystal growth: A joint morphological and atomic force microscopy analysis, Mater. Chem. Phys. **66**, 177–188 (2000)

3.102 E. Haitema, J.P. van der Eerden: Defect formation during crystal growth, J. Cryst. Growth **166**, 141–145 (1996)

3.103 N. Zaitseva, L. Carman, I. Smolsky, R. Torres, M. Yan: The effect of impurities and supersaturation on the rapid growth of KDP crystals, J. Cryst. Growth **204**, 512–524 (1999)

3.104 N. Zaitseva, L. Carman, I. Smolsky: Habit control during rapid growth of KDP and DKDP crystals, J. Cryst. Growth **241**, 363–373 (2002)

3.105 R. Rodriguez-Clemente, S. Veintemillas-Verdaguer, F. Rull-Perez: Mechanism of crystal growth from boiling water solutions of soluble inorganic salts, mainly KDP. In: *Morphology and Growth Unit of Crystals*, ed. by I. Sunagawa (Terra Scientific, Tokyo 1989) pp. 479–512

3.106 A.S. Myerson, A.F. Izmailov: The structure of supersaturated solutions. In: *Handbook of Crystal Growth. 1a Fundamentals*, ed. by D.T.J. Hurle (North-Holland/Elsevier, Amsterdam 1993) pp. 249–306

3.107 F. Abbona, R. Boistelle: Nucleation of struvite ($MgNH_4PO_4 \cdot 6H_2O$) single crystals and aggregates, Cryst. Res. Technol. **20**, 133–140 (1985)

3.108 I. Sunagawa: Morphology of minerals. In: *Morphology of Crystals. Part B*, ed. by I. Sunagawa (Terra Scientific, Tokyo 1987) pp. 509–587

3.109 P. Bennema: Analysis of crystal growth models for slightly supersaturated solutions, J. Cryst. Growth **1**, 278–286 (1967)

3.110 A.A. Chernov: The spiral growth of crystals, Sov. Phys. Usp. **4**, 116–148 (1961)

3.111 G.H. Gilmer, R. Ghez, N. Cabrera: An analysis of combined surface and volume diffusion processes in crystal growth, J. Cryst. Growth **8**, 79–93 (1971)

3.112 R. Kern: Crystal growth and adsorption. In: *Growth of Crystals*, Vol. 8, ed. by N.N. Sheftal (Consultant Bureau, New York 1969) pp. 3–23

3.113 C.H. Lin, N. Gabas, J.P. Canselier, N. Hiquily: Influence of additives on the growth morphology of γ-aminobutyric acid, J. Cryst. Growth **166**, 104–108 (1996)

3.114 S.D. Durbin, G. Feher: Simulation of lysozyme crystal growth by the Monte Carlo method, J. Cryst. Growth **110**, 41–51 (1991)

3.115 R. Boistelle: Survey of crystal habit modification in solution. In: *Industrial Crystallization*, ed. by J.W. Mullin (Plenum, New York 1975) pp. 203–214

3.116 R.J. Davey, J.W. Mullin: The effect of supersaturation on growth features on the {100} faces of ammonium dihydrogen phosphate crystals, J. Cryst. Growth **29**, 45–48 (1975)

3.117 A.A. Chernov, L.N. Rashkovic, A.A. Mkrtchan: Solution growth kinetics and mechanism: Prismatic face of ADP, J. Cryst. Growth **74**, 101–112 (1986)

3.118 A.A. Chernov, A.I. Malkin: Regular and irregular growth and dissolution of (101) faces under low supersaturation, J. Cryst. Growth **92**, 432–444 (1988)

3.119 K. Sangwal: On the mechanism of crystal growth from solutions, J. Cryst. Growth **192**, 200–214 (1998)

3.120 J. Prywer: Effect of supersaturation on evolution of crystal faces – Theoretical analysis, J. Cryst. Growth **289**, 630–638 (2006)

3.121 M. Kitamura: Controlling factor of polymorphism in crystallization process, J. Cryst. Growth **237–239**, 2205–2214 (2002)

3.122 M. Mirmehrabi, S. Rohani: Polymorphic behaviour and crystal habit of an anti-viral/HIV drug: Stavudine, Cryst. Growth Des. **6**, 141–149 (2006)

3.123 R. Bourne, R.J. Davey: The role of solvent-solute interactions in determining crystal growth mechanisms from solution. I. The surface entropy factor, J. Cryst. Growth **36**, 278–286 (1976)

3.124 P. Hartman: Le coté cristallographique de l'adsorption vu par le changement de faciès. In: *Adsorption et Croissance Cristalline* (CNRS, Paris 1965) pp. 479–506, in French

3.125 M. Lahav, L. Leiserowitz: The effect of solvent on crystal growth and morphology, Chem. Eng. Sci. **56**, 2245–2258 (2001)

3.126 S.D. Elwell, H.J. Scheel: *Crystal Growth from High-Temperature Solutions* (Academic, London 1975)

3.127 H.C. Zeng, L.C. Lim, H. Kumagai, M. Hirano: Effect of ambient water on crystal morphology and coloration of lead molybdate, J. Cryst. Growth **171**, 493–500 (1997)

3.128 J. Wang, C. Loose, J. Baxter, D. Cai, Y. Wang, J. Tom, J. Lepore: Growth promotion by H_2O in organic

solvent-selective isolation of a target polymorph, J. Cryst. Growth **283**, 469–478 (2005)
3.129 R. Boistelle: Crystal growth from non aqueous solutions. In: *Interfacial Analysis of Phase Transformations*, ed. by B. Mutaftschiev (Reidel, Dordrecht 1982) pp. 531–557
3.130 B. Bourne, R.J. Davey: Solvent effects in the growth of hexamethylene tetramine crystals. In: *Industrial Crystallization*, ed. by J.W. Mullin (Plenum, New York 1975) pp. 223–237
3.131 J.R. Bourne, R.J. Davey: The role of solvent-solute interactions in determining crystal growth mechanisms from solution. II. The growth kinetics of hexamethylene tetramine, J. Cryst. Growth **36**, 287–296 (1976)
3.132 H.-X. Cang, W.-D. Huang, Y.-U. Zhou: Effects of organic solvents on the morphology of the meta-nitroaniline crystal, J. Cryst. Growth **192**, 236–242 (1998)
3.133 C. Stoica, P. Verwer, H. Meekes, P.J.C.M. van Hoof, F.M. Karspersen, E. Vlieg: Understanding the effect of a solvent on the crystal habit, Cryst. Growth Des. **4**, 765–768 (2004)
3.134 P.J.C.M. van Hoof, M. Schoutsen, P. Bennema: Solvent effect on the roughening transition and wetting of n-paraffin crystals, J. Cryst. Growth **192**, 307–317 (1998)
3.135 W.S. Wang, M.D. Aggarwal, J. Choi, T. Gebre, A.D. Shields, B.G. Penn, D.O. Frazier: Solvent effects and polymorphic transformation of organic nonlinear optical crystal L-pyroglutamic acid in solution growth process. I. Solvent effects and growth morphology, J. Cryst. Growth **198/199**, 578–582 (1999)
3.136 H. Oosterhof, R.M. Geertman, G.J. Witkamp, G.M. van Rosmalen: The growth of sodium nitrate from mixtures of water and isopropoxyethanol, J. Cryst. Growth **198/199**, 754–759 (1999)
3.137 X. Holmbäck, Å.C. Rasmuson: Size and morphology of benzoic acid crystals produced by drowning-out crystallization, J. Cryst. Growth **198/199**, 780–788 (1999)
3.138 K.-S. Seo, C. Han, J.-H. Wee, J.-K. Park, J.-W. Ahn: Synthesis of calcium carbonate in a pure ethanol and aqueous ethanol solution as the solvent, J. Cryst. Growth **276**, 680–687 (2005)
3.139 W. Polak, K. Sangwal: Modelling the formation of solute clusters in aqueous solutions of ionic salts, J. Cryst. Growth **152**, 182–190 (1995)
3.140 R.E. Aigra, W.S. Graswinckel, W.J.P. van Enckevort, E. Vlieg: Alizarin crystals: An extreme case of solvent induced morphology change, J. Cryst. Growth **285**, 168–177 (2005)
3.141 J.H. ter Horst, R.M. Geertman, G.M. van Rosmalen: The effect of solvent on crystal morphology, J. Cryst. Growth **230**, 277–284 (2001)
3.142 M. Ledésert, J.C. Monier: Modification du faciès des cristaux de cyanure mercurique par adsorption spécifique de molécules CH_3OH. In: *Adsorption et Croissance Cristalline* (CNRS, Paris 1965) pp. 537–554, in French
3.143 B. Simon, A. Grassi, R. Boistelle: Cinétique de croissance de la face (110) de la paraffine $C_{36}H_{74}$ en solution. I. Croissance en milieu pur, J. Cryst. Growth **26**, 77–89 (1974), in French
3.144 H. Cano, N. Gabas, J.P. Canselier: Experimental study on the ibuprofen crystal growth morphology in solution, J. Cryst. Growth **224**, 335–341 (2001)
3.145 E. Buckley: *Crystal Growth* (Wiley, New York 1951) pp. 330–385
3.146 *Adsorption et Croissance Cristalline*, Colloques Internationaux du CNRS, No. 152 (CNRS, Paris 1965), in French
3.147 G.M. Van Rosmalen, P. Bennema: Characterization of additive performance on crystallization: Habit modification, J. Cryst. Growth **99**, 1053–1060 (1990)
3.148 S. Sarig: Fundamentals of aqueous solution growth. In: *Handbook of Crystal Growth. 2b*, ed. by D.T.J. Hurle (North-Holland/Elsevier, Amsterdam 1994) pp. 1217–1269
3.149 K. Sangwal: Effect of impurities on the processes of crystal growth, J. Cryst. Growth **128**, 1236–1244 (1993)
3.150 K. Sangwal: Effects of impurities on crystal growth processes, Prog. Cryst. Growth Charact. Mater. **32**, 3–43 (1996)
3.151 R.J. Davey: The control of crystal habit. In: *Industrial Crystallization*, ed. by E.J. de Jong, S.J. Jančić (North-Holland, Amsterdam 1979) pp. 169–183
3.152 G. Bliznakow: Die Kristalltracht und die Adsorption fremder Beimischungen, Fortschr. Min. **36**, 149–191 (1958), in German
3.153 N. Cabrera, D.A. Vermileya: The growth of crystals from solutions. In: *Growth and Perfection of Crystals*, ed. by R.H. Doremus, B.W. Roberts, D. Turnbull (Wiley, New York 1958) pp. 393–408
3.154 G.W. Sears: The effect of poisons on crystal growth. In: *Growth and Perfection of Crystals*, ed. by R.H. Doremus, B.W. Roberts, D. Turnbull (Wiley, New York 1958) pp. 441–444
3.155 N. Albon, W.J. Dunning: Growth of sucrose crystals: determination of edge energy from the effect of added impurity on rate of step advance, Acta Cryst. **15**, 474–478 (1962)
3.156 R.J. Davey: Adsorption of impurities at growth steps, J. Cryst. Growth **29**, 212–214 (1975)
3.157 N. Kubota, J.W. Mullin: A kinetic model for crystal growth from aqueous solution in the presence of impurity, J. Cryst. Growth **152**, 203–220 (1995)
3.158 N. Kubota, M. Yokota, J.W. Mullin: Supersaturation dependence of crystal growth in solutions in the presence of impurity, J. Cryst. Growth **182**, 86–94 (1997)
3.159 G. Sgualdino, D. Aquilano, A. Cincotti, L. Pastero, G. Vaccari: Face-by-face growth of sucrose crystals from aqueous solutions in the presence of raffinose. I. Experiments and kinetic-adsorption model, J. Cryst. Growth **292**, 92–103 (2006)

3.160 R.J. Davey: The effect of impurity adsorption on the kinetics of crystal growth from solution, J. Cryst. Growth **34**, 109–119 (1976)
3.161 R. Boistelle: Impurity adsorption in crystal growth from solution. In: *Interfacial Analysis of Phase Transformations*, ed. by B. Mutaftschiev (Reidel, Dordrecht 1982) pp. 621–638
3.162 T.A. Eremina, N.N. Eremin, V.A. Kuznetsov, T.M. Okhrimenko, N.G. Furmanova, E.P. Efremova, V.S. Urusov: Characterization of defects generated by di- and trivalent cations in the potassium-dihydrophosphate structure and their influence on growth kinetics and face morphology, Crystallogr. Rep. **47**, 576–585 (2002)
3.163 M. Rauls, K. Bartosch, M. Kind, S. Kuch, R. Lacmann, A. Mersmann: The influence of impurities on crystallization kinetics – A case study on ammonium sulphate, J. Cryst. Growth **213**, 116–128 (2000)
3.164 L.N. Balykov, M. Kitamura, I.L. Maksimov: Effect of kink contamination on habit of two-dimensional crystal during growth with edge diffusion, J. Cryst. Growth **275**, 617–623 (2005)
3.165 C. Sweegers, H.C. de Coninck, H. Meekes, W.J.P. van Enckevort, I.D.K. Hiralai, A. Rijkeboer: Morphology, evolution and other characteristic of gibbsite crystals grown from pure and impure aqueous sodium aluminate solutions, J. Cryst. Growth **233**, 567–582 (2001)
3.166 C. Li, L. Wu, W. Chen: The impurity effects on growth and physical properties of strontium nitrate crystals, Int. J. Mod. Phys. B **16**, 114–121 (2002)
3.167 E. Kirkova, M. Djarova, B. Donkova: Inclusion of isomorphous impurities during crystallization from solutions, Prog. Growth Charact. Mater. **32**, 111–134 (1996)
3.168 H. Füredi-Milhofer, S. Sarig: Interactions between polyelectrolytes and sparingly soluble salts, Prog. Growth Charact. Mater. **32**, 45–74 (1996)
3.169 T. Jung, W.-S. Kim, C.K. Chou: Crystal structure and morphology control of calcium oxalate using biopolymeric additives in crystallization, J. Cryst. Growth **279**, 154–162 (2005)
3.170 F. Abbona, M. Angela-Franchini, C. Croni-Bono, H.E. Lundager Madsen: Effect of ammonia excess on the crystal habit of $NiNH_4PO_4{\cdot}6H_2O$ (Ni-struvite), J. Cryst. Growth **43**, 256–260 (1994)
3.171 J.J. Lu, J. Ulrich: An improved prediction model of morphological modification of organic crystals induced by additives, Cryst. Res. Technol. **38**, 63–73 (2003)
3.172 G.K. Kirov, I. Vesselinov, Z. Cherneva: Condition of formation of calcite crystals of tabular and acute rhmbohedral habits, Krist. Tech. **7**, 497–509 (1972)
3.173 A. Millan, P. Bennema, A. Verbeeck, D. Bollen: Morphology of silver bromide crystals from KBr-AgBr-DMSO-water systems, J. Cryst. Growth **192**, 215–224 (1998)
3.174 S. Veintemillas-Verdaguer: Chemical aspects of the effect of impurities in crystal growth, Prog. Cryst. Charact. Mater. **32**, 75–109 (1996)
3.175 I.M. Byteva: Effects of pH and crystal holder speed on the growth of crystals of ammonium dihydrogen phosphate. In: *Growth of Crystals*, Vol. 3, ed. by A.V. Shubnikov, N.N. Sheftal (Consultants Bureau, New York 1962) pp. 213–216
3.176 L.N. Rashkovic, G.T. Moldazhanova: Growth kinetics and morphology of potassium phosphate crystal faces in solutions of varying acidity, J. Cryst. Growth **151**, 145–152 (1995)
3.177 M. Delfino, J.P. Dougherty, W.K. Zwicker, M.M. Choy: Solution growth and characterization of L(+) glutamic acid hydrochloride single crystals, J. Cryst. Growth **36**, 267–272 (1976)
3.178 E.V. Khamskii: Some problems of crystal habit modification. In: *Industrial Crystallization*, ed. by J.W. Mullin (Plenum, New York 1975) pp. 215–221
3.179 H.E. Lundager Madsen: Influence of magnetic field on the precipitation of some inorganic salts, J. Cryst. Growth **152**, 94–100 (1995)
3.180 V.E. Ivashchenko, V.V. Boldyrev, Y.A. Zakharov, T.P. Shakhtshneider, A.E. Ermakov, V.I. Krasheninin: The effect of magnetic field on the shape of etch pits of paracetamol crystals, Mater. Res. Innov. **5**, 214–218 (2002)
3.181 C. Zhong, L. Wang, N.I. Wakayama: Effect of a high magnetic field on protein crystal growth-magnetic field induced order in aqueous protein solutions, J. Cryst. Growth **233**, 561–566 (2001)
3.182 T. Kaito, S. Yanagiya, A. Mori, M. Kurumada, C. Kaito, T. Inoue: Effects of magnetic field on the gel growth of $PbBr_2$, J. Cryst. Growth **289**, 275–277 (2006)
3.183 J. Garside: Kinetics of crystallization from solution. In: *Crystal Growth and Materials*, ed. by E. Kaldis, H.J. Scheel (Elsevier, Amsterdam 1978) pp. 483–513
3.184 W.R. Wilcox: Influence of convection on the growth of crystals from solution, J. Cryst. Growth **65**, 133–142 (1983)
3.185 R. Boistelle, F. Abbona: Morphology, habit and growth of newberyite crystals ($MgHPO_4{\cdot}3H_2O$), J. Cryst. Growth **54**, 275–277 (1981)
3.186 F. Abbona, M. Franchini-Angela: Crystallization of calcium and magnesium phosphates from solution of low concentration, J. Cryst. Growth **104**, 661–671 (1990)

4. Generation and Propagation of Defects During Crystal Growth

Helmut Klapper

This chapter presents a review of the typical growth defects of crystals fully grown on (planar) habit faces, i. e., of crystals grown in all kinds of solutions, in supercooled melt (mainly low-melting organics) and in the vapor phase. To a smaller extent growth on rounded faces from the melt is also considered when this seems appropriate to bring out analogies or discuss results in a more general context. The origins and typical configurations of defects developing *during* growth and *after* growth are illustrated by a series of selected x-ray diffraction topographs (Lang technique) and, in a few cases, by optical photographs.

After an overview (Sect. 4.1) the review starts with the formation of inclusions (Sect. 4.2), which are the main origin of other growth defects such as dislocations and twins. Three kinds of inclusions are treated: foreign particles, liquid inclusions (of nutrient solution), and solute precipitates. Particular attention is directed to the regeneration of seed crystals into a fully facetted shape (*capping*), and inclusion formation due to improper hydrodynamics in the solution, especially for potassium dihydrogen phosphate (KDP).

Section 4.3 deals briefly with striations (treated in more detail in Chap. 6 of this Handbook) and more comprehensively with the different kinds of crystal regions grown on different growth faces: growth sectors, vicinal sectors, and facet sectors. These regions are usually differently perfect and possess more or less different physical properties, and the boundaries between them are frequently faulted internal surfaces of the crystal. Two subsections treat the optical anomalies of growth and vicinal sectors and the determination of the relative growth rates of neighboring growth faces from the orientation of their common sector boundary.

In Sect. 4.4 distinction is made between dislocations connected to and propagating with the growth interface (*growth dislocations*), and dislocations generated *behind* the growth front by plastic glide due to stress relaxation. The main sources of both types of dislocations are inclusions. In crystals grown on planar faces, growth dislocations are usually straight-lined and follow (frequently noncrystallographic) preferred directions depending on the Burgers vector, the growth direction, and the elastic constants of the crystal. These directions are explained by a minimum of the dislocation line energy per growth length, or equivalently by zero force exerted by the growth surface on the dislocation. Calculations based on anisotropic linear elasticity of a continuum confirm this approach. The influence of the discrete lattice structure and core energy on dislocation directions is discussed. Further subsections deal with Burgers vector determination by preferred directions, postgrowth movement of grown-in dislocations, generation of postgrowth dislocations, and the growth-promoting effect of edge dislocations.

Section 4.5 presents *twinning*, the main characteristics of twins and their boundaries, their generation by nucleation and by inclusions, their propagation with the growth front, and their growth-promoting effect. Postgrowth formation of twins by phase transitions and ferroelastic (mechanical) switching is briefly outlined. Finally, Sect. 4.6 compares the perfection of crystals (KDP and ammonium dihydrogen phosphate (ADP)) slowly and rapidly grown from solutions. It shows that the optical and structural quality of rapidly grown crystals is not inferior to that of slowly grown crystals, if particular precautions and growth conditions are met.

4.1 Overview

The present chapter mainly deals with growth defects in crystals fully grown on (planar) habit faces. To a smaller extent crystals grown on rounded faces from the melt are also considered when this seems to be appropriate to bring out analogies or discuss results in a more general context. Crystals grow on habit faces in solutions, supercooled melts, and vapor. A special feature of this growth method is that there is practically no temperature gradient inside the crystal, provided that facet growth occurs freely on the whole surface of the crystal (without contact with a container wall). This is also the case for growth in the supercooled melt: the crystallization heat released at the growing habit faces keeps these at the crystallization temperature – or at least close to it [4.1]. The absence of a temperature gradient, and thus of thermal stress, inside the crystal allows the development of defect structures according to first thermodynamical principles and their preservation in their as-grown geometries, unless thermal gradients are introduced by improper cooling to room temperature after growth. This particularly concerns dislocations in crystals growing in their plastic state. Dislocations are the essential elements of stress relaxation by plastic glide: they are generated, moved, and multiplied by stress. Thus – in the presence of thermal stress – it makes an essential difference whether crystals are grown in their brittle or their plastic state. From solution, crystals grow in the brittle or in the plastic state (depending on the specific mechanical properties at the growth temperature); from the melt, however, crystals *always* grow in the plastic state, because each material has a more or less extended plastic zone below its melting point. It will be shown that growth dislocations develop the same geometrical features in crystals grown on habit faces from solution in the brittle state and from supercooled melts in the plastic state, provided that thermal gradients are absent.

In this review the generation of defects at the interface and their propagation with the advancing growth front are considered separately. This is because certain defects formed by a growth disturbance (e.g., by inclusions) may *heal out* and do not continue into the further growing crystal, whereas other defects (dislocations, twins, and grain boundaries), once initiated, are *forced* to proceed with the interface despite growth under optimal conditions. These defects can only be eliminated by growing out at the sides of the crystal, e.g., during Czochralski pulling on interfaces which are convex toward the melt. Moreover, distinction is made between *defects always connected to the interface* (*growth defects*, especially *growth dislocations*) and *defects generated "behind" the growth front* (*postgrowth defects*). The latter defects may be formed already during the growth run, either by thermal stress or by precipitation. Furthermore, defect configurations may be preserved in their *as-grown* geometry or changed after growth (e.g., by *postgrowth* movement of dislocations).

Many experimental results and the majority of photographs presented in this review were obtained by growth experiments and x-ray topographic studies

(Lang technique) in the author's laboratories in Aachen and Bonn. Crystals were grown from aqueous and organic solutions, from supercooled melts (organics), and by Czochralski pulling (organics). The organic crystals were considered as low-melting *model substances* (melting points below 100 °C), chosen with the primary aim of studying the generation and propagation of growth defects in dependence on growth methods and varying growth conditions. The main characterization method, x-ray diffraction topography, is not treated here; the reader is referred to the reviews [4.2–5] in the literature and to Chap. 42 in this Handbook. More specialized x-ray topographic treatments are given for twinned crystals [4.6] and for organic crystals [4.7]. Earlier reviews on the generation and propagation of growth defects were published by the author [4.8–11].

4.2 Inclusions

Two categories of inclusions are distinguished according to their origin [4.12, 13]: *primary inclusions* are associated with the growth front, i. e., they arise during growth, whereas *secondary inclusions* are formed after growth. Primary inclusions are *key defects* because they are the source of other defects (dislocations, twins) which propagate with the growth front into the further growing crystal. Inclusions of both categories may form stress centers which give rise to dislocation loops or half-loops by plastic glide (stress relaxation). Among the primary inclusions, three kinds are distinguished:

- Foreign particles
- Solvent (liquid) inclusions in crystals grown from solutions
- Solute precipitations in crystals grown from impure or doped melts.

Secondary inclusions are precipitates of solute impurities (dopants) formed after growth in the solid state during slow cooling, annealing or processing of crystals which are grown at high temperatures. They are due to the supersaturation of solutes at temperatures below the temperature at which the crystal was grown. These solutes precipitate if their diffusion mobility is sufficiently high and not frozen-in (as is usually the case at room temperature). In the same way vacancies and self-interstitials may condense into dislocation loops and stacking faults, e.g., during processing silicon crystals for electronic applications (*swirl defects*, e.g., [4.14]).

Here we treat only primary inclusions. A very detailed theoretical and experimental treatment of the capture of inclusions during crystal growth is presented by *Chernov* and *Temkin* [4.15]. A similar study with particular consideration of crystallization pressure is reported by *Khaimov-Mal'kov* [4.16].

4.2.1 Foreign Particles

Foreign particles preexisting in the nutrient (solution, melt) increase the risk of (heterogeneous) nucleation. Their incorporation into the growing crystal, however, is often considered as not very critical due to the crystallization pressure [4.16] (*disjoining force* after *Chernov* and *Temkin* [4.15]) which repulses foreign particles from the growth interface. Nevertheless, particles coming into contact with the growth face may be incorporated, depending on the size and chemical/physical nature of the particles and on growth conditions such as stirring, growth rate, and supersaturation [4.15]. For example, potassium alum crystals can be grown inclusion free (as assessed by optical inspection and x-ray topography) from *old* (i. e., repeatedly used) unfiltered aqueous solutions containing many floating dust particles, provided that growth conditions (temperature control, stirring) are stable enough to avoid the formation of liquid inclusions (see below). On the other hand in crystals of benzil grown in *old* (repeatedly filled up) supercooled melts ($T_m = 96\,°C$), flocks of solid decomposition products floating in the melt are quite readily incorporated (unpublished observation by the author). In contrast to the solution growth of potassium alum, such benzil melts were not stirred, but thermal convection occurred due to the release of crystallization heat at the crystal surface [4.1]. In the latter case the incorporation seems to be favored by the higher viscosity and the lower agitation of the nutrient phase, and probably also by the chemical similarity (carbon) of the particles to the growing crystal.

Foreign solid inclusions are very common in minerals. In laboratory and industrial crystal growth they usually play a minor role because they are easily avoided by filtering of the nutrient before growth. If solid inclusions appear during the growth run, e.g., as abrasives of the stirring device, continuous filtering is advised. This has been demonstrated by *Zaitseva* et al. [4.17] for the rapid growth of huge potassium dihydrogen phosphate (KDP) and deuterated potassium dihydrogen phosphate (DKDP) crystals with linear sizes up to 55 cm: continuous filtering during the whole growth run considerably increased the optical quality and laser damage threshold of these crystals.

The intentional incorporation of particle inclusions for the study of the generation of dislocations is reported in Sect. 4.4.2. The intentional inclusion of oil drops during crystallization from solutions was studied by *Kliia* and *Sokolova* [4.18].

4.2.2 Solvent Inclusions

Solvent inclusions are very common in crystals grown by all variants of solution growth (aqueous and organic solvents, flux). Two origins are distinguished.

Faceting (*Capping*) of Rounded Surfaces

In general crystals grow from solutions with planar faces (habit faces), whereby faces with low surface energy grow slowly and determine the final morphology of the crystal (*Wulff theorem* [4.19]). If surfaces are rounded, (e.g., of the seed crystal or after redissolution), during first growth, facets of habit faces and (between them) terraces of these faces are formed. The facets become larger and the terraced regions grow out until a single edge between the two habit faces engaged is formed, as shown in Fig. 4.1 (*theorem of Herring* [4.20–23]; see also *growth on spheres* [4.24] and [4.25, p.130]). The *healed-out* regions often have the shape of caps (*capping region*). The growth on terraced surfaces favors the entrapment of solvent inclusions, which may lead in extreme cases to a spongy structure of the capping region. This usually happens during first growth on seed crystals which were rounded

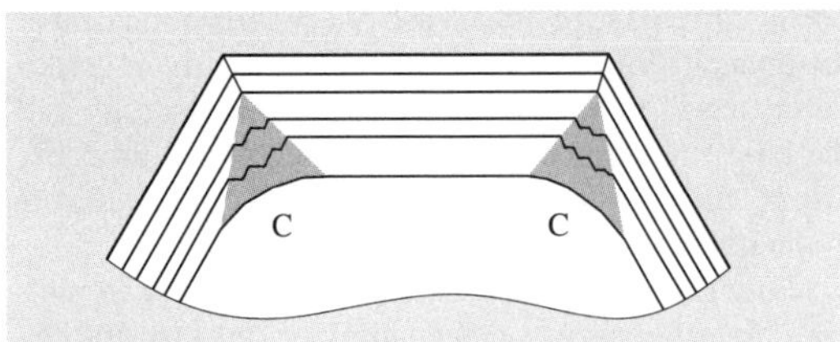

Fig. 4.1 Faceting and *capping* on rounded crystal surfaces. The shaded regions of terraced growth favor entrapment of liquid inclusions. They grow out and finally form the growth-sector boundary between the main habit faces

Fig. 4.2 A KDP crystal (length 45 mm) with {011} capping pyramid on a (001) seed plate

during a final etching (which is necessary in order to remove surface impurities and defects) before seeding-in. Therefore the zone of first growth around the seed crystals is usually more or less disturbed by liquid inclusions. These inclusions, however, can be largely avoided by a very slow (and thus time-consuming) growth under low supersaturation during the seed-faceting period.

A conspicuous example of capping is provided by potassium dihydrogen phosphate (KDP) grown in aqueous solution on (001) seed plates (Fig. 4.2). KDP develops habit faces {100} (tetragonal prism) and {011} (tetragonal dipyramid), but {001} is not a habit face. Thus in the first stage of growth on a (001) seed plate a spongy capping zone in the form of a tetragonal pyramid {011} over the seed plate as basis is formed, followed by clear further growth on {011} pyramid faces (Fig. 4.2). Detailed descriptions of this (001) capping process in KDP and ammonium dihydrogen phosphate (ADP) crystal growth are presented by *Zerfoss* and *Slawson* [4.12] and *Janssen-van Rosmalen* et al. [4.26].

Fluctuation of Growth Conditions (Growth Accidents)

A sufficiently strong change of growth condition (e.g., of supersaturation, stirring rate, stirring direction) may introduce – due to local variations of supersaturation – a (temporary) instability of growth faces: regions of retarded and promoted growth occur, leading to elevations and depressions on the growth face. Overhanging layers then spread over the depressions and close them, thus trapping nutrient solution. Usually a group of inclusions arranged in a plane parallel to the growth face is formed (*zonal inclusions*, Fig. 4.3). If all growth faces of the crystal are affected by the same growth disturbance, inclusions are formed on all faces. After stabilization of growth conditions and further clear growth the inclusions, which are visible by scattered light (if the crystal is transparent), reveal the shape of the crystal at the instant of the disturbance (*phantom crystal* in mineralogy). This is often observed when, after an accidental (or intentional) temporary redissolution, the crystal is rounded, so that during further growth refaceting with increased tendency for inclusion trapping occurs. Due to the capping effect, these inclusions are concentrated and most visible in the edge regions of the crystal [4.27] (Fig. 4.1).

The tendency to form solvent inclusions may strongly depend on the type of growth face $\{hkl\}$. In general the formation of solvent inclusions is favored on faces with high surface (attachment) energy, and thus high growth rate. An instructive example is provided by potassium alum growing from aqueous solution: fluctuations of growth conditions lead to pronounced liquid-inclusion entrapment on the smaller and fast-growing cube faces {100}, whereas the slow and morphologically dominant octahedron faces {111} resist the formation of inclusions even for strong changes of growth parameters [4.27]. There are also crystals for which certain growth faces trap tiny liquid inclusions despite controlled growth conditions. An example is shown in Fig. 4.4: The two pinacoid growth sectors of the crystal appear milky due to light scattering at solvent inclusions, whereas sectors of other faces are optically clear. The reason for the preferred inclusions trapping on certain growth faces is their specific surface structure, which favors the incorporation of solvent molecules and the formation of solvent bubbles. This phenomenon is known as *hourglass inclusions*, of which two types are distinguished: (1) face-specific preferential formation of liquid bubbles (which is the case in Fig. 4.4), and (2) preferential absorption of solvent and other foreign molecules as solid solution in the host crystal (without the formation of bubbles or solid precipitates). A well-known and frequently studied example of *hourglass inclusions* of type 2 is provided by potassium sulfate (e.g., *Buckley* [4.25, p. 415–420], *Vetter* et al. [4.28]; see also *dyeing crystals* [4.29, 30]). It is also repeatedly referred to in Sect. 4.3 of this chap-

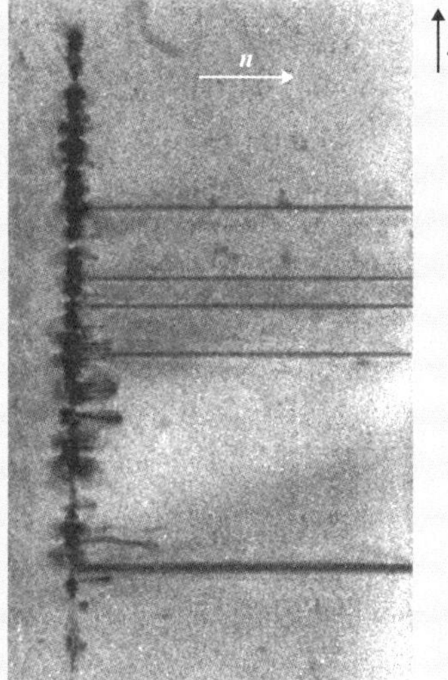

Fig. 4.3 Liquid (*zonal*) inclusions in solution-grown potassium alum (water), triggered on a (110) face by intentionally introduced redissolution due to a temporary increase of the solution by 1 °C. The original temperature (about 40 °C) was restored after a slight rounding of the crystal edges had appeared. *Arrow* ***n***: growth direction. A few edge dislocations originate from the inclusions. Section ($6\times12\,mm^2$) of an x-ray topograph of a 0.9 mm-thick (001) plate. Diffraction vector $\boldsymbol{g}(2\bar{2}0)$, Mo$K_\alpha$ radiation

Fig. 4.4 Preferred liquid-inclusion entrapment in the {201} pinacoid growth sectors of monoclinic Tutton salt $K_2Zn(SO_4)_2\cdot6H_2O$ grown from aqueous solution under well-controlled conditions (horizontal diameter 35 mm). The crystal plate, cut from a bulk crystal, contains the seed crystal with the nylon thread for suspension in the solution

ter. Often the preferred formation of liquid inclusions on certain growth faces can be largely suppressed by a proper pH value.

The hydrodynamics of the solution flow around the growing crystal may also play a significant role in the formation of liquid inclusions (e.g., *Chernov* et al. [4.31]). A particularly interesting example of this influence for the growth of KDP crystals is reported by *Janssen-Van Rosmalen* and *Bennema* [4.32], *Janssen-van Rosmalen* et al. [4.26], and *van Enckevort* et al. [4.33]. In their experiments the KDP crystals (shape: tetragonal prism {100} terminated on both side by dipyramid {011}) were mounted on a *tree* which rotated in the solution. The solution was flowing toward one (front) and away from the other (rear) pyramid. On pyramid faces on the trailing side, solvent inclusions are often formed in a quasiperiodic sequence (Fig. 4.5), whereas on the front side inclusions do not appear. This phenomenon is explained by the hydrodynamic situation at the rear-side pyramid face: in the wake *behind* the crystal a swirling region with no or strongly reduced liquid exchange with the bulk mother solution is formed. Thus the saturation decreases locally and growth is retarded compared with regions neighboring edges of the growth face. This leads to a depression in the growth face. After some distance of further growth this cavity is overgrown, forming a solvent inclusion. As shown in Fig. 4.5, this process is repeated several times in a quasiperiodic manner. A detailed study of this effect, including flow simulation experiments in a model system, is presented by *Janssen-Van Rosmalen* et al. [4.26], *Janssen-Van Rosmalen* and *Bennema* [4.32], and *van Enckevort* et al. [4.33], who also report that the formation of these inclusions is avoided by stronger stirring, in their experiment by faster rotation of the crystal tree. In any case, strong stirring smoothes out supersaturation differences on the growth face and thus may largely avoid interface instabilities. This is particularly significant in the solution growth of very large crystals where high saturation differences between the edge regions and the center of a growth face may occur. For the *rapid growth* of KDP (e.g., [4.34–36]) and other crystals very strong stirring is a prerequisite for inclusion-free growth.

Fig. 4.5 X-ray diffraction topograph of a (100)-plate (about 1.5 mm thick, about 50 mm high), cut from the rear side of a KDP crystal moved by rotation through the solution. Due to a closed wake of solution with reduced supersaturation behind the $(0\bar{1}1)$ growth face, liquid inclusions were repeatedly formed. They are the origin of numerous dislocations which grow out of the crystal at the side because the {010} prism faces practically do not grow. The dislocations in the triangular region above the capping zone belong to one of the growth sectors (101) or $(\bar{1}01)$. They are inclined and emerge out of the plate at their top ends. Diffraction vector $\boldsymbol{g}(020)$, AgK_α radiation

Finally a special type of liquid inclusions, so-called *hair inclusions*, is mentioned. These were, for example, observed by *Smolsky* et al. [4.37] in rapidly grown KDP; they consist of long hair-thin channels or strings of tiny bubbles filled with mother liquor. These pipes and strings are not arranged along the instantaneous growth front but form a more or less large angle with it, which indicates that they have proceeded with the growth front. Their origin is unclear for the most part, but in the case of KDP [4.37] it was shown by in situ atomic force microscopy of the growth face that at least some of them are triggered by tiny solid inclusions. As was shown by x-ray diffraction topography [4.37], these *hairs* are not correlated with dislocations. Thus they are different from the so-called *micropipes* (channels), frequently observed along the hexagonal axis of beryl [4.38] and silicon carbide [4.39–41]. These channels are the hollow cores of screw dislocations with large Burgers vectors.

Liquid inclusions play a significant role in mineralogy because they allow the reconstruction of the conditions of mineral formation. A comprehensive review is given by *Roedder* [4.43].

4.2.3 Solute Precipitates

A critical parameter in the growth of crystals from melts containing solute impurities or dopants is the effective distribution (segregation) coefficient k_{eff} of the impurity (dopant) between melt and crystal (see Chap. 6 in this Handbook). If $k_{eff} < 1$ (which is mostly the case), the solutes are rejected by the growing crystal, which leads to a higher solute concentration in the melt in front of the growth interface. The excess solute diffuses away from the interface. For high growth rates, however, the solute concentration may become supersaturated, leading to the precipitation of the solute and incorporation into the crystal during further growth. Thus the solute precipitation is an interplay between k_{eff}, the characteristic time scale of solute diffusion in the melt, and the growth rate.

If a solute precipitation is triggered by a short temporary increase of the growth rate, a solitary sheet of precipitations, marking the instantaneous growth front, will be formed. If, however, the high growth rate is permanent, the precipitations extend in the growth direction and form so-called *solute trails* (e.g., [4.44]). In more extreme development they lead to constitutional supercooling with cellular and dendritic growth (see Chap. 6 in this Handbook).

As examples the formation of solute inclusions during melt growth of organic crystals is shown in Figs. 4.6 and 4.7. The melts of organic materials contain considerable amounts of solute atmospheric gas which is precipitated as small bubbles. Figure 4.6 shows bubble precipitation in Czochralski crystals of benzophenone [$(C_6H_5)_2CO$, $T_m = 48\,°C$] induced by intentionally introduced changes of the growth conditions [4.42]. In the example of Fig. 4.6a, the pulling rate was temporarily increased from 0.4 to 0.6 mm/h, the other parameters remaining constant. This instantaneously led to bubble precipitation and reduction of the crystal diameter. The shape of the growth front at the instant of the intervention and its changes during further growth are clearly visible in dark-field light illumination. Note that the bubbles are also arranged in strings normal to the growth interfaces, corresponding to solute trails. After about 2 h the pulling rate was again reduced to its former value of 0.4 mm/h. Then the crystal adopted its former width and the bubble precipitation stopped. Figure 4.6b shows the effect of an intentionally introduced transition from Czochralski growth to supercooled-melt growth by lowering the temperature of the melt from about 1.5 °C above to about 1 °C below $T_m = 48\,°C$ and stopping the pull (but retaining the crystal rotation). The bubble precipitations reveal the transition of the concave (towards the melt) interface to a convex one. After the growth rate fell below the critical value, the bubble precipitation stopped. The crystal grew into the now supercooled melt and adopted a fully faceted shape at its end [4.42]. We have also observed that the (internal) surface of gas bubbles, precipitated in old and impure melts, were often covered with a layer of brownish material. This indicates that nongaseous impurities of the melt have been precipitated together with the solute gas.

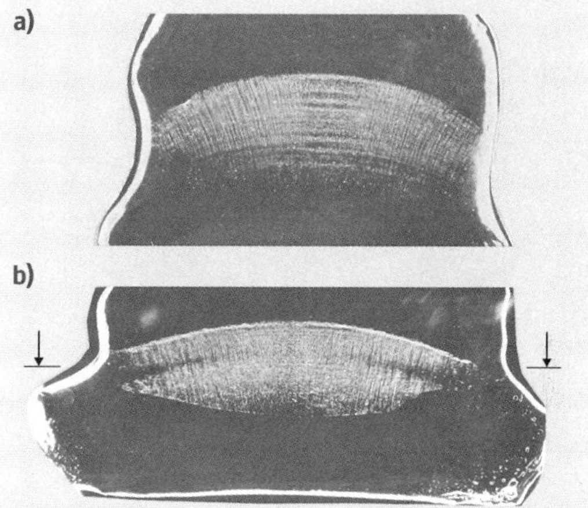

Fig. 4.6a,b Bubble precipitation in Czochralski benzophenone ($T_m = 48\,°C$) by an intentionally introduced change of growth parameters. Optical dark-field photographs of about 1.6 mm-thick plates. **(a)** Temporary increase of the pulling rate from 0.4 to 0.6 mm/h without change of other growth parameters. The diameter of the crystal is temporarily reduced from about 11 to 10 mm. **(b)** Transition from Czochralski growth to growth from supercooled melt by lowering the temperature from about 1.5 °C above to about 1 °C below T_m and stopping the pulling while retaining the rotation. The crystal grew into the now supercooled melt and became fully faceted. The *arrows* indicate the level of the melt after stopping the pulling (after [4.42])

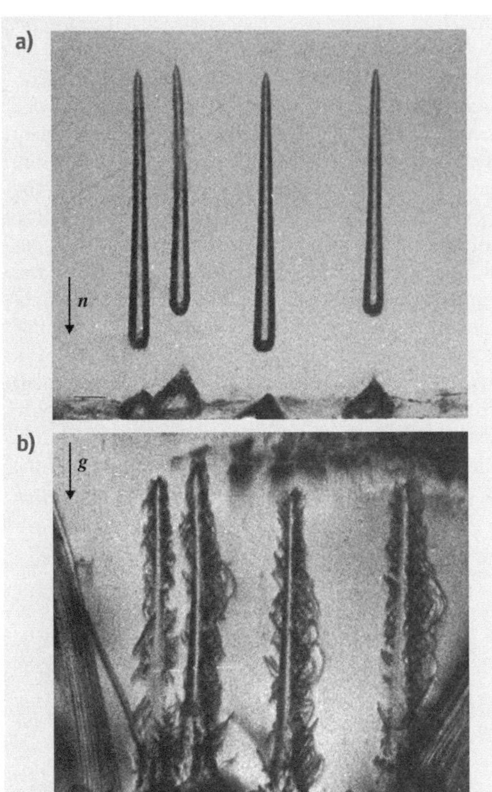

Fig. 4.7a,b Bubble channels filled with gas in a prism growth sector of trigonal benzil ($T_m = 96\,°C$) grown from supercooled melt ($5.6 \times 5.1\,mm^2$ section of a 2.2 mm-thick (0001) plate cut from the bulk crystal). Growth direction vertically downward. **(a)** Optical photograph. The channels originate from four gas bubbles sticking to the seed surface when starting growth. They have partially been filled by recrystallization within about 6 months after growth (see text). **(b)** X-ray topograph taken before the channels were closed. Note the glide-dislocation half-loops emitted from the channels (after [4.1]). Diffraction vector $\boldsymbol{g}(2\bar{2}00)$, Cu$K_{\alpha_1}$ radiation

Another instructive example is provided by Fig. 4.7, which shows a section of a benzil crystal ($(C_6H_5CO)_2$, $T_m = 96\,°C$) grown from slightly supercooled melt (ΔT about 1 °C) with four gas inclusions elongated normal to the planar growth face (i.e., in the growth direction) [4.1]. The inclusions started from small bubbles sticking to the bottom side of the seed crystal already during the seeding-in procedure. During further growth the growth rate was always below the critical rate for gas precipitation. Nevertheless, the bubbles advanced with the growth front and elongated and became even wider by collecting gas from the gas-rich zone in front of the growth face, thus leaving behind a gas-filled channel.

The photograph in Fig. 4.7a was taken about half a year after growth [4.1]. During this period the channels were partially filled by recrystallization, as can be recognized by the funnels in the surface at the bottom, indicating the former channel openings. Optically the recrystallized regions appear homogeneous except for a string of tiny scatterers (bubbles) aligned along the central axis of the former channel, revealed by optical dark-field observation. The mechanism of this recrystallization is not clear, but it probably occurred via sublimation, since the vapor pressure of benzil is relatively high, and the crystal was welded between two gastight plastic foils and, thus, stored in his own vapor.

Figure 4.7b shows an x-ray diffraction topograph taken shortly after growth was finished, when the channels were still open [4.1]. Numerous glide dislocations in the shape of half-loops emitted from the channel surface have formed (cf. also Sect. 4.4.6).

The entrapment of gas bubbles and their elongation into channels has been studied in detail by *Chernov* and *Temkin* [4.15], *Khaimov-Mal'kov* [4.16], and *Gegusin* and *Dziyuba* [4.45].

The relatively high concentration of solute gas in organic melts significantly limits the growth rate allowing bubble-free growth. Comparative growth experiments in outgassed supercooled melts under their own vapor showed that the growth rate for visually perfect growth could be increased by a factor of about three [4.46] compared with growth in gas-rich melts. In these experiments the growth rate (i.e., the supercooling), however, cannot be increased too much, because strong thermal upward convection in the melt (induced by the high release rate of heat of crystallization at crystal surfaces) leads to turbulent melt flow and serious defects (rugged growth) at the top end of the crystals.

Precipitation of solutes is a serious problem in melt growth of doped and mixed crystals. A detailed treatment is given in Chap. 6 in this Handbook. An instructive investigation of this effect, including an x-ray topographic study of dislocations formed around the precipitations, is reported by *Bardsley* et al. [4.44].

4.3 Striations and Growth Sectors

4.3.1 Striations

Striations are local variations of the impurity (dopant) concentration or of the crystal stoichiometry. They arise from fluctuations of growth conditions, such as changes of temperature, cooling rate, pressure, or convection in the solution or melt. These fluctuations lead to temporary changes in the growth rate, and thus to changes of the impurity incorporation. As a rule, they affect the whole growth front and thus form inhomogeneous layers parallel to the interface. The term *striations* is usually applied when the impurity layers appear in a (quasi)periodic sequence (Fig. 4.8a). If there are isolated layers, due to sporadic changes of growth conditions, often the term *growth bands* is used. In mineralogy, the term *growth zoning* is common. In crystals grown under rotation, strictly periodic *rotational striations*, correlated with the rotation rate, may occur. They are due to a nonuniform radial temperature distribution around the rotation axis, leading to slight changes of growth conditions (even with remelting) within a rotation period.

The impurities may be contaminants of the solution or of the melt, or incorporated solvent components. Striations are also formed by dopants intentionally introduced with the aim of tailoring specific physical properties of the crystals. The rate of incorporation depends on the impurity (dopant) species and is governed by their distribution (segregation) coefficient with regard to the crystal to be grown.

The regions of different impurities/dopants form layers coinciding with the instantaneous growth front. In crystals grown on habit faces (from solution or supercooled melt) they are planar, as shown in Figs. 4.8a and 4.9. The *intensity* of the striations, i. e., the concentration of impurities, may be considerably different in distinct growth sectors (Sect. 4.3.2). This is due to different surface structures of different growth faces which may facilitate or impede impurity capture. Symmetrically equivalent growth sectors show the same *intensity* of striations unless the growth conditions (e.g., solution

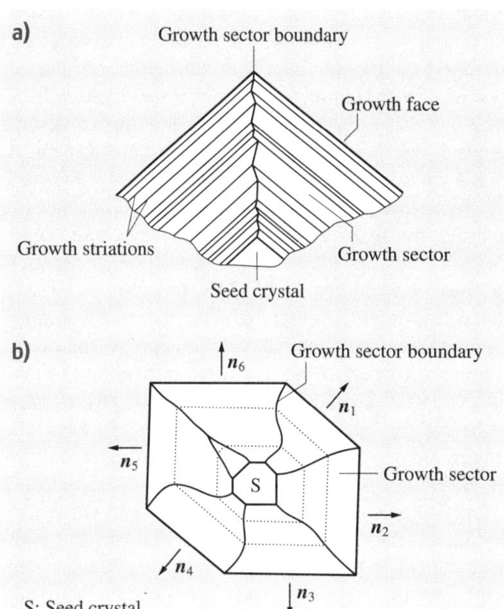

Fig. 4.8 (**a**) Growth striations and growth-sector boundary in a crystal grown on planar (habit) faces. The sector boundary is an internal surface formed by the movement of the edge joining the two faces during growth. It separates regions of different growth directions. (**b**) Division of a fully faceted crystal into growth sectors. The vectors $\boldsymbol{n}_i$ indicate the growth directions. *Dashed lines*: contours of the crystal at different stages of growth. One of the growth sectors has *grown out*

Fig. 4.9 X-ray topograph of a (0001) plate (about 8 mm diameter, 0.35 mm thick) cut out of a quartz-homeotypic gallium phosphate $GaPO_4$ crystal grown from high-temperature solution in phosphoric/sulfuric acid (after [4.47]), showing the triangular arrangement of growth sectors with pronounced striations. Growth-sector boundaries are visible by topographic contrast or by bends of the striations. Diffraction vector $\boldsymbol{g}(10\bar{1}0)$, AgK_α radiation

flow) at the corresponding growth faces are different. Conspicuous examples are the so-called hourglass growth patterns of crystals stained with organic dyes. *Staining of crystals* has been thoroughly studied by *Kahr* and *Guerney* [4.29] and *Kahr* and *Vasquez* [4.30]. Striations are often modified by growth hillocks (*vicinal pyramids*) as discussed in Sect. 4.3.3. An x-ray topographic study of the striation formation in the presence of vicinal pyramids in rapidly grown KDP crystals is presented by *Smolsky* et al. [4.48].

Crystals grown on rounded interfaces exhibit curved striations. An example of striations accompanied by tiny gas bubbles in a Czochralski crystal is shown in Fig. 4.10. Facets formed on rounded interfaces lead to regions (*facet sectors*, Sect. 4.3.4) with planar striations. The occurrence and *intensity* of these striations may be quite different from those of striations formed along curved interfaces. This is due to distinct growth modes with different distribution coefficients for rough growth on curved interfaces and growth on facets from supercooled melt (cf. Chap. 6 of this Handbook).

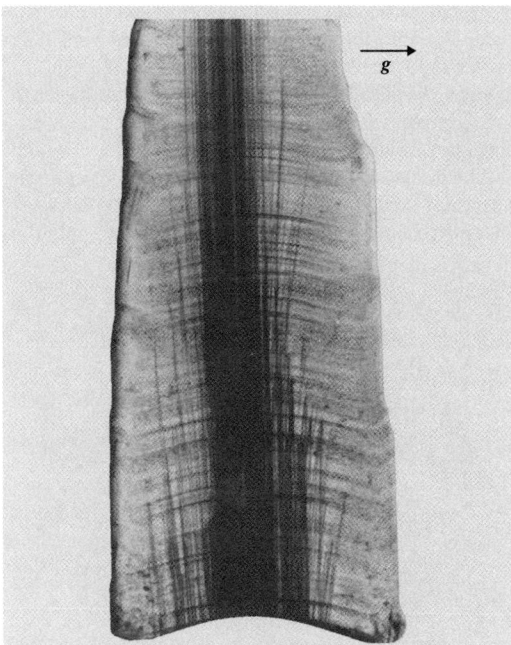

Fig. 4.10 Plate cut from the center of a Czochralski boule of orthorhombic salol ($T_m = 42\,°C$), about 1.3 mm thick; imaged length is about 40 mm. The growth striations marking the interface at different stages of growth contain tiny gas bubbles, many of which are sources of growth dislocations. Due to the concave interface the dislocations are focused toward the center of the boule. Due to this effect many dislocations enter the plate from above through the plate surface. Diffraction vector $\boldsymbol{g}(002)$, CuK_{α_1} radiation

In general, growth striations lead to local changes of physical properties (e.g., electric conductivity, optical birefringence). This is a major problem in the growth of doped crystals for sophisticated electronic and optical solid-state devices. This can be encountered by suppression of melt convection, e.g., by growth under microgravity [4.49] or by growth in magnetic fields [4.50, 51], which are treated in Chaps. 7 and 17 of this Handbook. An extensive treatment of the origin of striations and of recipes to largely avoid them is presented by *Scheel* [4.52].

4.3.2 Growth Sectors

Bulk crystals grow in all directions of space. Due to their structural and physical anisotropy, the types, distribution, and geometry of growth defects are distinct for different growth directions. This is pronounced in crystals grown from solutions and supercooled melts, which develop planar growth (habit) faces, and thus consist of regions (*growth sectors*) grown in discrete directions defined by the normals of the growth faces involved (Fig. 4.8b). Among all habit faces that are possible in principle, the *final* crystal usually exhibits only those faces which possess low surface (attachment) energies and thus – according to Wulff's theorem [4.19–21] – have low growth velocities. *Fast* faces with higher attachment energy grow out and vanish from the external morphology (cf. *Wulff–Herring construction* [4.20–23]). Thus the crystal may contain more growth sectors, usually in close neighborhood of the seed crystal, than are recognized from its final outward morphology (Fig. 4.8b).

Growth sectors are separated by growth-sector boundaries. These boundaries are internal surfaces over which the edges between neighboring faces have swept during growth. They are surfaces generated by the parallel movement of a straight line. When projected parallel to the edge (zone axis) of the two faces 1 and 2 involved, the boundary appears as a straight or somewhat curved line, the (local) direction of which depends on the (instantaneous) relative growth velocity v_1/v_2 of these faces (Sect. 4.3.6 and Fig. 4.14). If v_1/v_2 is constant, the line is straight (i. e., the boundary is planar); if v_1/v_2 fluctuates, the line is irregular, often zigzag-like, as sketched in Fig. 4.8a (i. e., the boundary is an irregularly waved internal surface).

Growth-sector boundaries and their surroundings may be perfect crystal regions. In many cases, however, they are fault surfaces which can be observed by etching, optical birefringence, and x-ray diffraction topography. The fault may be due to increased local impurity incorporation when growth layers on neighboring faces meet at their common edge, or due to slightly different lattice parameters in both sectors. The latter lead to a transition zone along the boundary with lattice distortions which can be detected by the methods mentioned above. An example of this case is shown in Fig. 4.11. Lattice distortions preferentially occur along boundaries between symmetrically *nonequivalent* faces, due to different incorporation of impurities which leads to slight differences of their d-values. Boundaries between symmetrically *equivalent* sectors are often strain free, but may be visible by the sharp bends of growth striations (if present), see Fig. 4.9. An illustrative example of the extraordinarily rich growth sectoring of natural beryl, revealed by x-ray topographic imaging of sector boundaries and striations, is presented by *Herres* and *Lang* [4.53]. For x-ray topographic characterization of faulted growth-sector boundaries as shift or tilt boundaries, see *Klapper* [4.7, 8, 54].

Fig. 4.11 (0001) Plate (about 14 mm diameter, 1 mm thick) of benzil grown from solution in xylene, containing faulted boundaries between the growth sectors shown in the drawing. The boundaries are inclined to the plate normal and appear as contrast bands with increased intensity at their emergence from the surface (increased strain due to stress relaxation at the surface). Some boundaries are invisible in the x-ray reflection used here. The plate tapers toward its edges, thus giving rise to *pendellösung* fringes. Some contrasts are due to surface damages. Diffraction vector $\boldsymbol{g}(\bar{2}020)$, Cu$K_{\alpha_1}$ radiation

The different incorporation of additives in different growth sectors is strikingly demonstrated in the so-called *dyeing of crystals* which goes back to Sénarmont (1808–1862) and was extensively studied in the last two decades by *Kahr* and coworkers (e.g., [4.29, 30]). They grew crystals from solutions with organic dye molecules as additives. The distinct incorporation of these molecules on different growth faces is conspicuously apparent from the different coloring of their growth sectors (see also Sect. 4.3.5). A similar study of coloring of the growth sectors of KDP with organic dyes is reported by *Maeda* et al. [4.55].

4.3.3 Vicinal Sectors

Another, less pronounced kind of sectoring frequently arises within the growth sectors treated above, due to growth hillocks (growth pyramids). These very flat *vicinal* pyramids, which are caused by dislocations emerging at their apex, often exhibit facets (*vicinal facets*) deviating by only very small angles from the main growth face. The facets are formed by terraces of growth layers, and their slopes depend on the step height and the widths of the terraces. On facets with different slope angles the incorporation of impurities is different. This leads to slightly distinct d-values of the regions grown on different vicinal facets (*vicinal sectors*). In analogy to growth sectors, the ridges of vicinal pyramids are termed *vicinal-sector boundaries*, which may be faulted surfaces. This also holds for the *valleys* between neighboring vicinal pyramids.

A detailed x-ray topographic study of vicinal sectors and their boundaries formed on {011} dipyramid faces and {100} prism faces of KDP and ADP was published by *Smolsky* et al. [4.56] and *Smolsky* and *Zaitseva* [4.57], who also coined the term *vicinal sector*. Atomic force microscope in situ investigation of the step structures of vicinal hillocks in relation to the Burgers vectors of unit and multiple unit height of the dislocations generating the hillocks is presented by *De Yoreo* et al. [4.58]. Pronounced triangular vicinal pyramids are observed on {111} octahedron faces of potassium alum (cf. the optical and x-ray topographic study by *Shtukenberg* et al. [4.59] and *Klapper* et al. [4.60]). Tetragonal vicinal pyramids generating faulted sector boundaries on {001} faces of tetragonal nickel sulfate hexahydrate have been studied by *van Enckevort* and *Klapper* [4.61]. Impurity incorporation on different slopes of vicinal hillocks on {111} faces of synthetic diamond has been investigated by *Kanda* et al. [4.62].

Vicinal sectors are usually accompanied by optical inhomogeneities (variation of refractive index, stress birefringence) which can be visualized by sensitive optical polarization means. These inhomogeneities, although usually small, can reduce the threshold for laser damage in high-power optical applications, e.g., in KDP crystals used for laser applications [4.63, 64]. Another method of optical visualizing vicinal sectors, by staining with organic dyes (chromophores), was studied in KDP by *Zaitseva* et al. [4.65]. The staining was due to different incorporation of the dye molecules on the distinct slopes of vicinal pyramids: the corresponding sectors appear with coloring of different strengths. A similar study is presented by *Bullard* et al. [4.66], who doped vicinal slopes of potassium hydrogen phthalate during growth from aqueous solution with luminescent organic molecules (fluorophores) and observed the slope pattern by luminescence microscopy. This staining and doping, however, develops during growth and is therefore not applicable to already grown crystals. It is also *destructive* insofar as it increases the degree of imperfection by addition of impurities (see also Sect. 4.3.5).

Part A | 4.3

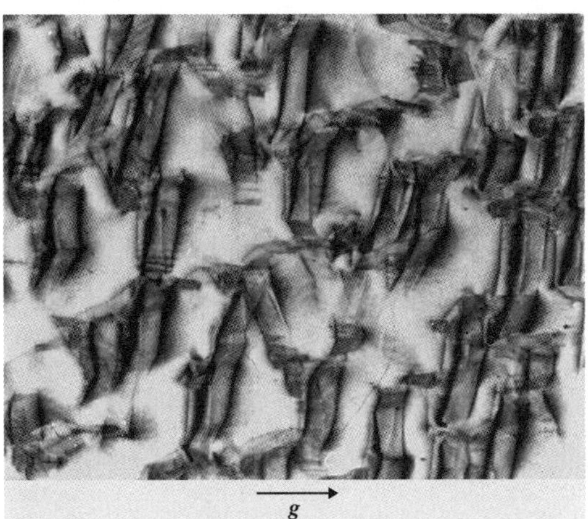

Fig. 4.12 X-ray topograph of a Z-plate of synthetic quartz cut from the growth sector of a strongly cobbled (0001) face (imaged section about 20×25 mm^2, about 1 mm thick, diffraction vector $\boldsymbol{g}(10\bar{1}1)$, Mo$K_\alpha$ radiation). The cobble mounds exhibited a small $\{10\bar{1}1\}$ rhombohedron facet at their steepest side. The planar defects shown in the topograph are formed by the *trajectory*, during growth, of the groove segments between the hills attached to the facets, and result from increased impurity entrapment. The topographic contrast is stronger at the outcrops of these faults at the plate surfaces (increased strain due to stress relaxation at the surface). Since always the same facet of the three symmetrically equivalent ones was formed, all faults have the same orientation

In this context another similar phenomenon, the very particular growth-cell formation on {0001} growth faces of synthetic (hydrothermal) quartz, is noteworthy. The (0001) face is not a habit face according to the Wulff theorem (therefore it never appears on natural crystals), but it is forced to appear when large (0001) seed plates are used in synthetic growth. During growth these faces usually develop a pronounced *cobble* texture (*Lang* and *Miuskov* [4.67]), consisting of rounded hills, which define conical or columnar regions inside the grown crystal. The boundaries between these sectors, defined by the trajectory of the grooves between the *cobbles* during growth, are mostly faulted due to the increased incorporation of impurities. This is particularly the case when habit facets appear on the side of the cobble hill (Fig. 4.12). Moreover, dislocations are trapped into the groove, and thus form part of these boundaries. X-ray topographic studies of these *impurity cell walls* in synthetic quartz, containing dislocations, are presented by *Lang* and *Miuskov* [4.68].

4.3.4 Facet Sectors

These sectors, formed during melt growth on rounded interfaces, are analogous to the growth sectors described above. They arise at the segments of the interface whose orientation coincides with that of a pronounced habit face. This habit face then appears as a planar facet. When growth proceeds, this facet defines inside the crystal a conical or cylindrical region (*facet sector*) whose perfection usually differs considerably from that of the crystal regions grown on the rounded interface (rough growth). This is due to the different modes of growth on rounded and planar interfaces: growth on the latter proceeds from supercooled melt and thus exhibits different (usually higher) incorporation of impurities (dopants) compared with growth on rounded interfaces. This is apparent from the usually much more pronounced striations in the facet sectors and the lattice distortions along their boundaries.

A favorite method to observe facet sectors and their perfection, in particular in garnets, is by polarized light (stress birefringence), e.g., *Schmidt* and *Weiss* [4.70] and *Cockayne* et al. [4.71]. An x-ray topography study is given by *Stacy* [4.72]. For a more detailed review on faceting, see Chap. 6 in this Handbook.

4.3.5 Optical Anomalies of Growth Sectors

As discussed above, different growth sectors usually contain different concentrations of impurities, or different deviations from stoichiometry, or – in mixed crystals – different compositional ordering. As a consequence, the physical properties are also – more or less – different in different growth sectors. This is conspicuously apparent from the staining of crystals with organic dyes ([4.29,30], and the end of Sect. 4.3.2). In symmetrically equivalent growth sectors the *magnitude* of the property changes is essentially the same, unless the growth conditions at the corresponding faces are different (e.g., due to different hydrodynamics). The different growth directions of equivalent faces, however, lead to different *orientations* of the growth-induced (additional) anisotropies of the properties also for equivalent sectors. This effect can be considered as the reduction of the symmetry of the basic crystal by superposition with the symmetry of the external influence of *crystal growth*, which is represented by a rotational symmetry with polar axis along the growth normal (*Curie principle*, e.g., [4.73, Chap. 3.2]). This *dissymmetry* influences, in principle, all properties, but it is very pronounced for optical refraction due to the high sensitivity of the refractive index to compositional variations and stress. An instructive example is presented in Fig. 4.13, showing the optical anomaly of a (K,NH_4)-alum mixed crystal which is basically cubic, and thus should be optically isotropic [4.59, 69]. Here the growth-induced birefringence (*optical anomaly*) is due to a partial ordering of K and NH_4 ions in the {111} sectors [4.59, 73].

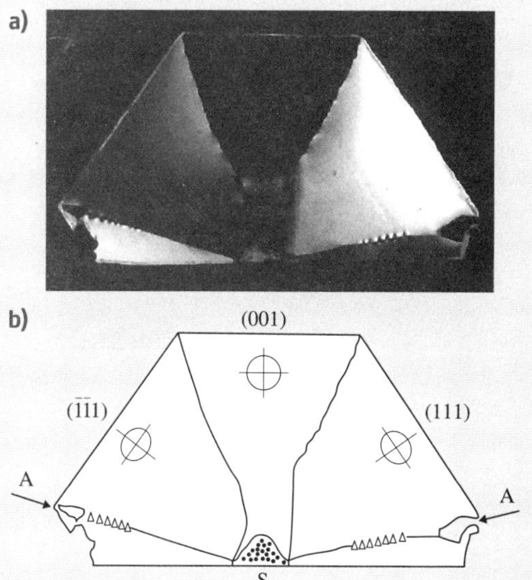

Fig. 4.13 (**a**) Optical anomaly of a cubic mixed (K,NH_4)-alum crystal grown from aqueous solution, as revealed by polarized light (crossed polarizers): (110) plate, 1 mm thick, horizontal width about 4 cm. (**b**) Sketch of growth sectors and their boundaries of the plate shown in (**a**). The {111} growth sectors are optically negative and approximately uniaxial, with their optic axes parallel to their growth directions ⟨111⟩ (after [4.59]). The (001) sector is nearly isotropic. Along the boundaries A between {111} sectors, small {110} growth sectors (resulting from small periodically appearing {110} facets) have formed during growth. S: seed crystal. After *Hahn* and *Klapper* [4.69, p. 393] (© 2003 IUCr)

Optical anomalies of crystals had already been observed and investigated in the early 19th century, as documented in the review by *von Brauns* (1891) [4.74]. A richly illustrated survey, including a historical overview, is presented by *Kahr* and *McBride* [4.75]. Very recently a monograph has been published by *Shtukenberg* et al. [4.73]. Finally it is mentioned that optical anomalies also occur in vicinal sectors and facet sectors. An interesting example of the former is given by *Zaitseva* et al. [4.65] for large-scale KDP crystals.

4.3.6 Growth-Sector Boundaries and Relative Growth Rates

As mentioned above, the orientation of the growth-sector boundary is dependent on the relative growth velocity of the two neighboring faces involved. This is illustrated in Fig. 4.14, from which the relation

$$v_1/v_2 = \cos\beta_1/\cos\beta_2\,, \quad \text{with } \beta_1+\beta_2=\alpha\,,$$

is easily derived, where α is the angle between the growth directions $\boldsymbol{n}_1$ and $\boldsymbol{n}_2$, and β_1 and β_2 are the angles between $\boldsymbol{n}_1$ and $\boldsymbol{n}_2$ and the growth-sector boundary. The boundary is straight when the relative growth rate is constant; it is curved when it changes. In the latter case the *local* relative growth velocity is derived

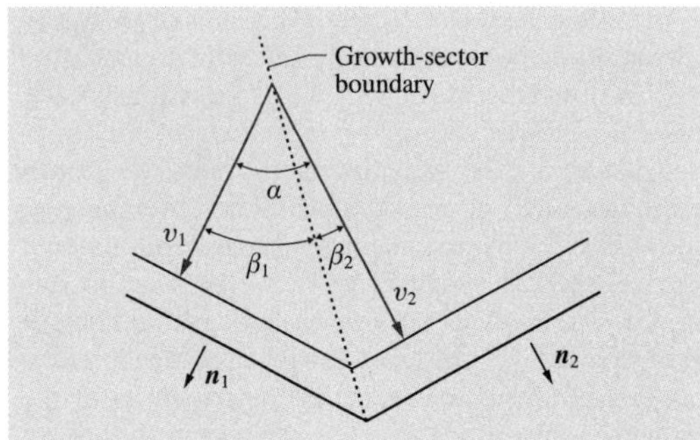

Fig. 4.14 Relation between growth velocities v_1 and v_2 of neighboring growth faces (growth directions $\boldsymbol{n}_1$ and $\boldsymbol{n}_2$) and the direction of the growth-sector boundary (*dotted line*)

from the angles β_1 and β_2 between $\boldsymbol{n}_1$ and $\boldsymbol{n}_2$ and the tangent plane to the sector boundary in the corresponding growth stage. Thus, the relative growth rates are easily reconstructed if the growth-sector boundary is visible.

An illustrative example, showing the strong changes of the growth velocities (*growth-rate dispersion*) due to fluctuations of growth conditions and defects is presented in Fig. 4.15a,b [4.60, 76], which shows x-ray topographs of a (110) plate cut from a potassium alum crystal (grown from aqueous solution by temperature lowering) which was subjected to temporary redissolution by a temporary temperature increase of about 1 °C [4.27]. The two boundaries between the central cube sector (001) and the two neighboring octahedron sectors $(\bar{1}11)$ and $(1\bar{1}1)$ are clearly depicted by kinematical contrast due to lattice distortions. (For the contrast variations in different x-ray reflections, see Chap. 42 in this Handbook.) Figure 4.15c outlines the shape of the crystal in different stages of growth, reconstructed from the course of the growth-sector boundaries (dotted line). Four regions (1–4) of different relative growth rates can be distinguished. In the first period, after seeding-in, the crystal was grown by continuous temperature decrease of about 0.3 °C/day until it reached the shape outlined by A–A–A–A. At this stage the temperature of the growth chamber was increased in one step by 1 °C. Due to the slow transfer of the temperature jump into the solution, redissolution started about half an hour later, recognized by the rounding of the crystal edges. Now the previous temperature and decrease rate were restored and growth continued as before. Due to this disturbance a layer of liquid inclusions covering a part of the (001) facet was formed, and its growth rate, relative to the neighboring {111} faces (the growth rate of which remained constant during the whole experiment), was strongly increased in region 2; in regions 3 and 4 it decreased again. From the angles β_1 and β_2 ($\alpha = 35.26°$) the relative growth rates were determined as

$$v(001)/v(111) = 1.0/5.6/1.7/0.8$$
$$\text{in growth intervals } 1/2/3/4$$

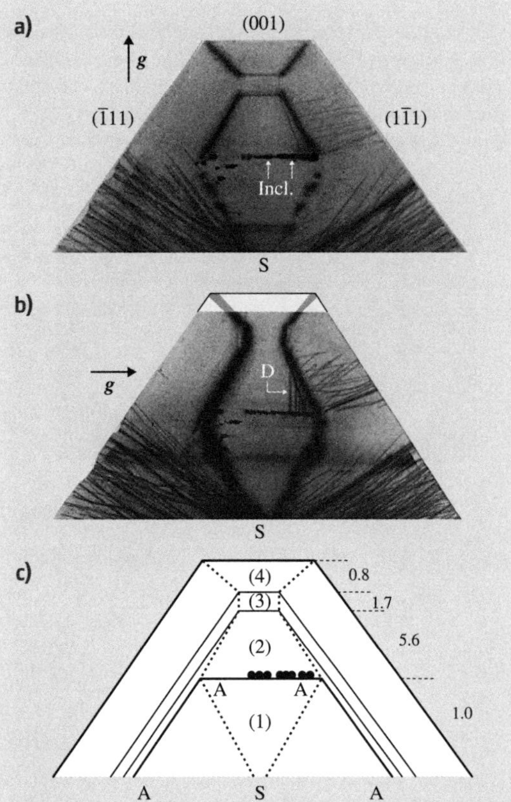

Fig. 4.15a–c Sections of topographs of a (110) plate cut from a potassium alum crystal subjected to a temporary redissolution (vertical extent 12 mm, reflections 004 (**a**) and 220 (**b**)). S: location of the seed (outside the section); Incl.: liquid inclusions; D: edge dislocations. (**c**) Illustration of the development of the crystal shape and of growth-sector boundaries (*dotted lines*) during growth. *Contour* A–A–A–A outlines the shape of the crystal at the time of redissolution. At the right side the relative growth rates $v(001)/v(111)$ of growth intervals 1–4 are given (after [4.76])

(averaged over the nearly equal left- and right-hand sides). The drastic increase in period 2 is obviously due to the dislocations D originating from the inclusions, and the retardation in periods 3 and 4 may arise from the elimination of these dislocation from the (001) face by bending at the growth-sector boundary into the (111) sector. Note that only part of the dislocations involved in this process is visible in the topographs (Fig. 4.15a,b) of the 1.4 mm-thick crystal cut, since the larger part of the (001) sector (with a basis of about $8\times8\,mm^2$ in growth stage A–A–A–A) is outside the cut and thus not recorded. In this context reference is made to similar and more detailed studies on dislocation-dependent growth rate dispersion of {100} and {110} growth faces of potassium alum by *Sherwood* and *Shiripathi* [4.77], *Bhat* et al. [4.78], and *Ristic* et al. [4.79]. An interesting output of their investigations is evidence for the growth-promoting role of pure edge dislocations (Sect. 4.4.7).

4.4 Dislocations

4.4.1 Growth Dislocations and Postgrowth Dislocations

Dislocations are generated during crystal growth, by plastic deformation and by the condensation of self-interstitials and vacancies. In the study of crystal growth defects it is useful to distinguish between two categories of dislocations:

1. Dislocations which are connected with the growth front and proceed with it during growth (*growth dislocations* or *grown-in dislocations*)
2. Dislocations which are generated *behind* the growth front, either still during the growth run or during cooling to room temperature (*postgrowth dislocations*), or later during processing or by improper handling.

The final arrangement of dislocations in a crystal at room temperature results from growth dislocations, postgrowth dislocations, and the movement, multiplication, and reactions of both after growth. Crystals grown at low temperatures (e.g., from aqueous solution) and in their brittle state usually contain dislocations in their original *as-grown* configuration, whereas in crystals grown at high temperatures, the original dislocation configurations may be drastically altered by dislocation movement, dislocation multiplication, and dislocation reactions. These processes, which may occur during the growth run (*behind* the growth front), are induced by thermal stress due to temperature gradients and, particularly in crystals grown at very high temperatures, by the absorption of interstitials and vacancies (*dislocation climb*).

In this chapter the formation and propagation of dislocations in crystals grown at low temperatures (below 100 °C) under zero or only low thermal gradients are treated. The development of dislocation configurations during growth from melt under high thermal gradients or during processing at elevated temperatures has been experimentally and theoretically studied by various authors (e.g., [4.80–83]) and is reviewed in Chap. 6 of this Handbook.

4.4.2 Sources of Growth Dislocations

For topological reasons dislocation lines cannot start or end in the interior of a perfect crystal. They either form closed loops, or they start from external and internal surfaces (e.g., grain boundaries), or from other defects with a *break* of the crystal lattice. In crystal growth, such defects may arise from all kinds of inclusions (e.g., foreign particles, liquid inclusions, bubbles, solute precipitates). When inclusions are overgrown and *closed* by growth layers, *lattice closure errors* may occur. These errors are the origin of growth dislocations which are connected to the growth front and propagate with it during further growth.

It is a very common observation that inclusions are the source of growth dislocations. Examples are shown in Figs. 4.3, 4.5, 4.10, etc. The appearance of dislocations *behind* an inclusion (viewed in the direction of growth) is correlated with its size: small inclusions emit only a few dislocations or are often dislocation free. Large inclusions ($> 50\,\mu m$) usually emit bundles of dislocations. In some cases, however, large inclusions (several millimeters in diameter) of mother solutions *without* dislocation generation have been observed (e.g., in the capping zone of KDP [4.84, 85]).

The generation of growth dislocations by foreign-particle inclusions has been experimentally studied by *Neuroth* [4.86] in crystals growing in aqueous solution (potassium alum) and in supercooled melt (benzophenone $(C_6H_5)_2CO$, $T_m = 48\,°C$; salol $C_{13}H_{10}O_3$, $T_m = 42\,°C$). A seed crystal is fixed to a support in such an orientation that a dominant growth face [octahedron (111) for cubic potassium alum, prism (110) for orthorhombic benzophenone, pinacoid (100) for orthorhombic salol] develops horizontally. After a sufficiently long distance of (visually) perfect growth, a small ball of solder (0.3–0.5 mm) is placed on the horizontal growth face, and growth continued without change of conditions. During the whole experiment the growth surface was observed with a microscope (long focal distance) in reflected light or by Michelson interferometry, both with videotape recording. After the deposition of the ball the face grows slowly as before without additional surface features as long as the ball is not covered by growth layers. During this period the crystal seems to *sink* into the growing crystal. In the moment when the ball is covered by growth layers, a conspicuous, fast extending growth hillock appears, emitting macrosteps from its apex. After some time of growth the originally single hillock splits into a group of diverging hillocks. This indicates that a bundle of dislocations, fanning out during growth, has been created by the ball inclusion (Fig. 4.16).

Fig. 4.16 X-ray topograph of a (010) plate (about 1.5 mm thick, width 21 mm) cut from a crystal of orthorhombic salol grown in supercooled melt. It contains a solder ball (diameter about 0.4 mm) dropped on a perfectly growing (100) facet (directed upward). Numerous dislocations were generated *behind* the ball. Dislocations of the fan propagating to the left are pure screw and exactly parallel to one of the prominent ⟨101⟩ edges which dominate the shape of the crystal (cf. *Deviations from Calculated Directions* (i) in Sect. 4.4.4 and Fig. 4.22). The (unresolved) dislocations of the vertical bundle have Burgers vectors [100] and [001]. Diffraction vector $\boldsymbol{g}(200)$, CuK_{α_1}

For the study of the dislocations associated with the inclusions, a plate containing the ball and the region *behind* it was cut out of the crystal and subjected to x-ray topography. Figure 4.16 shows that – in accordance with the observed surface pattern – numerous dislocations originate from the *back side* of the ball. Their density is partially too high to be resolved by this imaging method.

Similar experiments have been performed with mechanical in situ violation (puncturing, scratching) of an interface perfectly and steadily growing in solution, supercooled melt, and by Czochralski pulling [4.86]. Again, bundles of dislocations originate from these damages, which in solution growth frequently give rise to liquid inclusions. In plastic crystals (always the case in melt growth) the mechanical impact generates glide dislocations which emerge at the growth front and continue as growth dislocations. Similar experiments are reported by *Forty* ([4.87, esp. p. 23]). His review presents a rich collection of photographs of growth spirals and other surface patterns on growth faces of various crystals.

The formation of screw dislocations in thin plates of organic crystals during growth from solution and from the vapor has been studied in situ by Russian authors using (polarized) light microscopy with film recording [4.88–90]. Screw dislocations arise at reentrant corners between branches of dendrites [4.88] and by growth around intentionally introduced particles [4.89, 90]. In these cases the dislocations run through the lamellae and do not end inside the crystal. The mechanism of formation of lattice closure errors and of dislocations *behind* an inclusion on the nanometer scale is not yet fully understood, although simple models have been derived. An example is presented by *Dudley* et al. [4.91].

As pointed out in Sect. 4.4.2, in habit-face crystals inclusions preferentially arise in the regeneration zone of growth on rounded interfaces, in particular in the zone of first growth on a seed crystal. Moreover, dislocations and other defects (grain boundaries, twins) preexisting in the seed will continue into the growing crystal. Thus, the perfection of the seed as well as the seeding-in process are most crucial for the growth of perfect crystals. This holds for all methods of seeded growth, not only for habit-face crystals. That the regeneration zone around the seed crystal is the main source of dislocations is apparent from several topographs shown in this chapter (e.g. Figs. 4.19, 4.21 and 4.24). It is stressed that inclusions and dislocations can largely be avoided by very slow (and thus time-consuming) growth during the regeneration period of first growth on a perfect seed.

Finally it is emphasized that inclusions can also block already existing dislocations. This has been observed several times by the author, and reported in the literature. It frequently happens to dislocations in the seed crystal which are blocked by inclusions formed in the regeneration zone (*capping*) of first growth and do not enter the growing crystal [4.84, 85]. Thus provoking a capping zone by an intentionally introduced deviation of the seed surface from a habit face may be helpful for reducing the number of dislocations coming from the seed, but it implies also a considerable risk of generating *new* dislocations behind the inclusions. The blocking of growth dislocations by closed inclusions must obey the conservation law of Burgers vectors, as discussed below in Sect. 4.4.3.

4.4.3 Burgers Vectors, Dislocation Dipoles

The sum of the Burgers vectors of all dislocations originating from an inclusion fully embedded in an otherwise perfect crystal is zero [4.84]. This directly follows from Frank's conservation law of Burgers vectors (see textbooks on dislocations, e.g., [4.92–94]), which states that the sum of Burgers vectors $\boldsymbol{b}_i$ of all dislocation lines going into a dislocation node (i. e., with line direction *into* the node) is zero $\sum \boldsymbol{b}_i = 0$ (analogous to Kirchhoff's law of electrical currents). Another proof may be given via the Burgers-circuit definition of Burgers vectors (e.g., [4.92–94]): imagine a Burgers circuit parallel to the growth face in the perfect crystal region grown before the inclusion was formed. Now shift the circuit stepwise in the growth direction over the inclusion and the dislocation bundle behind it. No closure error of the circuit, which now encircles all dislocations, will arise during this (virtual) procedure: $\sum \boldsymbol{b}_i = 0$.

From this it immediately follows that a *single* dislocation cannot originate from an inclusion. If dislocations are formed, there must be at least two of them, with opposite Burgers vectors. This is often observed when the inclusions are very small. Two slightly diverging dislocation lines emanating from small, x-ray topographically invisible or nearly invisible inclusions were observed in KDP by *Fishman* [4.84]. Examples are presented in Fig. 4.19 (label A) for salol grown from supercooled melt. A few pairs of slightly diverging dislocations, starting from a point, can also be recognized in Fig. 4.10 of a Czochralski salol specimen. There are, however, many x-ray topographic observations of apparently only one dislocation line arising from an inclusion (e.g., in Fig. 4.3, where only one wider contrast line indicates the presence of more than one dislocation). In all these cases the single lines must represent pairs of two closely neighboring (x-ray topographically unresolvable) parallel dislocations with opposite Burgers vectors: a dislocation dipole. Such a dipole can alternatively also be considered as a single dislocation in the shape of a narrow hairpin with its (virtual) bend in the inclusion. In this approach the two branches of the *hairpin* have the same Burgers vector, but opposite directional line sense. Examples of a pure-screw and two pure-edge dislocation dipoles are shown in Fig. 4.17. The two branches of the dipoles attract each other and may annihilate if they come close enough together. This annihilation is possible for screw dislocation dipoles, and for edge dipoles if both edge dislocations of the latter are on the same glide plane. If they are located on

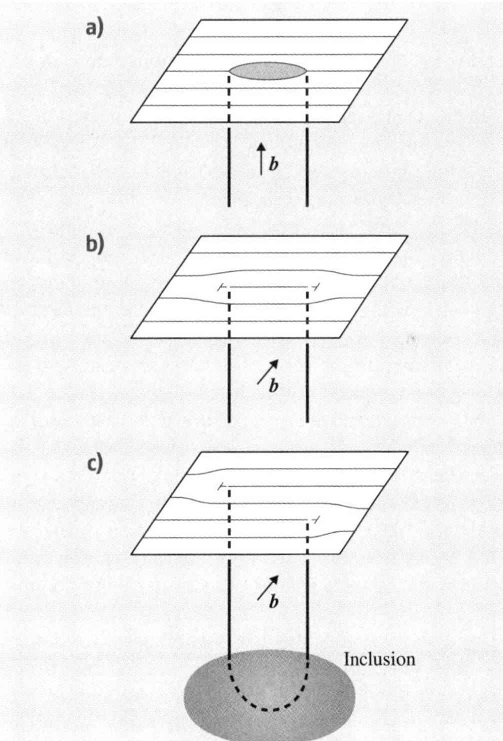

Fig. 4.17a–c Sketches of dislocation dipoles originating from an inclusion. Here a dipole is considered a single (hairpin) dislocation with Burgers vector $\boldsymbol{b}$ but opposite line direction sense of the two branches. (**a**) Pure-screw dipole; (**b**,**c**): pure-edge dipoles

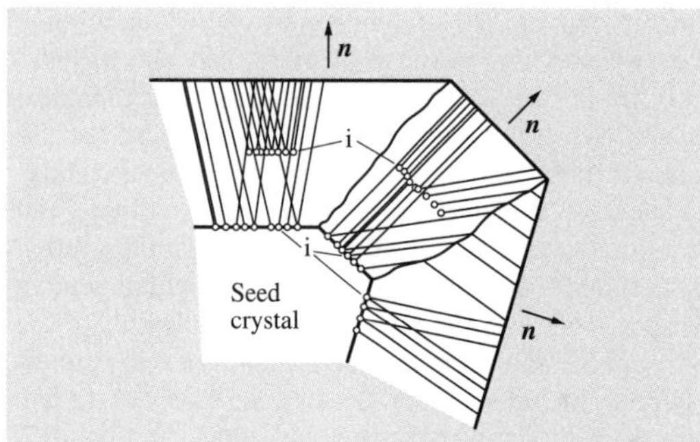

Fig. 4.18 Typical geometry of growth dislocations in crystals grown on habit faces. The different preferred directions of dislocations lines within one growth sector result from different Burgers vectors. These directions abruptly change their directions when they penetrate a growth-sector boundary, i. e., when, during growth, their outcrops shift over the edge from one face to the other (i: Inclusions)

different glide planes, the two dislocations can approach each other to a minimum separation, where they have prismatic character and form the edges of stripes of inserted or missing lattice planes (Fig. 4.17b). A model of the formation of a screw dislocation *behind* an inclusion is presented by *Dudley* et al. [4.91].

During x-ray topographic studies of growth dislocations it is often observed that *only one* contrast line originates from an inclusion, indicating – at first sight – a single dislocation (Fig. 4.3). Here the question arises, how by x-ray topography a dislocation dipole can be distinguished from a single dislocation with the same Burgers vector. If the dipole dislocations are sufficiently separated, they are resolved as two lines or appear as a broader contrast line. However, since the strain fields of the two dislocation have opposite signs and subtract each other, the resultant strain may also be smaller and less extended, if the separation of the two dislocations is small. Thus a dipole may appear on x-ray topographs with similar or even narrower contrast than a single dislocation with the same Burgers vector, and distinction of the two is often not immediately possible. An example is given by *van Enckevort* and *Klapper* [4.61, Fig. 11a], where a single contrast line represents a screw dislocation dipole in nickel sulfate hexahydrate, as is proven by the bulge of one of the two dipole arms. In the same crystal the presence of two closely neighboring etch pits at the apices of growth pyramids and slip traces, indicating the escape of one of the two screw dislocations from the hillock center, has been observed. Dislocation dipoles are also formed during plastic flow of crystals, when the movement of glide dislocations is locally blocked by obstacles (inclusions, jogs; [4.95]). Examples are given in Sect. 4.4.5 (Figs. 4.25 and 4.26).

4.4.4 Propagation of Growth Dislocations

Characteristic Configurations, Theory of Preferred Direction

A dislocation line ending on a growth face will proceed with that face [4.7, 8, 96]. Its direction depends on the shape and orientation of the growth face and its Burgers vector. As shown in Sect. 4.3.2 and Fig. 4.8b, crystal growing on planar (habit) faces consist of growth sectors belonging to different growth faces (different growth directions $\boldsymbol{n}$). This leads – under ideal conditions (i. e., stress-free growth) – to a characteristic configuration of growth dislocations which is illustrated in Fig. 4.18. The dislocations start from inclusions and propagate as straight lines with directions $\boldsymbol{l}$ usually close to, and frequently parallel to, the growth direction of the sector in which they lie. They usually exhibit sharply defined, often *noncrystallographically preferred*, directions $\boldsymbol{l}_0$ which depend on the growth direction $\boldsymbol{n}$ and the Burgers vector $\boldsymbol{b}$: $\boldsymbol{l}_0 = \boldsymbol{l}_0(\boldsymbol{n}, \boldsymbol{b})$. The dependence of the preferred direction $\boldsymbol{l}_0$ on the growth direction $\boldsymbol{n}$ becomes strikingly apparent when the dislocations penetrate growth-sector boundaries. This implies an abrupt change of the growth direction: the dislocation lines undergo an abrupt change of their preferred direction $\boldsymbol{l}_0$ (*refraction* of dislocation lines). An example is shown in Fig. 4.19.

These preferred direction of growth dislocations are explained by two approaches [4.7, 8, 96]:

1. *Minimum-energy theorem* (Fig. 4.20a). The dislocation lines adopt a direction $\boldsymbol{l}$ (unit vector) for which its energy within any growth layer is a minimum. For a growth layer of unit thickness $d = 1$ this can be expressed as

   $$E/\cos\alpha = \text{minimum}\,,$$

   where $E = E(\boldsymbol{l}, \boldsymbol{b}, c_{ij})$ is the elastic energy (strain energy) per unit length of the dislocation line (c_{ij} are the elastic constants of the crystal) and α is the angle between $\boldsymbol{n}$ and $\boldsymbol{l}$. The factor $1/\cos\alpha$ accounts for the length of the dislocation line in the layer.
2. *Zero-force theorem* (Fig. 4.20b). A dislocation line emerging at the surface experiences an *image* force $\mathrm{d}\boldsymbol{F}$ which depends on the angle α between the dislocation direction $\boldsymbol{l}$ and the surface normal $\boldsymbol{n}$, and on its distance r from the surface. At the surface this

force is infinitely large. According to *Lothe* [4.97] there exists always a direction $\boldsymbol{l}_0$ for which this force is zero for the dislocation line segments at any depth below the surface. It is plausible that, during growth, a dislocation emerging at the growth face follows this direction of zero force.

Using the formula

$$\mathrm{d}F = -\frac{1}{r}\left(\frac{\partial E}{\partial \alpha} + E\tan\alpha\right)\mathrm{d}\boldsymbol{l}$$

(*Lothe* theorem [4.97]), it can be shown that both approaches lead to the same preferred directions.

Verification of the Minimum-Energy Approach

The strain energy [4.7,8,96] per unit length of a straight dislocation line is given by (see textbooks on disloca-

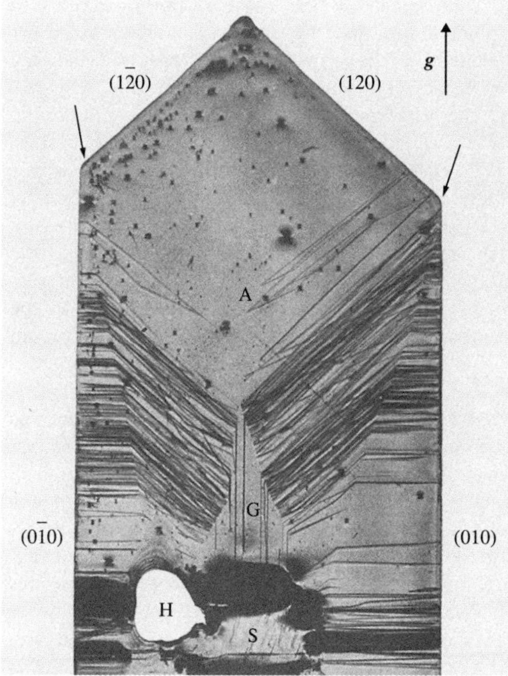

Fig. 4.19 Section of a (001) plate (horizontal width about 26 mm, thickness 1.5 mm) cut from a salol crystal grown from supercooled melt. The dislocation lines change their preferred directions when they penetrate the boundaries (*arrows*) from the {120} sectors into the {010} growth sectors (*refraction* of dislocation lines). A: Dislocation pairs originating from tiny inclusions. *Dots*: surface damages. Diffraction vector $\boldsymbol{g}(200)$, CuK_α

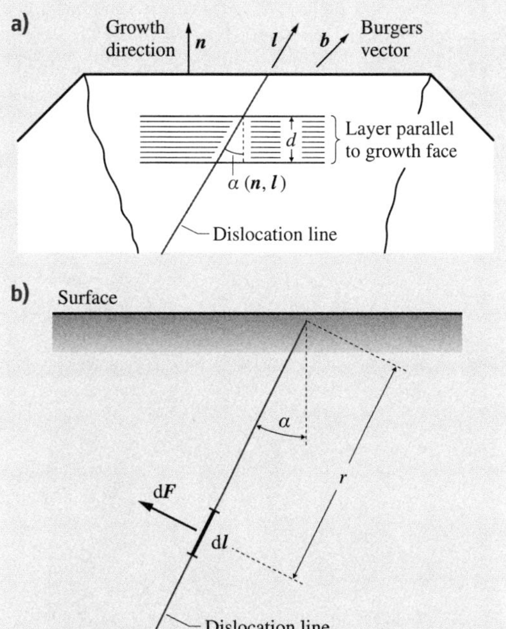

Fig. 4.20 (**a**) Derivation of the energy of a straight dislocation line within a layer parallel to the growth face. (**b**) Illustration of the force d$\boldsymbol{F}$ exerted by the crystal surface on a line segment d$\boldsymbol{l}$ of a straight dislocation line emerging at the surface (theorem of *Lothe* [4.97])

tions, e.g., [4.92–94])

$$E = \frac{Kb^2}{4\pi}\ln\left(\frac{R}{r_0}\right),$$

with $K = K(\boldsymbol{l}, \boldsymbol{l}_\mathrm{b}, c_{ij})$ the so-called *energy factor* of a straight dislocation line, R the outer cutoff radius, r_0 the inner cutoff radius, and b the modulus of the Burgers vector $\boldsymbol{b}$. The energy factor K describes the variation of the strain energy with the direction $\boldsymbol{l}$ of the dislocation line. It also depends on the Burgers vector direction $\boldsymbol{l}_\mathrm{b}$ and on the elastic constants c_{ij} of the crystal. The inner cutoff radius defines the limit until which the linear elasticity theory is applicable, and it corresponds to the radius of the dislocation core.

Since the logarithmic term and the core energy are, in general, not accessible to a numerical calculation, the variation of strain energy E of a given dislocation with Burgers vector $\boldsymbol{b}$ with direction $\boldsymbol{l}$ is – in a certain approximation – considered as proportional to the energy factor $K(\boldsymbol{l}, \boldsymbol{l}_\mathrm{b}, c_{ij})$, assuming the logarithmic term to be independent of $\boldsymbol{l}$. The energy factor has been calculated using the theory of dislocations in elastically anisotropic crystals developed by *Eshelby* et al. [4.98]. Since, in general, analytic solutions are not possible, numerical calculations have been performed using the program DISLOC, accounting for the elastic anisotropy of any symmetry down to the triclinic case [4.96]. Figure 4.21a,b shows a comparison of observed and calculated preferred directions of dislocations with four different Burgers vectors in the (011) growth sector of KDP [4.99]. The agreement is excellent with deviations of 3–6°, except for dislocation 4, the observed directions of which scatter by ±5° around a direction deviating by about 20° from the calculated one. This may be due to the very flat minimum of K, which makes the minimum-energy directions more subject to other influences such as surface features and core-energy variations (see *Deviations from Calculated Directions*). Similar comparisons have been carried out for various crystals grown on planar faces from solutions and supercooled melts: benzil, $(C_6H_5CO)_2$ [4.54, 100]; thiourea, $(NH_2)_2CS$ [4.101]; lithium formate monohydrate $HCOOLi \cdot H_2O$ [4.102]; ammonium hydrogen oxalate hemihydrate, $NH_4HC_2O_4 \cdot \frac{1}{2}H_2O$ [4.103]; and zinc oxide, ZnO [4.96]. In general, the agreement of observed and calculated directions is satisfactory and confirms the validity of the above theorems. It is pointed out that the preferred directions are independent of the growth method, provided that the growth faces (growth sectors) are the same. This has been demonstrated for benzil grown *in solution* in xylene and *in supercooled melt* ([4.1], [4.7, p. 138, Fig. 17]). Moreover, basal growth dislocations (Burgers vectors $\boldsymbol{b} = \langle 100 \rangle$) in prism sectors of hydrothermally grown hexagonal (wurtzite-type) ZnO crystals [4.104] show the same minimum-energy configuration as the corresponding growth dislocations in the prism sectors of benzil [4.96]. This similarity is due to the hexagonal lattice, the same prism growth sector, and

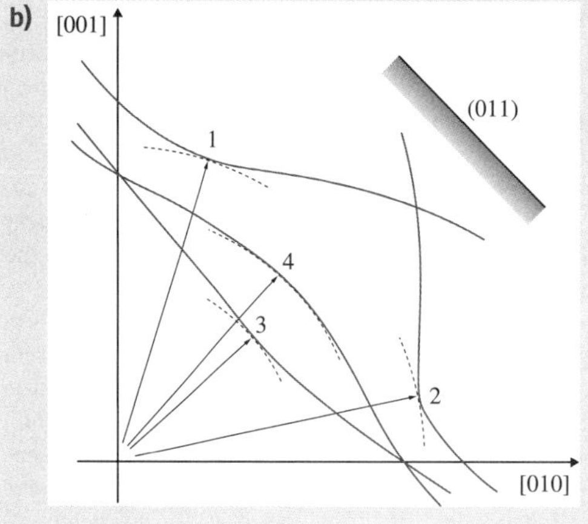

Fig. 4.21 **(a)** X-ray topograph ($\boldsymbol{g}$(020), Ag K_α radiation) of a KDP (100) plate (horizontal width ≈ 28 mm, thickness ≈ 1.5 mm), showing bundles of dislocations with noncrystallographically preferred directions emanating from small liquid inclusions (especially in the right-hand (011) growth sector) and from the capping region. The Burgers vectors of these dislocations can be recognized from their preferred directions. In addition, growth bands and features due to vicinal effects are visible. ABC: Boundary between (011) and (0$\bar{1}$1) growth sectors. **(b)** Plots of calculated energies $E/\cos\alpha$ (arbitrary units) of dislocations with Burgers vectors $\boldsymbol{b} = [001]$ (1), $\boldsymbol{b} = [100]$ (2), $\boldsymbol{b} = [011]$ (3), and $\boldsymbol{b} = [0\bar{1}1]$ (4) in growth sector (011) of KDP (polar coordinates: the energies are given by the length of the radius vector to the *curves*). The preferred directions of minimum $E/\cos\alpha$ are represented by *arrows*. The *dashed lines* are circles with radii equal to the minimum values of $E/\cos\alpha$. Note the close coincidence to observed **(a)** and calculated **(b)** dislocation line directions ◄

the same Burgers vectors of the dislocations in both cases.

An interesting experimental study of preferred dislocation directions in synthetic quartz is presented by *Alter* and *Voigt* [4.105]. They cut (0001) plates (Z-plates) out of different growth sectors of previously grown highly perfect quartz crystals and used them as seed plates for further growth experiments. The seed plates contained growth dislocations following the preferred directions typical for the sector from which they were cut. Since growth now proceeded on the (0001) face, the dislocations of the seed continued into the growing crystals with preferred directions typical of the Z-sector, exhibiting sharp bends of up to 90° at their transition from the seed into the grown crystal. This is instructively shown by x-ray topography [4.105].

On account of the factor $1/\cos\alpha$ in the energy term above, the preferred directions of growth dislocations are mostly normal or nearly normal to the (local) growth face. In some cases of planar interfaces, however, deviations from the growth normal of up to 30° have been observed, in agreement with the calculations.

For interfaces with convex curvature (e.g., in Czochralski growth) the dislocation lines, propagating more or less normal to the growth front, diverge and grow out of the crystal boule through its side faces. For concave interfaces the dislocation lines are focused into the center of the crystal boule (Fig. 4.10). Trajectories of growth dislocations in Czochralski gadolinium gallium garnet (GGG) have been calculated and compared with observed ones by *Schmidt* and *Weiss* [4.70]. The curvature of the interface has been taken into account by performing the calculations stepwise in small increments, leading to curved dislocation trajectories. Again the agreement is satisfactory. Moreover, this allowed assignment of Burgers vectors to the different dislocation trajectories which were observed optically with polarized light.

In 1997 and following years, preferred dislocation directions and their bending when penetrating a growth-sector boundary were observed by transmission electron microscopy in GaN grown by metalorganic vapor-phase epitaxy (MOVPE) using the epitaxial lateral overgrowth (ELO) technique [4.106–110]. The GaN hexagonal pyramids $\{11\bar{2}2\}$ growing through the windows in the mask are in the first stage topped by the (0001) basal plane, which during further growth becomes smaller and finally vanishes. Thus growth dislocations propagating normal to the (0001) facet penetrate the boundary to a $\{11\bar{2}2\}$ sector and are bent by about 90° into the preferred direction in this sector [4.109]. By this process the number of *threading* dislocations is drastically reduced. Very recently the bending of dislocations in growth-sector boundaries has also been used for the elimination of threading dislocations by *aspect-ratio trapping* in Ge selectively grown in submicron trenches on Si substrates [4.111]. Similarly, dislocations are eliminated from prism growth sectors of rapidly grown KDP crystals ([4.112, cf. Fig. 4]).

The above theory of preferred dislocation directions does not allow a dislocation line to proceed along a growth-sector boundary, as is sometimes discussed: the line would emerge on the edge constituting the boundary and thus be in a labile position. If, however, the edge is a narrow facet, the dislocation line can lie in its sector and appear to proceed along the boundary, but the probability of *breaking out* into one of the adjacent sectors would be high. *Ester* and *Halfpenny* [4.113] and *Ester* et al. [4.114] have observed in potassium hydrogen phthalate V-shaped pairs of dislocations originating from growth-sector boundaries, with the two *arms* of the *V*, following sharply defined directions, in the two adjacent sectors. This is in accordance with the minimum-energy concept.

Deviations from Calculated Directions

Although the agreement of observed and calculated directions of growth dislocations is in general satisfactory, frequently discrepancies are found. The reasons for this may be insufficient approximation in the model on which the calculations are based, or by influences of other defects or particular surface relief. The above calculations are based on linear anisotropic elasticity of the continuum and do not account for the discrete structure of the crystals, the dislocation core energy or (in piezoelectric crystals) for electrical contributions.

The following three causes have been found to affect preferred directions:

i. *Discrete lattice structure of the crystals and the neglect of the core energy*
   It is frequently observed that growth dislocations are exactly parallel to a low-index lattice direction (mostly a symmetry direction), although calculations suggest another (usually noncrystallographic) direction. This seems to happen in cases of pronounced Peierls energies, where the dislocations tend to align along the Peierls *energy valleys* (see textbooks on dislocations, e.g., [4.92–94]). It is plausible that dislocations will favor directions parallel to closed-packed directions, strong bond chains, structural channels, and planes of

pronounced cleavage. A dislocation line inclined to such a direction will consist of line segments along this direction, (i. e., lie in the Peierls valley) and kinks or jog across the Peierls potential barrier [4.93, pp. 229–234]. This strongly affects the core energy, which varies considerably with direction, particularly in the neighborhood of the structurally pronounced directions, for which it has a minimum. This is the same effect that leads to those favored directions of, for example, 60° dislocations in diamond- and sphalerite-structure crystals.

An example of the influence of the lattice structure in organic salol grown in supercooled melt and by Czochralski pulling is given in Figs. 4.16 and 4.22, where dislocations with Burgers vectors $\boldsymbol{b} = \langle 101 \rangle$ align along directions parallel to the prominent crystal edges $\langle 101 \rangle$, independent of the growth directions. A detailed discussion of the competing influences of core energy and elastic strain-field energy on dislocation directions, based on observations in solution-grown orthorhombic ammonium hydrogen oxalate hemihydrate, is presented by *Klapper* and *Küppers* [4.103].

ii. *Long-range stress*
Long-range stress arising near the growth face, e.g., due to other growth disturbances (neighboring inclusions, impurity layers), exerts forces on a dislocation and may locally change the zero-force direction, thus leading to curved dislocations lines.

iii. *Surface relief of the growth face*
Macrosteps sweeping over the dislocation outcrops lead to macroscopic (rounded) kinks in the dislocation lines. A particular influence is exerted by growth hillocks (vicinal pyramids, Sect. 4.3.3), the slopes of which possess an orientation somewhat different from that of the main face. Dislocations emerging from the vicinal slopes exhibit preferred directions different from those outcropping on the main face. Similar to the *refraction* of dislocation lines at growth-sector boundaries (Fig. 4.19), dislocation lines change their directions when they pass through an intervicinal boundary, i. e., when their outcrops shift from one vicinal facet to another. The influence of the surface relief, in particular of vicinal slopes, on the course of dislocations has been studied by *Smolsky* and *Rudneva* [4.116] and *Smolsky* et al. [4.48].

An example of the influence of vicinal pyramids on the course of growth dislocation lines is presented in Fig. 4.23. Here the dislocation lines in the right and left {011} sectors of KDP change their directions when they pass through the boundaries appearing in the topograph as dark and white contrast bands roughly parallel to the traces of the {011} faces. These boundaries are assumed to result from the competition of different growth pyramids. A dominating vicinal pyramid may be *overrun* by another pyramid which now takes over the dominating role. Since this overflowing of one growth pyramid over another can take place within a short growth period, the boundaries between regions grown on one or the other pyramid are practically parallel to the *main* growth face. Due to the different inclinations of the pyramid facets, the preferred directions of dislocations ending on these slopes are different.

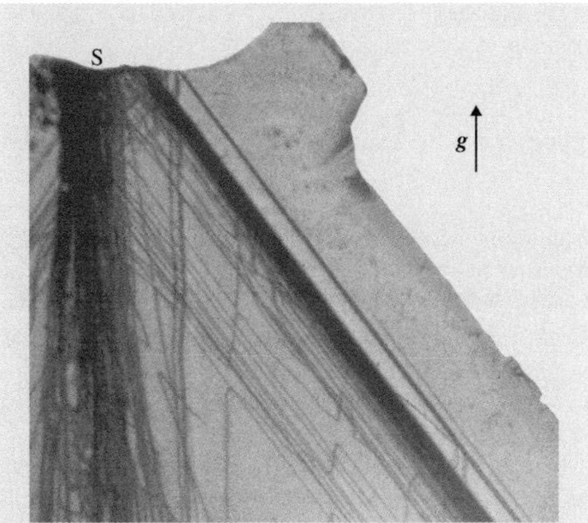

Fig. 4.22 Section (about $11 \times 11\,\text{mm}^2$) of a (010) plate (about 1.3 mm thick) cut from the cone region of a Czochralski salol crystal (pulling direction [100] upward; S: seed crystal, broken off). The topograph shows many straight dislocation lines ($\boldsymbol{b} = [101]$) exactly parallel to $\langle 101 \rangle$, which is the most prominent morphological edge of the crystal (Fig. 4.16). This direction is enforced by the discrete lattice structure, because the minimum-energy continuum approach suggests directions close to the normal to the concave interface. Furthermore, reactions (segment-wise annihilation) with a few *vertical* dislocations are recognized (after [4.115]). Diffraction vector $\boldsymbol{g}(200)$, CuK_{α_1}

Determination of Burgers Vectors

In x-ray topography, Burgers vectors are usually determined with the aid of the $\boldsymbol{g} \cdot \boldsymbol{b}$ criterion (*visibility rules*, see Chap. 42 of this Handbook). This requires the imaging of dislocations in several, at least two, different reflections, and sometimes an unambiguous determination is not possible with this method. Since the preferred directions of dislocation lines in a growth sector are characteristic of the Burgers vector, the observation of such directions may provide information on the Burgers vector direction. This should be particularly successful in cases where the dislocations penetrate a sector boundary, so that preferred directions of the same dislocation in different growth sectors are observed.

Depending on the knowledge of the elastic constants and the availability of a computer program the following three options are considered:

a) If in a crystal the Burgers vectors and corresponding preferential directions of dislocations in a growth sector have been determined by application of the visibility rules, then in further x-ray topographic studies of the same crystal species the Burgers vectors of these dislocations can be identified by their preferred direction $\boldsymbol{l}_0$ without necessarily taking exposures in different reflections.
b) If the elastic constants of the crystal under investigation are known and if a computer program is available, the directions $\boldsymbol{l}_0$ may be calculated for various Burgers vectors and growth sectors and compared with the observed ones. This may be particularly helpful in cases where unambiguous identification of the Burgers vector direction by the visibility rules is not possible.
c) Apart from these possibilities, the following three general statements, derived under the assumption that the energy factor K_s of a pure-screw dislocation is minimal and that K_e of a pure-edge dislocation is maximal (which is usually the case), may be useful [4.103]:

1. If $\boldsymbol{b}||\boldsymbol{n}$, then $\boldsymbol{l}_0||\boldsymbol{n}$ (pure-screw dislocation normal to the growth face)
2. If $\boldsymbol{b}\perp\boldsymbol{n}$, then $\boldsymbol{l}_0||\boldsymbol{n}$ (pure-edge dislocation normal to the growth face)
3. If $\boldsymbol{b}$ is inclined to $\boldsymbol{n}$, then $\boldsymbol{l}_0$ lies between $\boldsymbol{n}$ and $\boldsymbol{b}$.

In case 1 the energy per unit growth length $E = K/\cos\alpha$ (see *Characteristic Configurations, Theory of Preferred Direction*) has a steep minimum, because both K and $1/\cos\alpha$ (i. e., the length of the dislocation line in the unit growth layer) have a minimum along $\boldsymbol{n}$, whereas in case 2 K is maximal and $1/\cos\alpha$ is minimal. In the latter case $\boldsymbol{l}_0$ is parallel to $\boldsymbol{n}$ only if the decrease of K with increasing α is overcompensated by the increase of the length $1/\cos\alpha$ (which is mostly the case), leading to a flat minimum of E along $\boldsymbol{n}$. In case 3 the minima of K (along $\boldsymbol{b}$) and of $1/\cos\alpha$ (along $\boldsymbol{n}$) have different directions, thus the minimum direction $\boldsymbol{l}_0$ of E lies between them (i. e., it deviates from the growth direction $\boldsymbol{n}$ towards the Burgers vector $\boldsymbol{b}$).

These three rules have proved to be obeyed in most cases studied [4.54, 99–103]. Exceptions are provided by those dislocations whose directions are predominantly influenced by the discrete lattice structure. An application and detailed discussion of these rules for Burgers vectors of dislocations in orthorhombic ammonium hydrogen oxalate hemihydrate is reported in [4.103, p. 502].

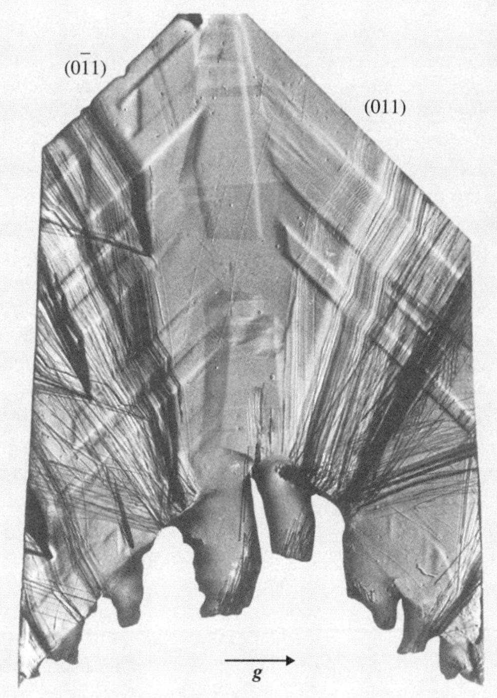

Fig. 4.23 X-ray topograph ($\boldsymbol{g}$(020), AgK_α radiation) of a (100) plate of KDP (horizontal width 26 mm, about 2 mm thick) showing growth dislocations in two {011} growth sectors. The preferred directions are modified by the varying vicinal surface relief on the {011} growth faces

4.4.5 Postgrowth Movement and Reactions of Dislocations

The as-grown geometry of dislocations with preferential directions, described in the previous sections, may be more or less drastically changed by thermal stress during cooling to room temperature [4.7–11]. Crystals grown in their brittle state from solution are rather insensitive in this respect. An example is benzil grown at about 40 °C from a solution in xylene by slow solvent evaporation (Fig. 4.24a): its growth dislocations are sharply straight-lined with preferred directions [4.7, 54, 100]. Even the thermal stress, which occurred by uncontrolled cooling and which induced a crack, did not change the grown-in dislocation configuration.

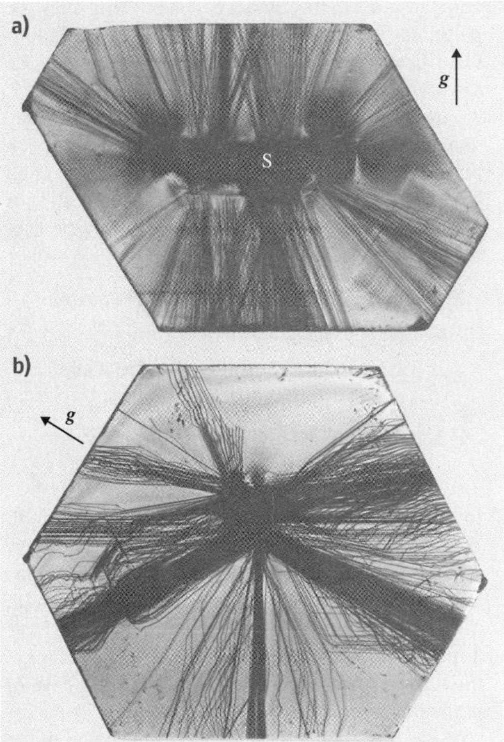

Fig. 4.24 **(a)** (0001) Plate (horizontal width 28 mm, thickness 1.2 mm) cut from a trigonal benzil crystal grown in its brittle state at about 40 °C from a solution in xylene by solvent evaporation. S: seed crystal, containing a crack which was formed at the end of the growth run. (The crack tips on the right and left sides of the seed induce extended long-range strain.) All dislocations are straight-lined with preferred directions. Diffraction vector $\boldsymbol{g}(20\bar{2}0)$, Cu$K_{\alpha_1}$. **(b)** (0001) Plate (horizontal width about 35 mm, thickness 1.2 mm) cut from a crystal grown in its plastic state from a supercooled melt ($T_m = 96$ °C). Numerous growth dislocations start from the surface of the rather strongly disturbed seed S, and many of them show postgrowth movement. $\boldsymbol{g}(02\bar{2}0)$, Cu$K_{\alpha_1}$

Crystals growing from their melt are always in the plastic state, and therefore their dislocations are subject to plastic glide even under small stress. As already mentioned in Sect. 4.1, gradient-free growth is possible from supercooled melt [4.1], and grown-in dislocation configurations can be preserved if the crystal is cooled very slowly to room temperature. This has been shown for benzil grown from slightly supercooled melt ($T_m = 96$ °C) and carefully cooled through its plastic zone into its brittle state: the grown-in dislocations are straight-lined and exhibit essentially the same geometry as that in crystals grown from solution ([4.1] and [4.7, Fig. 17]).

If stress arises during growth from the melt or during cooling through the plastic zone, the grown-in dislocations move and adopt a more or less irregular arrangement. This is shown in Fig. 4.24b, which presents a topograph of benzil grown from the melt with supercooling of about 0.5 °C. It shows numerous growth dislocations originating from the strongly disturbed zone of first growth on the seed crystal. Many of them follow irregular courses. They often exhibit straight-line segments along directions ⟨100⟩ (which are twofold-symmetry axes in the basal plane (0001) of the trigonal crystal), indicating the influence of the discrete lattice structure, and pinning points where dislocation movement was locally stopped by obstacles. Line elements parallel to the growth faces are also present. They cannot have formed in this orientation at the growth face, because the strong force at the surface would have pushed parallel segments out of the crystal or rotated them into an orientation of low or zero force (see the Lothe theorem above). This argues that the growth dislocations must have changed their course after growth, probably when the crystal was taken out of the melt. For this procedure the growth chamber was shortly opened whereby cool air flowed against the crystal. There are, however, also straight dislocations (pure-edge) normal to the growth faces, still following the preferred directions suggested by the minimum-energy theory.

Figure 4.25 shows, among others, four growth dislocations (arrows a) in orthorhombic benzophenone, grown in supercooled melt, which exhibit line segments with sharply defined noncrystallographic preferred direction (mainly screw character, Burgers vector [001] vertical) and segments subjected to postgrowth movement. The latter consist of several bows generated by dislocation movement which was locally hindered by pinning points. The *horizontal* line segments d are pure-edge dislocation dipoles (Sect. 4.4.3). Note that these dipole segments, consisting of two closely neighboring edge dislocations with opposite Burgers vectors, exhibit mostly about the same and in some cases wider and stronger x-ray topographic contrast, compared with the single dislocation lines. Similar observations of post-growth movement of growth dislocations with pinning points forming dipoles were made for sodium chlorate

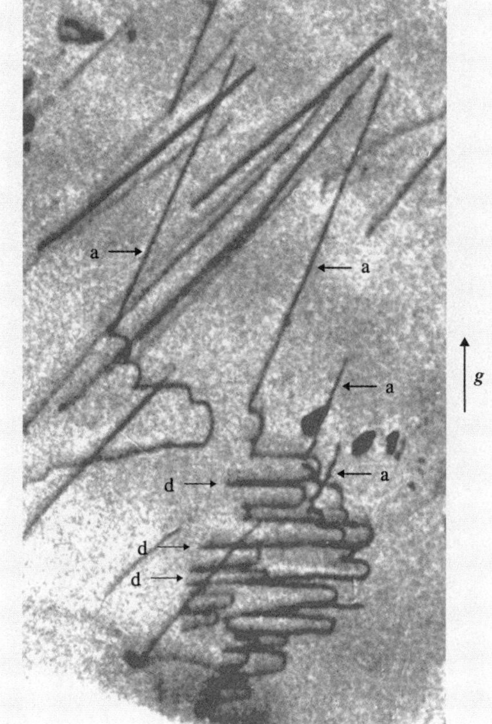

Fig. 4.25 Section (about 4×7 mm^2) of a (110) plate (about 1.5 mm thick) of benzophenone grown from supercooled melt. The *arrows* a mark four growth dislocation in a (111) sector showing sharply defined noncrystallographically preferred directions (predominantly screw character) in their upper, and postgrowth movement with pinning stops in their lower, parts. Burgers vector $\boldsymbol{b} = [001]$ vertical. The horizontal segments are pure-edge and form a few dislocation dipoles (*arrows* d). The dislocations emerge through the plate surfaces. $\boldsymbol{g}(002)$, CuK_{α_1}

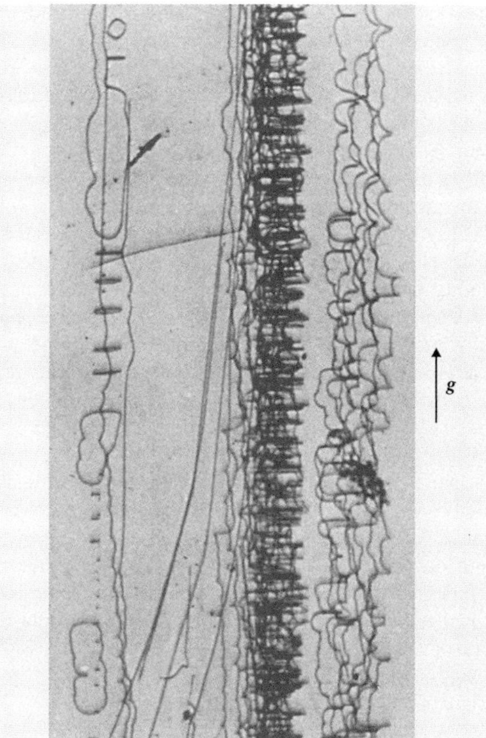

Fig. 4.26 Section (3.8×8.0 mm^2) of a plate (thickness about 1.2 mm) cut from a Czochralski crystal (pulling direction upward) of benzophenone ($T_m = 48\,°C$), showing growth dislocation after postgrowth movement. The pinning points of one of the dislocations at the *right-hand side* mark its originally straight course. At the *left side* some dislocations have partially annihilated and formed closed loops. $\boldsymbol{g}(002)$, CuK_{α_1}

$NaClO_3$ and lithium ammonium sulfate $LiNH_4SO_4$. This finding is remarkable insofar as these two crystals appear, by microhardness indentations, as highly brittle, forming cracks even under lowest indentation loads. This indicates that in these crystals fast plastic flow is impeded by high friction but slow dislocation movement (creep) is possible.

An even more drastic change of growth dislocations is presented in Fig. 4.26, which shows a section of a plate cut out of the center of a Czochralski-grown benzophenone crystal (temperature of the melt about 1.5 °C above $T_m = 48\,°C$, pulling rate 20 mm/day). During pulling the crystal was cooled by blowing air of about 35 °C against the rotating crystal in order to obtain a crystal diameter of 25 mm [4.42]. The topograph shows growth dislocations which originate from the zone of first growth on the seed crystal (outside the top of the figure). A few of the dislocations have retained their original course (straight or slightly curved lines), but most of them have suffered postgrowth movement (which may have occurred already during the pulling process): they consist of a series of bow-shaped segments connected at pinning points (right-hand side of Fig. 4.26). The pinning points are aligned along straight lines marking the original position of the dislocations. On the left-hand side of Fig. 4.26 the changes are even more drastic: some dislocations have partially annihilated, leaving behind only a few large and several small dislocation loops. Postgrowth reactions of growth dislocations forming dislocation nodes in Czochralski salol have been studied by *Neuroth* and *Klapper* [4.115].

Interestingly, the dislocations in the Czochralski salol of Fig. 4.10 have preserved their growth configuration with straight lines roughly normal to the growth front, despite the thermal stress which is always present in this growth technique. This may be due to the low growth temperature ($T_m = 42\,°C$) and the narrow plastic zone of salol below the melting point.

4.4.6 Postgrowth Dislocations

Dislocations formed in the interior of already grown crystal without connection to the growth front or other surfaces must be closed loops [4.7–11]. It is practically impossible to generate closed loops in a perfect crystal by stress, since the stress required for such processes would be extremely high. Inclusions, however, usually represent stress centers and form internal surfaces in the crystal. The stress in the crystal around the inclusions is relieved by the emission of concentric dislocations loops or – more frequently – of dislocation half-loops. The half-loops are – strictly speaking – also closed loops with a virtual closing line element inside the inclusion. Half-loops can also generate growth dislocations: if stress is built up around an inclusion just incorporated and still close to the growth front, half-loops emitted from the inclusion may *break through* to the growth interface, whereby each half-loop forms two separate dislocation lines with opposite Burgers vectors propagating with the growth front. Dislocation half-loops emitted from bubble inclusions and from decomposition particles in benzil grown from supercooled melt are shown in Fig. 4.27a,b. Examples of half-loops in solution-grown crystals, revealed by x-ray topography, are given in [4.9] (sodium chlorate) and [4.117] (tetraoxane). The latter study shows the successive emission of half-loops from inclusions and their splitting into two separate dislocations when reaching the crystal surface. A very peculiar kind of dislocations loops is found in octadecanol crystals grown from xy-

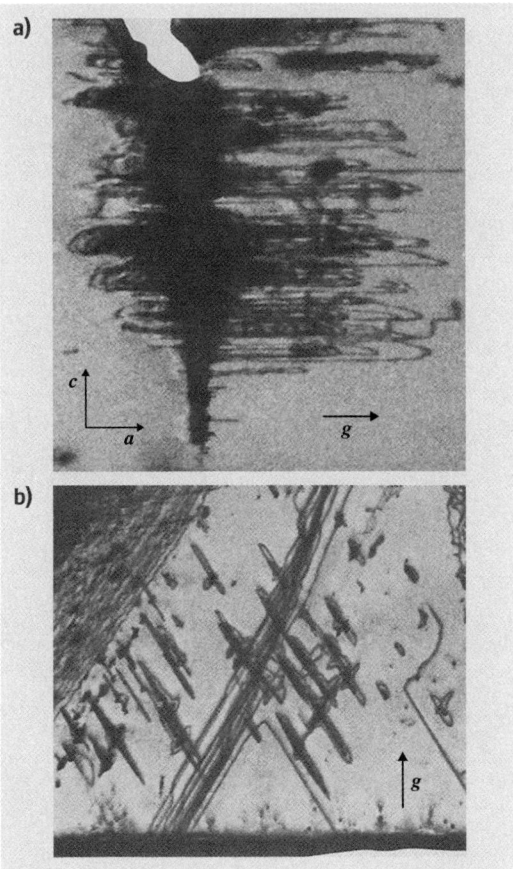

Fig. 4.27 **(a)** Postgrowth dislocation half-loops in benzil (grown from supercooled melt) emitted from a trail of bubbles. Section $4.5 \times 5\,mm^2$ of a $(01\bar{1}0)$ plate (after [4.1]). The dislocations belong to the $\{01\bar{1}0\}\langle100\rangle$ glide system and are pure-screw in their horizontal segments. $\boldsymbol{g}(20\bar{2}0)$, CuK_{α_1}. **(b)** Section (about $9.5 \times 8\,mm^2$) of a (0001) plate of benzil grown from supercooled melt, showing dislocation loops of glide system $(0001)\langle100\rangle$ emitted from small inclusions. The loops are elongated in the direction of their (symmetrically equivalent) Burgers vectors $\boldsymbol{b} = [100]$ and [110]; the loops with the third equivalent Burgers vector [010] are x-ray topographically *extinct* in the reflection used. $\boldsymbol{g}(02\bar{2}0)$, CuK_{α_1}

a)

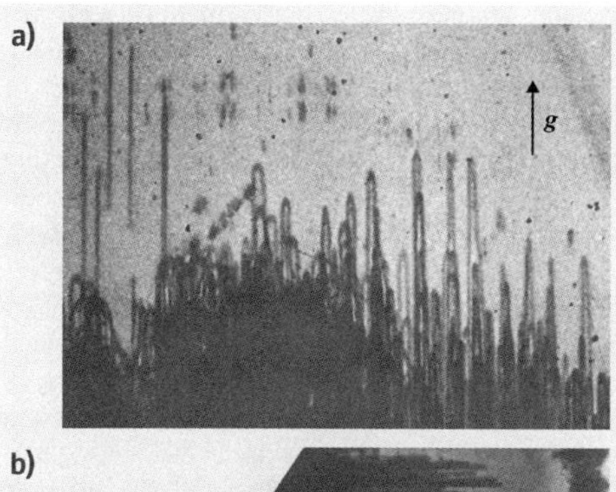

b)

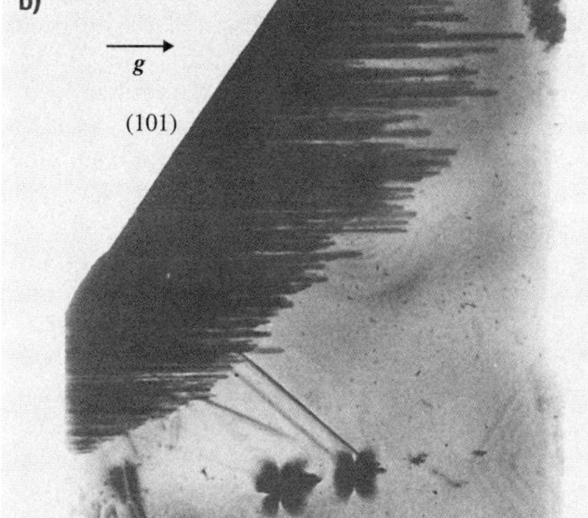

Fig. 4.28a,b Forests of hairpin dislocations (screw dipoles) of the easy glide system (100)[001] of orthorhombic thiourea, $(NH_2)_2CS$, grown from a water–methanol solution [4.119]. The *hairpins* have invaded the crystal through the {101} growth faces due to improper handling after growth. Both topographs $\boldsymbol{g}(001)$, MoK_α. **(a)** Plate (section about $4\times6\,mm^2$) parallel to glide plane (100), Burgers vector [001] vertical. *Upper-left corner*: growth dislocations penetrating the plate. **(b)** (010) Plate ($7\times9\,mm^2$); the *hairpins* are projected edge-on. *Bottom*: three growth dislocations, liquid inclusions

lene solution: columns of prismatic loops are punched out from inclusions [4.118].

Postgrowth dislocations may also arise from the (external) crystal surface. This is common for growth in contact with a container wall (e.g., Bridgman melt growth). Dislocations are generated by stress due to the different thermal expansion of crystal and container wall or by sticking of the crystal to the wall. An illustrative example is presented in [4.7, Figs. 32 and 33]: it shows numerous glide dislocations in cleavage lamellae of 2,3-dimethylnaphthalene (grown by the Bridgman method), which arise from the seed region, from the contact with the growth ampoule, from internal stress centers, and from the point of the cleavage impact. In crystals which are plastic at room temperature glide dislocations may invade through the surface by improper handling after growth, e.g., by mechanical impacts (mechanical polishing, scratches). This is illustrated in Fig. 4.28a,b for orthorhombic thiourea grown from a water–methanol solution. Numerous hairpin dislocations (dipoles) with pure-screw character of their long branches originate from surface damages and extend a few millimeters on the main glide plane (which is also a pronounced cleavage plane) into previously perfect crystal regions [4.119]. In Fig. 4.28b these *hairpins* are projected *edge-on*. The invasion of glide dislocations from the specimen surface into the bulk quite frequently occurs during high-temperature processing of crystals for solid-state electronics and optics (e.g., [4.82]).

4.4.7 The Growth-Promoting Role of *Edge* Dislocations

Since the pioneering works of *Frank* [4.120, 121] and *Burton*, *Cabrera*, and *Frank* (BCF theory) [4.122] it is well established that screw dislocations play a decisive role in growth at low supersaturations by forming persistent step sources in the form of spirals or concentric loops. The latter are generated by pairs of screw dislocations of *opposite* Burgers vectors (screw dislocation dipoles), which separately form spiral hills of *opposite handedness*. These hills fuse into a single growth pyramid with self-perpetuating concentric steps, emitted from the dislocation pair, if the distance between the two dislocations is larger than the critical radius ρ_c of the two-dimensional nucleus for stable growth. For a distance smaller than ρ_c a growth-promoting pyramid is not formed. In contrast to this, a group of N screw dislocations with the *same sign* of Burgers vectors (*same handedness* of growth spirals) increases the growth activity of the face on which they terminate by a factor up to N, if these dislocations are separated by distances smaller than ρ_c (*cooperating spirals*). For distances larger than ρ_c the growth activity corresponds to that of a single screw dislocation (*noncooperating spirals*). These findings hold for all dislocations, normal or inclined to the growth surface, which posses a screw component normal to the growth face [4.122].

Later it was recognized that pure-edge dislocations are capable of generating growth spirals also, e.g., a pure-edge dislocation, emerging at an angle of about 45° in such a way that its Burgers vector also forms an angle of 45° with the surface, will generate a growth spiral. This has been shown in models by *Strunk* [4.123] and *Ming* [4.124]. In their reasoning, dislocations with Burgers vector *parallel* to the growth surface, e.g., edge dislocations *emerging perpendicularly* from the growth face, were expected *not* to form persistent growth centers promoting growth. There is, however, much experimental evidence proving the opposite. For example, pure-edge dislocations generated by inclusions on {100} cube faces of potassium alum, growing from aqueous solution, drastically increase the growth rate of the {100} face (from which they perpendicularly emerge) in the moment of their appearance. The {100} growth rate is reduced again, when the edge dislocations shift during growth over a crystal edge on a neighboring {111} face (i. e., when they penetrate the boundary into the {111} growth sector [4.76–79], cf. Sect. 4.3.6 and Fig. 4.15). Detailed studies of this growth-promoting effect of edge dislocations in potassium alum are reported in [4.77–79].

The above results show that the term *screw dislocation mechanism* for the formation of growth spirals is not adequate. *Bauser* and *Strunk* [4.125] introduced the terms *longitudinal step source* for growth centers with a Burgers vector component normal to the growth face, and *transverse step source* for growth centers with zero component normal to the growth face. *Frank* [4.126] suggests the terms *rampant step source* and *couchant step source*, respectively, but a final choice of terms has not emerged in this matter.

The probably first demonstration of the growth-promoting effect of *transverse step sources*, based on a reliable Burgers vector determination, was provided by *Bauser* and *Strunk* [4.125]. They observed, by liquid-phase epitaxial growth on a (100) facet of GaAs, step patterns with well-defined growth centers. High-voltage transmission electron microscopy (TEM) showed that each of these centers was associated with a single dislocation. Burgers vector analysis revealed that, besides hill dislocations *with* Burgers vector components normal to the growth face (*longitudinal* step centers), also those *without* a normal component (*transverse* step centers) occur. Though a statistical analysis is lacking, both types of growth centers seem – in this case – to occur with comparable frequency and activity. *Frank* [4.126] expects that the *transverse growth center* is active only for increased supersaturation approaching the critical value necessary for two-dimensional nucleation, whereas the *longitudinal center* works already for very low supersaturations.

The origin of the activity of transverse step sources is not yet clear. *Bauser* and *Strunk* ([4.125] and references therein) assume that the dislocation is split into two partials (spanning a stacking fault) with a surface step formed between their endpoints on the surface. The growth-promoting activity of this defect arrangement has been demonstrated by *Ming* [4.124]. This approach, however, can only hold for crystals capable of stacking faults. *Frank* [4.126] suggests that surface stress around the dislocation outcrop leads to increased local adsorption of atoms, and *Giling* and *Dam* [4.127] postulate a local roughening of the surface with increased growth, forming a hill which emits from its periphery concentric step rings over the otherwise flat growth face.

4.5 Twinning

4.5.1 Introductory Notes

A twin is a frequently occurring aggregate or intergrowth of two or more crystals of the same species (same chemical composition and crystal structure) with a defined *crystallographic orientation relation* (determining the *orientation states* of the *twin components*), which in mineralogy is called the *twin law*. Besides the twin law, the boundary between twin domains (the *contact relation*) plays a decisive role: twins occur in those crystals in which boundaries of low energy can be formed. This has been proven quantitatively by *Gottschalk* et al. [4.128], who showed that the ease and frequency of the formation of (111) spinel twins in the sphalerite (zincblende) structure of III–V semiconductor compounds (GaAs, InP, etc.) is correlated to the (111) stacking-fault (twin-boundary) energy (see also [4.69, p. 422]). There is, however, also a theory which states that the boundary energy is of minor importance and that kinetic influences play a decisive role in the formation of twins [4.129]. This approach, however, is critically discussed [4.130, 131].

Historically, the concept of twinning was developed in mineralogy due to the rich occurrence of morphologically prominent twins in minerals. In the 1920s another approach, independent of mineralogy, was developed in physics with the investigation of ferroelectricity and ferroelasticity. The spontaneous electric polarization and spontaneous mechanical strain occurring in these *ferroic* crystals lead to ferroelectric and ferroelastic domains which are twin domains as defined in mineralogy. Though the concepts of *twinning* and *domain structures* (as the physical approach is now called) deal with the same phenomenon, there are (apart from a different nomenclature) certain differences: in contrast to *twinning*, the *domain structure* approach requires a (real or hypothetic) *crystallographic parent symmetry* (supergroup), from which the *orientation states* can be derived by a real or hypothetic phase transition. Therefore, this approach cannot be applied to many growth and deformation twins. A comparison of the concepts of *twinning* and of *domain structures* is given by *Janovec* et al. [4.132] and *Janovec* and *Přívratská* [4.133].

Twins are mainly classified by morphological features (dovetail, contact, penetration, sector, polysynthetic twins, etc.), by their genetic origin (growth, transformation, mechanical or deformation twins) and by their lattice coincidence features: *merohedral* twins (full, three-dimensional lattice coincidence), which are also called *twins with parallel axes*, and *nonmerohedral* twins (two or one-dimensional lattice coincidence), also called *twins with inclined axes*. Reviews on twinning are presented in many textbooks of mineralogy and crystallography (e.g., [4.134, 135]). A comprehensive treatment is provided by *Hahn* and *Klapper* [4.69, especially section *Growth twinning* p. 412–414]. A survey on x-ray topographic characterization of twinned crystals is given by *Klapper* [4.6]. In the present section only growth twins and twins generated in the cooling period after growth (postgrowth twins) are treated, whereby emphasis is placed on crystals grown from solutions on habit faces. A treatment of twinning in crystals grown from the melt is presented by in Chap. 6 of this Handbook.

4.5.2 Twin Boundaries

The shape and arrangement of twin domains is essentially governed by the twin boundaries and their preferred orientations. As mentioned in the previous subsection, twin boundaries are contact faces with good structural fit of the twin partners involved, i. e., they are internal surfaces of low energy. Accordingly they are, as a rule, low-index and structurally densely packed lattice planes common to both twin partners. For reflection twins and twins by a twofold axis, the twin mirror plane or the plane normal to the twofold twin axis is always an energetically favored contact plane because along this boundary the corresponding lattice planes of the two partners match perfectly. There is, however, a basic difference between *twins with parallel axes* (three-dimensional lattice coincidence: *merohedral* twins) and *twins with inclined axes* (two- or one-dimensional lattice coincidence: *nonmerohedral* twins [4.69, p. 422] [4.132, 133]). In the first case the lattices of the twin components match perfectly along any twin boundary for any twin law. In the latter case the boundary along the common lattice plane of perfect matching is usually strictly adopted. Deviations from strict planarity of these boundaries may occur by steps which form the *twinning dislocations*. These dislocations play an important role in the plastic deformation of crystals by twin formation.

For twins with parallel axes (full lattice coincidence, e.g., for inversion twins) the lattices of the twin components match perfectly along any arbitrary (also curved) boundary. Thus, from this lattice aspect alone, arbitrarily oriented boundaries are expected to occur. A prominent example is the Dauphiné twinning of quartz, which occurs as growth as well as transformation and mechanical (ferrobielastic) twinning: Dauphiné twin boundaries usually follow (at least macroscopically) arbitrary and curved surfaces, whereby low-index boundary segments sometimes also occur [4.136, pp. 75–99]. As a rule, however, low-index twin contact planes are favored also in twins with full lattice coincidence (merohedral twins). An example is again provided by quartz: Brazil twins (twin law: inversion), which are exclusively growth twins, develop boundaries along low-index planes, preferably parallel to prism and rhombohedron faces [4.136, Fig. 61].

A special case is given by the spinel twins of cubic crystals with twin law *reflection plane (111)* or *twofold axis along [111]*. They occur as growth twins in technologically significant crystals with diamond structure (Si, Ge), sphalerite (zincblende) structure (e.g., ZnS, GaAs, InP, and CdTe [4.128, 137]), and sodium chloride structure (photographic materials AgCl and AgBr [4.138–142]). They are characterized by a partial lattice coincidence of 1/3 of the lattice points (so-called *Σ3 twins*), which form a hexagonal sublattice of the cubic lattice, with the hexagonal axis along the threefold axis [111] common to both twin components. Preferred twin boundaries are planes {111} and $\{11\bar{2}\}$ (Fig. 4.29), which are twin reflection planes (in centrosymmetric crystals such as diamond, Si, Ge, AgBr) or planes

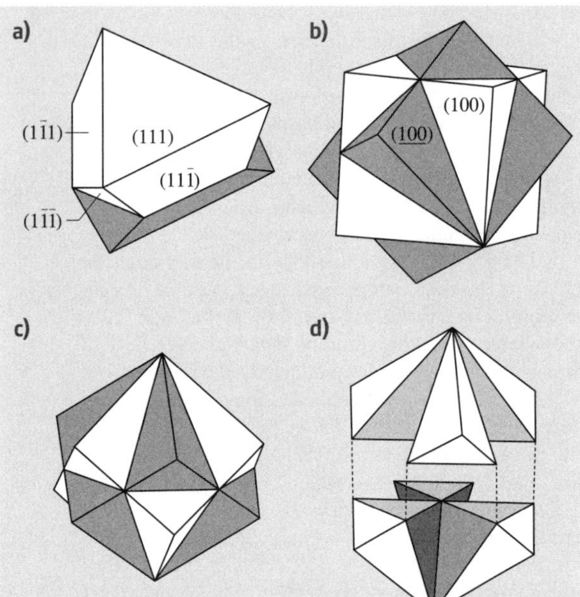

Fig. 4.29a–d Spinel twin of cubic crystals with twin mirror plane (111) or twofold twin axis [111] (or alternatively ±60° rotation around [111]). The domains of the two orientation states are shown *white* and *shaded*. (**a**) Contact twin with (111) contact plane (two twin components). (**b**), (**c**) penetration twin (idealized) with one (111) and three $\{11\bar{2}\}$ contact planes (12 twin components, 6 of each orientation state) in two different views: (**b**) with one [001] axis vertical, (**c**) with the threefold-symmetry axis [111] common to the two orientation states and coinciding with the twofold twin axis vertical. (**d**) Skeleton of the six components (exploded along [111]) of the *shaded* orientation state of (**c**). The components are connected along one [111] and three ⟨110⟩ edges meeting in the center

normal to twofold twin axes (in noncentrosymmetric crystals of sphalerite structure, e.g., ZnS, GaAs).

4.5.3 Formation of Twins During Growth

Formation During Nucleation of the Crystal

In many cases, twins are formed already during the first stage of spontaneous nucleation, possibly before the subcritical nucleus reaches the critical size necessary for stable growth. This formation is strongly evidenced for penetration and sector twins, where all domains are of similar size and originate from one common, well-defined *point* in the center of the twinned crystal, which marks the location of the spontaneous nucleus (Figs. 4.29 and 4.30). Other prominent examples are the penetration twins (two orientation states) in quartz-homeotypic gallium orthophosphate ($GaPO_4$ [4.47]) and in rhombohedral crystals such as corundum (sapphire, Al_2O_3 [4.143]) or iron borate ($FeBO_3$, calcite structure [4.6, p. 390]), and sector twins of pseudo-hexagonal crystals such as lithium ammonium sulfate (Fig. 4.30), potassium sulfate [4.69, p. 408], and aragonite $CaCO_3$ (Fig. 4.31), which all form twins with three orientation states. The origin of twinning by nucleation must also be assumed for contact twins (Fig. 4.29a), if both partners of the twin have roughly the same size, or

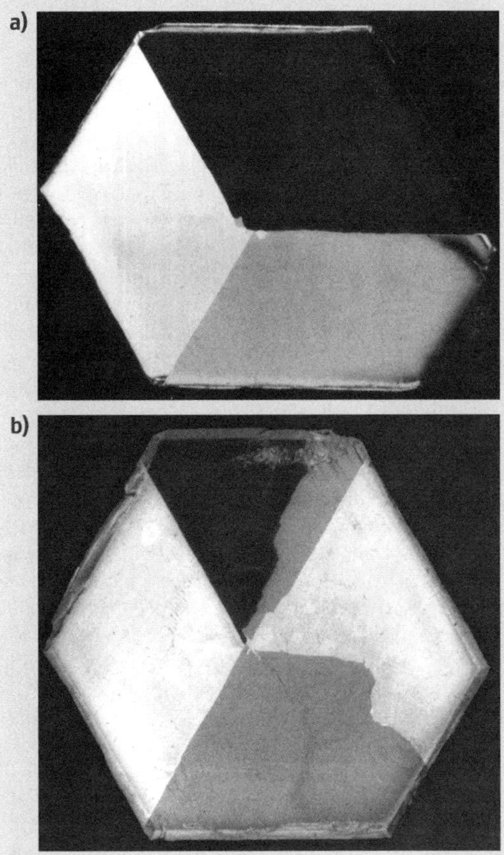

Fig. 4.30a,b Photographs of (001) plates (about 20 mm diameter, about 1 mm thick) of orthorhombic pseudo-hexagonal $LiNH_4SO_4$ between crossed polarizers, showing sector growth twins. (**a**) Nearly regular threefold sector twin (three orientation states, three twin components). (**b**) Irregular sector twin (three orientation states, but five twin components). After [4.69, p. 413]

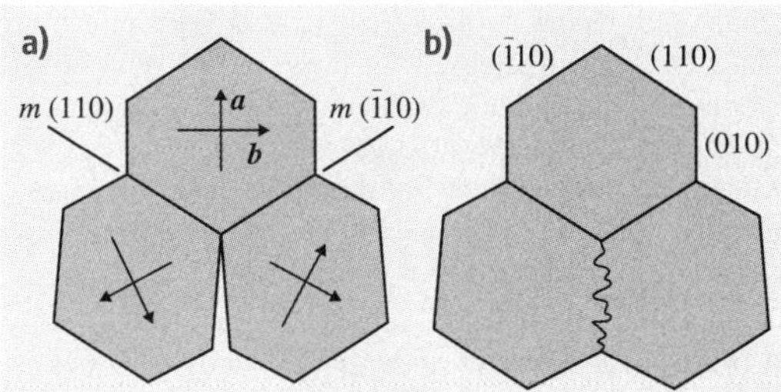

Fig. 4.31 (**a**) Triple growth twin of orthorhombic pseudo-hexagonal aragonite, $CaCO_3$. The three twin components are related by two symmetrically equivalent mirror planes (110) and $(\bar{1}10)$. Due to a relatively high deviation from the hexagonal metrics, a gap of 11.4° should be formed. In actual crystal the gap is usually closed, as shown in (**b**), leading to a strongly disturbed irregular twin boundary

if all spontaneously nucleated crystals in one batch are twinned. For example, *all* crystals of monoclinic lithium hydrogen succinate precipitated from aqueous solution are, without exception, dove-tail contact twins.

An approach to twin formation during crystal nucleation has been advanced by *Senechal* [4.144]. She proposes that the crystal nucleus first formed has a symmetry which is not compatible with the lattice of the (macroscopic) crystal. This symmetry may even be noncrystallographic. It is assumed that, after the nucleus has reached a critical size beyond which the translation symmetry becomes decisive, the nucleus collapses into a twinned crystal with domains of lower symmetry and continues to grow as a twin. This idea of twin formation from noncrystallographic nuclei has been substantiated experimentally by high-resolution TEM (HRTEM) investigations of nanocrystalline diamond-type and face-centered cubic (fcc) crystals, such as Ge, Ag, and Ni [4.145].

Twin Formation by Inclusions

As they are for dislocations (Sect. 4.4.2), inclusions are frequently sources of twins. It is assumed that a nucleus in twinned orientation forms at the inclusion and proceeds in this orientation during further growth. An instructive experimental key study of this process is presented by *Sunagawa* et al. [4.146] for Dauphiné and Brazil twins in synthetic amethyst quartz. Amethyst quartz contains much more Dauphiné and, to a still higher extent, Brazil growth twins than normal (colorless) quartz. The higher frequency of twin formation is doubtlessly due to its relatively high content of iron [4.147]. *Sunagawa* et al. [4.146, and references therein] grew amethyst quartz hydrothermally in various solutions containing ferric iron, on amethyst seed plates of various orientations. They studied the twinning on as-grown faces, on etched growth faces, and on cut surfaces by light-microscopic methods. In all cases the Dauphiné and Brazil twins originated from solid inclusions containing iron (probably as goethite, FeOOH). In some cases the Brazil twins are associated with dislocations originating from the same inclusion.

A twin component originating from an inclusion may have the following shapes, depending on the preferred orientation of the twin boundary:

a) It forms a conical (pyramidal) insert, embedded in the *mother* crystal, with its apex in the inclusion and widening in the growth direction. This shape occurs, e.g., in quartz for Dauphiné as well as Brazil growth twins, whereby the surface of the twin inserts (i.e., the twin boundary) is rounded for the Dauphiné and faceted by low-index habit planes for the Brazil twin [4.136, Fig. 61]. In amethyst quartz the inserts are lamellae preferably parallel to the major rhombohedron faces.
b) For crystals with a pronounced preference of a low-index planar twin boundary, the twin insert is a band-shaped lamella originating from the inclusion and proceeding with the growing crystal. An example is given by hexagonal potassium lithium sulfate ($KLiSO_4$, merohedral reflection twin) grown from aqueous solution [4.148].

For the generation of twinning in melt-grown crystals, the reader is referred to Chap. 6 in this Handbook.

Propagation of Twin Boundaries

In contrast to a twin lamella, a (single) twin boundary cannot end *within* an otherwise perfect crystal. Therefore, a twin boundary emerging on a growth face must proceed with it during growth, whereby it should obey the theorem of minimum-energy orientation, postulated in Sect. 4.4 for growth dislocations. This means that the orientation of the twin boundary results from the competition to minimize the area of its surface within a growth layer (i.e., by orienting toward 90° with the grown face) and to minimize the boundary energy per unit area. Since, as a rule, the boundary energy has a pronounced and sharp minimum along a low-index lattice plane (similar to dislocations with pronounced Peierls energy, cf. *Deviations from Calculated Directions*), the boundary will follow this plane even for small angles between the twin contact plane and the growth face.

In mineralogy it is frequently observed that a twin boundary coincides with a large-area prominent growth face. A famous example is the polysynthetic lamellar twinning of the triclinic feldspar albite ([4.69, p. 410]). This indicates that a two-dimensional nucleus in twin position has formed on the growth face of the (previously untwinned) crystal [4.149, pp. 472–475]. Obviously, this process is triggered by defects in the growth face (e.g., by impurities, inclusions). If the twin nucleus spreads out over the growth face, the twin boundary coincides with the growth face. This mechanism seems to be possible only for twin boundaries of low energy, since the boundary energy of the large interface has to be supplied in one step, i. e., during spreading out of one single growth layer in twin position. The energy contribution of the first layer, however, may be quite small, because the structural alteration across the interface occurs only for second-nearest and more distant neighbors, so that the full boundary energy of the twin accumulates only after the deposition of a sufficiently thick package of growth layers. This is particularly the case if long-range (e.g., ionic) interactions are present. Thus, from an energetic point of view, the formation of twin boundaries coinciding with growth faces – after a two-dimensional nucleus in twin orientation has formed – appears probable. According to *Hartmann* [4.150], this kind of twin formation can only occur on *flat* or, with lower probability, on *stepped* faces (F- and S-faces, respectively). The critical step is the formation of the two-dimensional twin nucleus on these faces.

The formation of twin boundaries coinciding with the *actual* growth face, however, seems to be rather the exception, although in special cases, e.g., in polysynthetic lamellar twinning of the albite feldspar, it happens several times within the same crystal. It is pointed out that, for example, the planar low-index boundaries of Brazil twins in quartz *do not* develop parallel to prism or rhombohedron growth faces on which the twins are nucleated, although these faces are preferred (low-energetic) twin contact planes [4.136, 146]. This feature is obviously due to the special character of Brazil twins: they are merohedral inversion twins with full three-dimensional lattice coincidence.

Finally, boundaries of twins with more than two orientation states are considered. Such twins frequently occur in orthorhombic pseudohexagonal crystals (e.g., K_2SO_4, $LiNH_4SO_4$, aragonite $CaCO_3$) and consist mostly of domains of three orientation states (triple twins) related by three equivalent twin mirror planes or twofold twin axes. An example is shown in Fig. 4.30a. The crystals usually grow as *sector twins* starting from a common nucleus, with three or six twin components. In the latter case pairs of opposite sectors belong to the same orientation state. The twin boundaries mostly coincide with the twin mirror planes, but deviations leading to irregular boundaries may also occur (Fig. 4.30b). A twin boundary may coincide with a growth-sector boundary but usually does not (Fig. 4.30). An x-ray topographic study of pseudohexagonal growth twinning in $LiNH_4SO_4$ is presented by *Docherty* et al. [4.151].

For the sector-twin boundaries of pseudohexagonal crystals the angle $\gamma = 2\arctan(b/a)$ (where a and b are the orthorhombic lattice parameters of the pseudohexagonal plane) plays an important role. For an exact fit of the sectors $\gamma = 120°$ is required. Small misfits, e.g., $\gamma = 119.6°$ for orthorhombic lithium ammonium sulfate [4.151] (Fig. 4.30), are tolerated for triple twins without further disturbances, but often induce cracks when the twins grow to a larger size. An example of extreme sectorial misfit occurs in orthorhombic aragonite $CaCO_3$ ($\gamma = 116.2°$), which would generate a gap of 11.4° from the 360° closure. In this case the triple twin usually exhibits two boundaries coinciding quite perfectly with the two twin mirror planes {110}, whereas the third is irregular and strongly disturbed, thus closing the angular gap (Fig. 4.31b). A particularly interesting example is provided by nanocrystals of germanium nucleated in an amorphous Ge film deposited from the vapor on a NaCl cleavage plane (HRTEM study by *Hofmeister* [4.145]): pseudopentagonal sector twins by repeated (111) twin reflections and (111) twin contact planes are formed. The angle of one sector is theoretically 70.5° (the supplement to the tetrahedral angle 109.5°) and five of these angles leads to a gap of about 7.5°. This gap is compensated by slight widening of the sectors either by distortion or by the formation of stacking faults within a sector, whereby the five (111) twin boundaries of this fivefold sector twin remain perfect ([4.145] and [4.69, p. 439]).

4.5.4 Growth-Promoting Effect of Twin Boundaries

The growth-promoting influence of twin boundaries forming reentrant edges was already noticed by mineralogists in the 1890s and has later been described in detail by *Buerger* [4.149] and *Hartmann* [4.150]. This *twin-plane reentrant-edge effect* (TPRE effect after [4.137]) can occur only in faceted crystals and is pronounced for growth faces adjoining a reentrant edge, which provides a self-perpetuating step source [4.137, 152–156]. It leads to crystals laterally extended in directions parallel to the twin interface (compared with untwinned crystal) and has also been observed for organic crystals of orthorhombic *n*-alkanes and paraffins [4.157, 158]. The TPRE effect has a particularly strong impact on the morphology of cubic crystals twinned by the spinel law: whereas untwinned cubic crystal exhibits an isometric shape, (111)- and $(11\bar{2})$-twinned crystals grow as plates parallel to the twin plane. The growth of (111) and $(11\bar{2})$ platelets of cubic elemental and compound semiconductors from metal solutions using the TPRE effect has been studied in detail by *Faust* and *John* [4.137, and references therein]. Of particular significance for photographic products are tabular crystals of cubic silver halogenides, such as AgBr, AgCl, and AgI. The formation of tabular AgBr crystals by the TPRE mechanism is discussed by *Jagannathan* et al. [4.138, 139], *Bögels* et al. [4.140, 141], and more recently *Lee* et al. [4.156] using the hard-sphere model of *Ming* and *Sunagawa* [4.152, 153]. These authors also discuss the effect of two or more *parallel* twin planes (i. e., of twin lamellae).

A more drastic morphological change of cubic crystals is provided by two or more *intersecting* twin planes (*cross-twinning*). In this case the crystals grow as needles parallel to the intersection of the twin planes. Examples are presented and discussed by *Wagner* [4.159], *Hamilton* and *Seidensticker* [4.160] (critically reviewed by *van de Waal* [4.161]) for Ge (dendrites in $\langle 11\bar{2}\rangle$ direction), and *Bögels* et al. [4.142] for AgBr and AgCl (needles in $\langle 110\rangle$ directions).

4.5.5 Formation of Twins after Growth

There are two causes for the formation of twins during cooling to room temperature after growth: phase transitions and ferroelastic switching [4.133].

Phase transition: Crystals often can only be grown at elevated temperatures, where they crystallize in another (*high-temperature*) phase of usually higher symmetry than they adopt at room temperature. On cooling below the transition temperature, twin domains are formed, whereby the lost symmetry elements of the mother phase act as twin elements (twin laws) relating the twin domains. Examples: lithium niobate $LiBO_3$ is grown from the melt at $T_m = 1275\,°C$ and undergoes a paraelectric–ferroelectric transition at T_c about 1140 °C, whereby ferroelectric domains are formed. The high-temperature superconductor $YBa_2Cu_3O_{7-\delta}$, usually grown from flux, is subject to a transition at about 750 °C from the tetragonal parent phase into an orthorhombic modification whereby it develops two nearly orthogonal systems of twin lamellae parallel to the two {110} mirror planes of the tetragonal mother phase, lost in the transition [4.162]. These lamellae are ferroelastic and can be changed or even removed (detwinning) by mechanical stress. An example related to mineralogy is provided by quartz, which is stable in its trigonal phase below 575 °C (α-quartz, point group 32) and hexagonal (β-quartz, point group 622) above this temperature. The transition from the hexagonal to the trigonal modification invariably leads to the formation of Dauphiné twinning (due to the loss of the twofold axis in the sixfold axis). The shape and arrangement of twin domains generated by crystal growth or by phase transformation (e.g., for Dauphiné twins) is quite different. In mineralogy these different features of twin textures are helpful for the determination of the conditions of mineral formation.

Ferroelastic switching: The twin domains of ferroelastic crystals switch by mechanical stress from one orientation state into the other. This occurs quite easily at elevated temperatures where the coercitive stress is strongly reduced. Thus, in a ferroelastic crystal grown without twinning, twin domains may be introduced by stress developed during cooling to room temperature. Here again inclusions forming stress centers are the main reason. The ferroelastic twin pattern is often correlated to the growth defects of the crystal, e.g., with growth striations [4.163, Fig. 100], [4.6, p. 379]. Postgrowth formation of twins by ferroelastic switching has been observed in ammonium sulfate $(NH_4)_2SO_4$ [4.151] and Rochelle salt [4.163, p. 184], both grown from aqueous solution.

4.6 Perfection of Crystals Grown Rapidly from Solution

Until the 1980s the opinion prevailed among crystal growers that highly perfect crystals could only be obtained from solution by very slow growth. This has been disproved by the pioneering work of *Zaitseva*, *Smolsky*, and *Rashkovich* [4.34, 57], who showed that, by well-devised construction of the growth apparatus and careful pretreatment and strong stirring of the solution, large KDP crystals of high perfection can be grown from aqueous solution. Their work was the basis of the growth of huge, highly perfect KDP and DKDP crystals of up to 60 cm edge lengths under high supersaturation within a time period of only a few weeks (e.g., [4.17, 35, 36, 164]). The rapid growth of KDP and ADP is described in detail in Chap. 22 of this Handbook. In the present section the growth defects of a slowly grown KDP and a rapidly grown ADP crystal, revealed by x-ray topography, are compared. In addition, the rapid growth of potassium alum, mixed (K,NH_4)-alum, and sodium chlorate [4.165] is reported.

In recent years, various studies of the perfection of rapidly grown KDP crystals have been published, e.g., [4.17,37,48,58]. The most striking difference between slowly and rapidly grown KDP crystals appears in their morphology: crystal grown slowly develops the shape of an *obelisk*, because under low supersaturation, the {100} prism faces are inhibited by impurities so that the crystals grow nearly exclusively on {101} pyramid faces. In order to obtain bulky crystals, large (001) seed plates are used, which lead to a large capping zone (Figs. 4.2, 4.21a, and 4.23). In highly supersaturated and strongly stirred solutions, however, the {110} prism and {101} pyramid faces have essentially the same growth velocity. This allows growth to be started with small *point seeds* with a correspondingly small capping zone. The proper shape and mounting of the point seed and its impact on the habit of the grown crystal has been studied by *Zaitseva* et al. [4.112]. The typical morphology of slowly and rapidly grown KDP and ADP crystals, the development of their growth sectors, and the arrangement of dislocations is shown in Fig. 4.32.

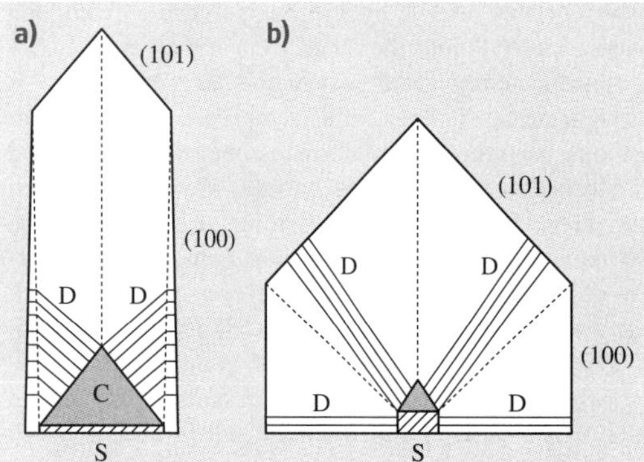

Fig. 4.32a,b Typical morphology and dislocation distribution of KDP and ADP crystal **(a)** slowly grown on a (001) seed plate S and **(b)** rapidly grown on a *point seed* S. D: dislocation lines, C: strongly disturbed regeneration (*capping*) zone (Figs. 4.21 and 4.23). In **(b)** this zone is small. *Dashed lines*: growth-sector boundaries. Note that here another (equivalent) labeling of faces is chosen than in Figs. 4.5, 4.21, and 4.23: prism {100}, dipyramid {101}

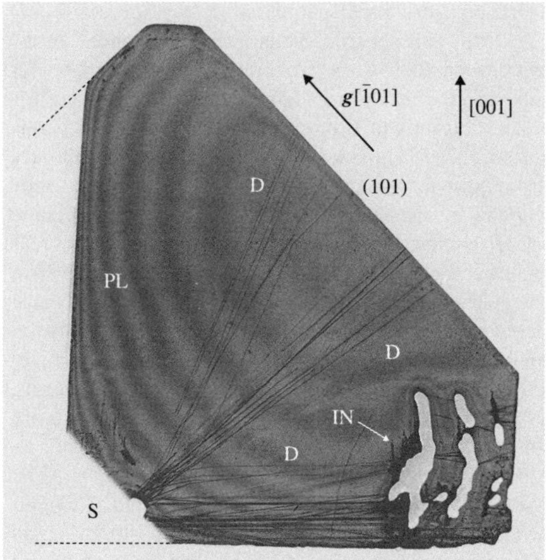

Fig. 4.33 X-ray topograph (MoK_α radiation, reflection $\bar{1}$01) of a (010) plate (about 2.5 mm thick, horizontal extension about 60 mm, tetragonal axis vertical) cut from an ADP crystal rapidly grown within 10 days (including the seeding-in process and regeneration period) to a size with ⟨100⟩ edge lengths of about 110 mm. S: Location of the seed. The crystal has grown with about the same growth rates (about 15 mm/day in the end phase) on all {100} and {101} faces. Since the plate was too large for the x-ray topography crystal holder, a part was cut off at the left side. The crystal is highly perfect, as indicated by the x-ray *Pendellösung* fringes (rounded soft contrast bands PL) of constant plate thickness. Only a few dislocations D originate from a liquid inclusion at the point seed (which was out of the x-ray beam). IN: Large liquid inclusions which have formed on the right (100) face due to growth-face instabilities intentionally provoked by decreasing the temperature by 4 °C in one step at 50 °C after about 40 mm of perfect growth. Note that these inclusions have formed on only one of the four prism faces: the four {101} faces stayed also inclusion free. Reflection **g**($\bar{1}$01), MoK_α. After [4.165]

The typical growth defects and their distribution in slowly and rapidly grown crystals are apparent from the x-ray topographs shown in Figs. 4.5, 4.21a, 4.23, and 4.33. The crystals shown in the former three figures were grown in the 1970s in the laboratory of *Bennema* (Delft and Nijmegen, The Netherlands) in a three-vessel growth system [4.26] at constant temperature of about 30 °C and supersaturation of about 5.4% with growth rate of about 1 mm/day on the {101} faces. The crystals were mounted on a tree which rotated in the solution. They contain inclusions, inhomogeneous impurity distributions (in the form of striations, growth, and vicinal sectors), and dislocations. As pointed out already by *Zerfoss* [4.12], *Janssen-van Rosmalen* et al. [4.26, 32], and *van Enckevort* et al. [4.33], liquid inclusions, in particular the quasiperiodic liquid inclusions shown in Fig. 4.5, can be avoided by strong stirring.

An x-ray topograph of an ADP crystal rapidly grown in the author's laboratory is shown in Fig. 4.33 [4.165]. The crystal was grown on a point seed in a sealed 10 l tank by temperature decrease from 55 to about 45 °C with about the same growth rates on {100} prism and {101} pyramid faces in 10 days to a width of 110 mm along the ⟨100⟩ edges and a height of about 80 mm. It was rotated in the solution on a platform at about 50 cycles/min with reversal of the rotation every 30 s. The rate of temperature decrease was 0.1–1 °C in the generation period and was then increased stepwise to 4 °C/day, dependent on the size of the crystal. It is obvious from Fig. 4.33 that this crystal is much more homogeneous and perfect than the crystals shown in Figs. 4.5, 4.21a, and 4.23. Only a few dislocations originate from the point seed into the prism and pyramid growth sectors. Due to growth face instabilities, large liquid inclusions have formed on the right prism face, but have healed out perfectly during further growth. Interestingly these inclusions block more growth dislocations than they generate. Convincing proof of the homogeneity of this crystal is the appearance of x-ray *Pendellösung fringes*, which result from the slight thickness decrease toward the edges of the plate. Neighboring (010) crystal plates, not cutting though the seed and through the inclusions, are dislocation free.

Rapid crystal growth of very large, highly perfect crystals from aqueous solution has been reported only for KDP, its deuterated variant DKDP, and for ADP [4.36]. In order to check the applicability of this method to other materials, growth experiments have been carried out with some other compounds usually used in student laboratory courses: pure and mixed potassium and ammonium alums, sodium chlorate, and nickel sulfate hexahydrate [4.165]. The growth procedure was essentially the same as reported above for ADP. Apart from some failures due to spontaneous nucleation, large crystals of these materials with edge lengths up to 12 cm could be successfully grown within 7–10 days with growth rates up to 25 mm/day. Optical inspection and x-ray topographic studies indicate that the optical and structural perfection is high and not at all inferior to the quality of corresponding traditionally grown crystals. These results indicate that the rapid growth method is applicable to the growth of many crystals from solution.

References

4.1 T. Scheffen-Lauenroth, H. Klapper, R.A. Becker: Growth and perfection of organic crystals from undercooled melt, J. Cryst. Growth **55**, 557–570 (1981)

4.2 A.R. Lang: Techniques and interpretation in x-ray topography. In: *Diffraction and Imaging Techniques in Materials Science*, 2nd edn., ed. by S. Amelinckx, R. Gevers, J. Van Landuyt (North-Holland, Amsterdam 1978) pp. 623–714

4.3 A. Authier: X-ray and neutron topography of solution-grown crystals. In: *Crystal Growth and Materials (ECCG-1 Zürich)*, ed. by E. Kaldis, H.J. Scheel (North-Holland, Amsterdam 1976) pp. 516–548

4.4 B.K. Tanner: *X-ray Diffraction Topography* (Pergamon, Oxford 1976)

4.5 A.R. Lang: Topography. In: *Internat. Tables for Crystallography, International Union of Crystallography, Vol. C*, ed. by A.J.C. Wilson (Kluwer Academic, Dordrecht 1995) pp. 113–123

4.6 H. Klapper: X-ray topography of twinned crystals. In: *Progress in Crystal Growth and Characterization*, Vol. 14, ed. by P. Krishna (Pergamon, Oxford 1987) pp. 367–401

4.7 H. Klapper: X-ray topography of organic crystals. In: *Crystals: Growth, Properties and Characterization*, Vol. 13, ed. by N. Karl (Springer, Berlin, Heidelberg 1991) pp. 109–162

4.8 H. Klapper: Defects in non-metal crystals. In: *Characterization of Crystal Growth Defects by X-ray Methods*, ed. by B.K. Tanner, D.K. Bowen (Plenum, New York 1980) pp. 133–160

4.9 H. Klapper: X-ray diffraction topography: Application to crystal growth and plastic deformation. In: *X-Ray and Neutron Dynamical Diffraction: Theory and Applications*, Proc. NATO ASI, Erice 1996, NATO Science Series B, Physics Vol. 357, ed. by A. Authier, S. Logomarsino, B.K. Tanner (Plenum Press, New York 1996) p. 167–177

4.10 H. Klapper: Structural defects and methods of their detection. In: *Materials Science Forum*, Vol. 276–277, ed. by R. Fornari, C. Paorici (Trans Tech, Switzerland 1998) pp. 291–306
4.11 H. Klapper: Generation and propagation of dislocations during crystal growth, Mater. Chem. Phys. **66**, 101–109 (2000)
4.12 S. Zerfoss, S.I. Slawson: Origin of authigenic inclusions in synthetic crystals, Am. Mineral. **41**, 598–607 (1956)
4.13 G. Laemmlein: Sekundäre Flüssigkeitseinschlüsse in Mineralien, Z. Kristallogr. **71**, 237–256 (1929), in German
4.14 A.R.J. de Kock: Effect of growth conditions on semiconductor crystal quality. In: *Crystal Growth and Materials (ECCG-1 Zürich)*, ed. by E. Kaldis, H.J. Scheel (North Holland, Amsterdam 1976) pp. 661–703
4.15 A.A. Chernov, D.E. Temkin: Capture of inclusions in crystal growth. In: *1976 Crystal Growth and Materials (ECCG-1 Zürich)*, ed. by E. Kaldis, H.J. Scheel (North Holland, Amsterdam 1977) pp. 4–77, esp. 53–54
4.16 V.Y. Khaimov-Mal'kov: (a) The thermodynamics of crystallisation pressure; (b) Experimental measurement of crystallization pressure; (c) The growth conditions of crystals in contact with large obstacles. In: *Growth of Crystals*, Vol. 2, ed. A.V. Shubnikov, N.N. Sheftal (Consultants Bureau Inc., New York 1959) pp. 3–13 (a), 14–19 (b), 20–28 (c)
4.17 N. Zaitseva, J. Atherton, R. Rozsa, L. Carman, I. Smolsky, M. Runkel, R. Ryon, L. James: Design and benefits of continuous filtration in rapid growth of large KDP and DKDP crystals, J. Cryst. Growth **197**, 911–920 (1999)
4.18 M.O. Kliia, I.G. Sokolova: The absorption of droplets of emulsion by a growing crystal during crystallization from solutions, Sov. Phys. Crystallogr. **3**, 217–221 (1958)
4.19 G. Wulff: Zur Frage der Geschwindigkeit des Wachstums und der Auflösung der Krystallflächen, Z. Kristallogr. **34**, 449–530 (1901), esp. 512–530, in German
4.20 R.F. Strickland-Constable: *Kinetics and Mechanisms of Crystallisation* (Academic, London, New York 1968) pp. 76–84
4.21 P. Bennema: Generalized Herring treatment of the equilibrium form. In: *Crystal growth: An introduction*, North-Holland Series in Crystal Growth I, ed. by P. Hartman (North-Holland, Amsterdam 1973), pp. 342–357
4.22 C. Herring: Some theorems on the free energies of crystal surfaces, Phys. Rev. **82**, 87–93 (1951)
4.23 C. Herring: The use of classical macroscopic concepts in surface energy problems. In: *Structure and Properties of Solid Surfaces*, ed. by R.G. Gromer, C.S. Smith (University of Chicago Press, Chicago 1953) pp. 5–72
4.24 W. Schnoor: Über das Wachstum von Auflösungskörpern und Kugeln aus Steinsalz, Z. Kristallogr. **68**, 1–14 (1928), in German
4.25 H.E. Buckley: *Crystal Growth* (Wiley, London, New York 1961)
4.26 R. Janssen-van Rosmalen, W.H. van der Linden, E. Dobinga, D. Visser: The influence of the hydrodynamic environment on the growth and the formation of liquid inclusions in large potassium hydrogen phosphate crystals, Krist. Tech. **13**, 17–28 (1978)
4.27 A. Faber: *Röntgentopographische Untersuchungen von Wachstumsstörungen durch alternierende Temperaturgradienten im Kali-Alaun. Studienarbeit* (Inst. f. Kristallographie, RWTH Aachen 1980), in German
4.28 W.M. Vetter, H. Totsuka, M. Dudley, B. Kahr: The perfection and defect structures of organic hourglass inclusion K_2SO_4 crystals, J. Cryst. Growth **241**, 498–506 (2002)
4.29 B. Kahr, R.W. Guerney: Dyeing crystals, Chem. Rev. **101**, 893–953 (2001)
4.30 B. Kahr, L. Vasquez: Painting crystals, Cryst. Eng. Commun. **4**, 514–516 (2002)
4.31 A.A. Chernov, G.Y. Kuznetsov, I.L. Smol'skii, V.N. Rozhanski: Hydrodynamic effects during ADP growth from aqueous solutions in the kinetic regime, Sov. Phys. Crystallogr. **31**, 705–709 (1986)
4.32 R. Janssen-van Rosmalen, P. Bennema: The role of hydrodynamics and supersaturation in the formation of fluid inclusions in KDP, J. Cryst. Growth **42**, 224–227 (1977)
4.33 W.J.P. van Enckevort, R. Janssen-van Rosmalen, H. Klapper, W.H. van der Linden: Growth phenomena of KDP crystals in relation to the internal structure, J. Cryst. Growth **60**, 67–78 (1982)
4.34 N.P. Zaitseva, I.L. Smolsky, L.N. Rashkovich: Study of rapid growth of KDP crystals by temperature lowering, Sov. Phys. Crystallogr. **36**, 113–115 (1991)
4.35 N.P. Zaitseva, J.J. De Yoreo, M.R. Dehaven, R.L. Vital, K.E. Montgomery, M. Richardson, L.J. Atherton: Rapid growth of large-scale (40–55 cm) KH_2PO_4 crystals, J. Cryst. Growth **180**, 255–262 (1997)
4.36 N. Zaitseva, L. Carman: Rapid Growth of KDP-type Crystals, Progr. Cryst. Growth Charact. Mater. **43**, 1–118 (2001)
4.37 I. Smolsky, J.J. de Yoreo, N.P. Zaitseva, J.D. Lee, T.A. Land, E.B. Rudneva: Oriented liquid inclusions in KDP crystals, J. Cryst. Growth **169**, 741–745 (1996)
4.38 E. Scandale, A. Zarka: Sur l'origine des canaux dans les cristaux, J. Appl. Cryst. **15**, 417–422 (1982), in French
4.39 X.R. Huang, M. Dudley, W.M. Vetter, W. Huang, S. Wang, C.H. Carter Jr.: Direct evidence of micropipe-related pure superscrew dislocations in SiC, Appl. Phys. Lett. **74**, 353–355 (1999)

4.40 J. Heindl, H.P. Strunk, V.D. Heydemann, G. Pensl: Micropipes: Hollow tubes in silicon carbide, Phys. Status Solidi (a) **162**, 251–262 (1997)

4.41 H.P. Strunk, W. Dorsch, J. Heindl: The nature of micropipes in 6H-SiC single crystals, Adv. Eng. Mater. **2**, 386–389 (2000)

4.42 Th. Scheffen-Lauenroth: *Czochralski-Züchtung und Perfektion organischer Kristalle*. Diplomarbeit (Inst. f. Kristallographie, RWTH Aachen 1983), in German

4.43 E. Roedder: Fluid inclusions. In: *Reviews in Mineralogy*, Vol. 12, ed. by P.H. Ribbe (Mineralogical Society of America, BookCrafters, Inc., Chelsea 1984)

4.44 W. Bardsley, D.T.J. Hurle, M. Hart, A.R. Lang: Structural and chemical inhomogeneities in germanium single crystals grown under conditions of constitutional supercooling, J. Cryst. Growth **49**, 612–690 (1980)

4.45 J.E. Gegusin, A.S. Dziyuba: Gas evolution and the capture of gas bubbles at the crystallization front when growing crystals from the melt, Sov. Phys. Crystallogr. **22**, 197–199 (1977)

4.46 M. Göbbels: *Züchtung organischer Molekülkristalle aus entgasten unterkühlten Schmelzen*. Studienarbeit (Inst. f. Kristallographie, RWTH Aachen), in German

4.47 G. Engel, H. Klapper, P. Krempl, H. Mang: Growth-twinning in quartz-homeotypic gallium orthophosphate crystals, J. Cryst. Growth **94**, 597–606 (1989)

4.48 I.L. Smolsky, A.E. Voloshin, N.P. Zaitseva, E.B. Rudneva, H. Klapper: X-ray topographic study of striation formation in layer growth of crystals from solution, Philos. Trans. Math. Phys. Eng. Sci. **357**, 2631–2649 (1999)

4.49 T. Nishinaga, P. Ge, C. Huo, J. He, T. Nakamura: Melt growth of striation and etch-pit free GaSb under microgravity, J. Cryst. Growth **174**, 96–100 (1997)

4.50 P. Dold: Czochralski growth of doped germanium with an applied rotating magnetic field, Cryst. Res. Technol. **38**, 659–668 (2003)

4.51 P. Rudolph: Travelling magnetic fields applied to bulk crystal growth from the melt: The step from basic research to industrial scale, J. Cryst. Growth **310**, 1298–1306 (2008)

4.52 H. Scheel: Theoretical and experimental solutions of the striation problem. In: *Crystal Growth Technology*, ed. by H.J. Scheel, T. Fukuda (Wiley, New York 2003), Chap. 4

4.53 N. Herres, A.R. Lang: X-ray topography of natural beryl using synchroton and conventional sources, J. Appl. Cryst. **16**, 47–56 (1983)

4.54 H. Klapper: Röntgentopographische Untersuchungen von Gitterstörungen in Benzil-Einkristallen, J. Cryst. Growth **10**, 13–25 (1971), in German

4.55 K. Maeda, A. Sonoda, H. Miki, Y. Asakuma, K. Fukui: Synergy of organic dyes for KDP crystal growth, Cryst. Res. Technol. **39**, 1006–1013 (2004)

4.56 I.L. Smol'skii, A.A. Chernov, G.Y. Kutznetsov, V.F. Parvov, V.N. Rozhanskii: Vicinal sectoriality in growth sectors of {011} faces of ADP crystals, Sov. Phys. Crystallogr. **30**, 563–567 (1985)

4.57 I.L.. Smol'skii, N.P. Zaitseva: Characteristic defects and imperfections in KDP crystals grown at high rates. In: *Growth of Crystals*, Vol. 19, ed. by E.I. Givargizov, S.A. Grinberg (Plenum, New York 1995) pp. 173–185

4.58 J.J. De Yoreo, T.A. Land, L.N. Rashkovich, T.A. Onischenko, J.D. Lee, O.V. Monovskii, N.P. Zaitseva: The effect of dislocation cores on growth hillock vicinality and normal growth rates of KDP {101} surfaces, J. Cryst. Growth **182**, 442–460 (1997)

4.59 A.G. Shtukenberg, Y.O. Punin, E. Haegele, H. Klapper: On the origin of inhomogeneity of anomalous birefringence in mixed crystals: An example of alums, Phys. Chem. Miner. **28**, 665–674 (2001)

4.60 H. Klapper, R.A. Becker, D. Schmiemann, A. Faber: Growth-sector boundaries and growth-rate dispersion in potassium alum crystals, Cryst. Res. Technol. **37**, 747–757 (2002)

4.61 W.J.P. Van Enckevort, H. Klapper: Observation of growth steps with full and half unit-cell heights on the {001} faces of $NiSO_4 \cdot 6H_2O$ in relation to the defect structure, J. Cryst. Growth **80**, 91–103 (1987)

4.62 H. Kanda, M. Akaishi, S. Yamaoka: Impurity distribution among vicinal slopes of growth spirals developing on the {111} faces of synthetic diamonds, J. Cryst. Growth **108**, 421–424 (1991)

4.63 J.J. De Yoreo, Z.U. Rek, N.P. Zaitseva, B.W. Woods: Sources of optical distortion in rapidly grown crystals of KH_2PO_4, J. Cryst. Growth **166**, 291–297 (1996)

4.64 K. Fujioka, S. Matsuo, T. Kanabe, H. Fujita, M. Nakjatsuka: Optical properties of rapidly grown KDP crystals improved by thermal conditioning, J. Cryst. Growth **181**, 265–271 (1997)

4.65 N. Zaitseva, L. Carman, I. Smolsky, R. Torres, M. Yan: The effect of impurities and supersaturation on the rapid growth of KDP crystals, J. Cryst. Growth **204**, 512–524 (1999)

4.66 T. Bullard, M. Kurimoto, S. Avagyan, S.H. Jang, B. Kahr: Luminescence imaging of growth hillocks in potassium hydrogen phthalate, ACA Transaction **39**, 62–72 (2004)

4.67 A.R. Lang, V.F. Miuskov: Dislocations and fault surfaces in synthetic quartz, J. Appl. Phys. **38**, 2477–2483 (1967), esp. p. 2482

4.68 A.R. Lang, V.F. Miuskov: Defects in natural and synthetic quartz. In: *Growth of Crystals*, Vol. 7, ed. by N.N. Sheftal (Consultants Bureau, New York 1969) pp. 112–123, esp. p. 122

4.69 T. Hahn, H. Klapper: Twinning of crystals. In: *International Tables for Crystallography*, Vol. D (Kluwer Academic, Dordrecht 2003) pp. 393–448

4.70 W. Schmidt, R. Weiss: Dislocation propagation in Czochralski grown gadolinium gallium garnet, J. Cryst. Growth **43**, 515–525 (1978)

4.71 B. Cockayne, J.M. Roslington, A.W. Vere: Microscopic strain in facetted regions of garnet crystals, J. Mater. Sci. **8**, 382–384 (1973)

4.72 W.T. Stacy: Dislocations, facet regions and grown striations in garnet substrates and layers, J. Cryst. Growth **24/25**, 137–143 (1974)

4.73 A. Shtukenberg, Y. Punin, B. Kahr: Optically anomalous crystals. In: *Springer Series in Solid State Science* (Springer, Berlin, Heidelberg 2007)

4.74 R. von Brauns: *Die optischen Anomalien der Krystalle* (S. Hirzel, Leipzig 1891), in German

4.75 B. Kahr, J.M. McBride: Optically anomalous crystals, Angew. Chem. Int. Ed. **31**, 1–26 (1992)

4.76 H. Klapper: Reconstruction of the growth history of crystals by analysis of growth defects. In: *Crystal Growth of Technologically Important Electronic Materials*, ed. by K. Byrappa, T. Ohachi, H. Klapper, R. Fornari (Allied Publishers PVT, New Delhi 2003)

4.77 J.N. Sherwood, T. Shiripathi: Evidence for the role of pure edge dislocations in crystal growth, J. Cryst. Growth **88**, 358–364 (1988)

4.78 H.L. Bhat, R.I. Ristic, J.N. Sherwood, T. Shiripathi: Dislocation characterization in crystal of potash alum grown by seeded solution growth und conditions of low supersaturation, J. Cryst. Growth **121**, 709–716 (1992)

4.79 R.I. Ristic, B. Shekunov, J.N. Sherwood: Long and short period growth rate variations in potash alum, J. Cryst. Growth **160**, 330–336 (1996)

4.80 E. Billig: Some defects in crystals grown from the melt I: Defects caused by thermal stresses, Proc. R. Soc. Lond. A **235**, 37–55 (1956)

4.81 V.L. Indenbom: Ein Beitrag zur Entstehung von Spannungen und Versetzungen beim Kristallwachstum, Kristall und Technik **14**, 493–507 (1979), in German

4.82 P. Möck: Comparison of experiments and theories for plastic deformation in thermally processed GaAs wafers, Cryst. Res. Technol. **35**, 529–540 (2000)

4.83 P. Rudolph: Dislocation cell structures in melt-grown semiconductor compound crystals, Cryst. Res. Technol. **40**, 7–20 (2005)

4.84 Y.M. Fishman: X-ray topographic study of the dislocations produced in potassium dihydrogen phosphate crystals by growth from solution, Sov. Phys. Crystallogr. **17**, 524–527 (1972)

4.85 G. Dhanaraj, M. Dudley, D. Bliss, M. Callahan, M. Harris: Growth and process induced dislocations in zinc oxide crystals, J. Cryst. Growth **297**, 74–79 (2006)

4.86 G. Neuroth: Der Einfluß von Einschlußbildung und mechanischer Verletzung auf das Wachstum und die Perfektion von Kristallen. Ph.D. Thesis (University of Bonn, Bonn 1996), (Shaker, Aachen 1996), in German

4.87 A.J. Forty: Direct observation of dislocations in crystals, Adv. Phys. **3**, 1–25 (1954)

4.88 G.G. Lemmlein, E.D. Dukova: Formation of screw dislocations in the growth process of a crystal, Sov. Phys. Crystallogr. **1**, 269–273 (1956)

4.89 M.I. Kozlovskii: Formation of screw dislocations in the growth of a crystal around solid particles, Sov. Phys. Crystallogr. **3**, 205–211 (1958/60)

4.90 M.I. Kozlovskii: Formation of screw dislocations at the junction of two layers spreading over the surface of a crystal, Sov. Phys. Crystallogr. **3**, 236–238 (1958/60)

4.91 M. Dudley, X.R. Huang, W. Huang, A. Powell, S. Wang, P. Neudeck, M. Skowronski: The mechanism of micropipe nucleation at inclusions in silicon carbide, Appl. Phys. Lett. **75**, 784–786 (1999)

4.92 W.T. Read: *Dislocations in Crystals* (McGraw-Hill, New York 1953) p. 47

4.93 D. Hull: *Introduction to Dislocations*, Vol. 2 (Pergamon, Oxford 1975) pp. 229–235

4.94 J.P. Hirth, J. Lothe: *Theory of Dislocations* (McGraw-Hill, New York 1968)

4.95 J. Weertmann, J.R. Weertmann: *Elementary Dislocation Theory* (Macmillan, New York 1964) p. 137

4.96 H. Klapper: Vorzugsrichtungen eingewachsener Versetzungen in lösungsgezüchteten Kristallen. Habilitation Thesis, (Technical University (RWTH) Aachen 1975), in German

4.97 J. Lothe: Force on dislocations emerging at free surfaces, Phys. Nor. **2**, 154–157 (1967)

4.98 J.B. Eshelby, W.T. Read, W. Shockley: Anisotropic elasticity with applications to dislocation theory, Acta Metall. **1**, 251–259 (1953)

4.99 H. Klapper, Y.M. Fishman, V.G. Lutsau: Elastic energy and line directions of grown-in dislocations in KDP crystals, Phys. Status Solidi (a) **21**, 115–121 (1974)

4.100 H. Klapper: Elastische Energie und Vorzugsrichtungen geradliniger Versetzungen in aus der Lösung gewachsenen organischen Kristallen. I. Benzil, Phys. Status Solidi (a) **14**, 99–106 (1972), in German

4.101 H. Klapper: Elastische Energie und Vorzugsrichtungen geradliniger Versetzungen in aus der Lösung gewachsenen organischen Kristallen. II. Thioharnstoff, Phys. Status Solidi (a) **14**, 443–451 (1972), in German

4.102 H. Klapper: Röntgentopographische Untersuchungen am Lithiumformiat-Monohydrat, Z. Naturforsch. **28a**, 614–622 (1973), in German

4.103 H. Klapper, H. Küppers: Directions of dislocation lines in crystals of ammonium hydrogen oxalate hemihydrate grown from solution, Acta Cryst. A **29**, 495–503 (1973), (correction: read $K/\cos\alpha$ instead of $K\cos\alpha$)

4.104 D.F. Croxall, R.C.C. Ward, C.A. Wallace, R.C. Kell: Hydrothermal growth and investigation of Li-doped zinc oxide crystals of high purity and perfection, J. Cryst. Growth **22**, 117–124 (1974)

4.105 U. Alter, G. Voigt: Direction change of dislocations on passing a growth-sector boundary in quartz crystals, Cryst. Res. Technol. **19**, 1619–1623 (1984)

4.106 A. Sakai, H. Sunakawa, A. Usui: Defect structure in selectively grown GaN films with low threading dislocation density, Appl. Phys. Lett. **71**, 2259–2261 (1997)

4.107 Z. Liliental-Weber, M. Benamara, W. Snider, J. Washburn, J. Park, P.A. Grudowski, C.J. Eiting, R.D. Dupuis: TEM study of defects in laterally overgrown GaN layers, MRS Internet J. Nitride Semicond. Res. **4s1**, 4.6 (1999)

4.108 H. Sunakawa, A. Kimura, A. Usui: Self-organized propagation of dislocations in GaN films during epitaxial lateral overgrowth, Appl. Phys. Lett. **76**, 442–444 (2000)

4.109 P. Venégues, B. Beaumont, V. Bousquet, M. Vaille, P. Gibart: Reduction mechanisms of defect densities in GaN using one- or two-step epitaxial lateral overgrowth methods, J. Appl. Phys. **87**, 4175–4181 (2000)

4.110 S. Gradezcak, P. Stadelman, V. Wagner, M. Ilegems: Bending of dislocations in GaN during epitaxial lateral overgrowth, Appl. Phys. Lett. **85**, 4648–4650 (2004)

4.111 J. Bai, J.-S. Park, Z. Cheng, M. Curtin, B. Adekore, M. Carroll, A. Lochtefeld, M. Dudley: Study of the defect elimination mechanism in aspect ratio trapping Ge growth, Appl. Phys. Lett. **90**, 101902 (2007)

4.112 N. Zaitseva, L. Carman, I. Smolsky: Habit control during rapid growth of KDP and DKDP crystals, J. Cryst. Growth **241**, 363–373 (2002)

4.113 G.R. Ester, P.J. Halfpenny: An investigation of growth-induced defects in crystals of potassium hydrogen phthalate, Philos. Mag. A **79**, 593–608 (1999)

4.114 G.R. Ester, R. Price, P.J. Halfpenny: The relationship between crystal growth and defect structure: A study of potassium hydrogen phthalate using x-ray topography and atomic force microscopy, J. Phys. D: Appl. Phys. **32**, A128–A132 (1999)

4.115 G. Neuroth, H. Klapper: Dislocation reactions in Czochralski-grown salol crystals, Z. Kristallogr. **209**, 216–220 (1994)

4.116 I.L. Smolsky, E.B. Rudneva: Effect of the surface morphology on the grown-in dislocation orientations in KDP crystals, Phys. Status Solidi (a) **141**, 99–107 (1994)

4.117 T. Watanabe, K. Izumi: Growth and perfection of tetraoxane crystals, J. Cryst. Growth **46**, 747–756 (1979)

4.118 K. Izumi: Lattice defects in normal alcohol crystals, Jpn. J. Appl. Phys. **16**, 2103–2108 (1977)

4.119 H. Klapper: Röntgentopographische Untersuchungen der Defektstrukturen im Thioharnstoff, J. Cryst. Growth **15**, 281–287 (1972), in German

4.120 F.C. Frank: The influence of dislocations on crystal growth, Disc. Faraday Soc. **5**, 48–54 (1949), and 66–68

4.121 C.F. Frank: Crystal growth and dislocations, Adv. Phys. **1**, 91–109 (1952)

4.122 W.K. Burton, N. Cabrera, F.C. Frank: The growth of crystals and the equilibrium structure of their surfaces, Philos. Trans. R. Soc. Lond. A **243**, 299–358 (1951), especially Part II, pp. 310–323

4.123 H.P. Strunk: Edge dislocation may cause growth spirals, J. Cryst. Growth **160**, 184–185 (1996)

4.124 N.-B. Ming: Defect mechanism of crystal growth and their kinetics, J. Cryst. Growth **128**, 104–112 (1993)

4.125 E. Bauser, H. Strunk: Analysis of dislocations creating monomolecular growth steps, J. Cryst. Growth **51**, 362–366 (1981)

4.126 F.C. Frank: "Edge" dislocations as crystal growth sources, J. Cryst. Growth **51**, 367–368 (1981)

4.127 L.J. Giling, B. Dam: A "rough heart" model for "edge" dislocations which act as persistent growth sources, J. Cryst. Growth **67**, 400–403 (1984)

4.128 H. Gottschalk, G. Patzer, H. Alexander: Stacking-fault energy and ionicity of cubic III–V compounds, Phys. Status Solidi (a) **45**, 207–217 (1978)

4.129 T.W. Donnelly: Kinetic considerations in the genesis of growth twinning, Am. Mineral. **52**, 1–12 (1967)

4.130 H. Carstens: Kinetic consideration in the genesis of growth twinning: A discussion, Am. Mineral. **53**, 342–344 (1968)

4.131 T.W. Donnelly: Kinetic consideration in the genesis of growth twins: A reply, Am. Mineral. **53**, 344–346 (1968)

4.132 V. Janovec, T. Hahn, H. Klapper: Twinning and domain structures. In: *International Tables for Crystallography*, Vol. D (Kluwer, Dordrecht 2003) pp. 377–378

4.133 V. Janovec, J. Přívratská: Domain structures. In: *International Tables for Crystallography*, Vol. D (Kluwer, Dordrecht 2003) pp. 449–505

4.134 F.D. Bloss: *Crystallography and Crystal Chemistry* (Rinehart & Winston, New York 1971) pp. 324–338

4.135 C. Giacovazzo (Ed.): *Fundamentals of Crystallography* (University Press, Oxford 1992) pp. 80–87, and 133–140

4.136 C. Frondel: Silica minerals. In: *The System of Mineralogy*, Vol. III, 7th edn. (Wiley, New York 1962) pp. 75–99

4.137 J.W. Faust Jr., H.F. John: The growth of semiconductor crystals from solution using the twin-plane reentrant-edge mechanism, J. Phys. Chem. Solids **25**, 1407–1415 (1964)

4.138 R. Jagannathan, R.V. Mehta, J.A. Timmons, D.L. Black: Anisotropic growth of twinned cubic crystals, Phys. Rev. B **48**, 13261–13265 (1993)

4.139 R. Jagannathan, R.V. Mehta, J.A. Timmons, D.L. Black: Reply to comment on anisotropic growth of twinned cubic crystals, Phys. Rev. B **51**, 8655 (1995), following the comment by B.W. van de Waal, Phys. Rev. B **51**, 8653–8654 (1995)

4.140 G. Bögels, T.M. Pot, H. Meekes, P. Bennema, D. Bollen: Side-face structure for lateral growth of tabular silver bromide crystals, Acta Cryst. A **53**, 84–94 (1997)

4.141 G. Bögels, H. Meekes, P. Bennema, D. Bollen: The role of {100} side faces for lateral growth of tabular

silver bromide crystals, J. Cryst. Growth **191**, 446–456 (1998)
4.142 G. Bögels, J.G. Buijnsters, S.A.C. Verhaegen, H. Meekes, P. Bennema, D. Bollen: Morphology and growth mechanism of multiply twinned AgBr and AgCl needle crystals, J. Cryst. Growth **203**, 554–563 (1999)
4.143 C.A. Wallace, E.A.D. White: The morphology and twinning of solution-grown corundum crystals. In: *Crystal Growth*, ed. by H.S. Peiser (Pergamon, Oxford 1967) pp. 431–435, supplement to Phys. Chem. Solids
4.144 M. Senechal: The genesis of growth twins, Sov. Phys. Crystallogr. **25**, 520–524 (1980)
4.145 H. Hofmeister: Forty years study of fivefold twinned structures in small particles and thin films, Cryst. Res. Technol. **33**, 3–25 (1998), especially Sect. 4
4.146 I. Sunagawa, L. Taijing, V.S. Balitsky: Generation of Brazil and Dauphiné twins in synthetic amethyst, Phys. Chem. Miner. **17**, 320–325 (1990)
4.147 A.C. MacLaren, D.R. Pitkethly: The twinning microstructure and growth of amethyst quartz, Phys. Chem. Miner. **8**, 128–135 (1982)
4.148 H. Klapper, T. Hahn, S.J. Chung: Optical, pyroelectric and x-ray topographic studies of twin domains and twin boundaries in $KLiSO_4$, Acta Cryst. B **43**, 147–159 (1987)
4.149 M.J. Buerger: The genesis of twin crystals, Am. Mineral. **30**, 469–482 (1945)
4.150 P. Hartmann: On the morphology of growth twins, Z. Kristallogr. **107**, 225–237 (1956)
4.151 R. Docherty, A. El-Korashi, H.-D. Jennissen, H. Klapper, K.J. Roberts, T. Scheffen-Lauenroth: Synchroton Laue topographic studies of pseudo-hexagonal twinning, J. Appl. Cryst. **21**, 406–415 (1988)
4.152 N.-B. Ming, I. Sunagawa: Twin lamellae as possible self-perpetuating steps sources, J. Cryst. Growth **87**, 13–17 (1988)
4.153 N.-B. Ming, K. Tsukamato, I. Sunagawa, A.A. Chernov: Stacking faults as self-perpetuating step sources, J. Cryst. Growth **91**, 11–19 (1988)
4.154 H. Li, X.-D. Peng, N.-B. Ming: Re-entrant corner mechanism of fcc crystal growth of A-type twin lamella: The Monte-Carlo simulation approach, J. Cryst. Growth **139**, 129–133 (1994)
4.155 H. Li, N.-B. Ming: Growth mechanism and kinetics on re-entrant corner and twin lamellae in a fcc crystal, J. Cryst. Growth **152**, 228–234 (1995)
4.156 R.-W. Lee, U.-J. Chung, N.M. Hsang, D.-Y. Kim: Growth process of the ridge-trough faces of a twinned crystal, Acta Cryst. A **61**, 405–410 (2005)
4.157 R. Boistelle, D. Aquilano: Interaction energies at twin boundaries and effects of the dihedral re-entrant and salient angles on the grown morphology of twinned crystals, Acta Cryst. A **34**, 406–413 (1978)
4.158 I.M. Dawson: The study of crystal growth with the electron microscope II. The observation of growth steps in the paraffin n-hexane, Proc. R. Soc. Lond. A **214**, 72–79 (1952)
4.159 R.S. Wagner: On the growth of Ge dendrites, Acta Metal. **8**, 57–60 (1960)
4.160 D.R. Hamilton, R.G. Seidensticker: Propagation mechanism of germanium dendrites, J. Appl. Phys. **31**, 1165–1168 (1960)
4.161 B. van de Waal: Cross-twinning model of fcc crystal growth, J. Cryst. Growth **158**, 153–165 (1996)
4.162 G. Roth, D. Ewert, G. Heger, M. Hervieu, C. Michel, B. Raveau, B. D'Yvoire, A. Revcolevschi: Phase transformation and microtwinning in crystals of the high-Tc superconductor $YBa_2Cu_3O_{8-x}$, Z. Phys. B **69**, 21–27 (1987)
4.163 I.S. Zheludev: Crystallography and spontaneous polarisation. In: *Physics of Crystalline Dielectrics*, Vol 1 (Plenum Press, New York 1971)
4.164 M. Nakatsuka, K. Fujioka, T. Kanabe, H. Fujita: Rapid growth of over 50 mm/day of water-soluble KDP crystal, J. Cryst. Growth **171**, 531–537 (1997)
4.165 H. Klapper, I.L. Smolsky, A.E. Haegele: Rapid growth from solution. In: *Crystal Growth of Technologically Important Electronic Materials*, ed. by K. Byrappa, T. Ohachi, H. Klapper, R. Fornari (Allied Publishers PVT, New Delhi 2003)

5. Single Crystals Grown Under Unconstrained Conditions

Ichiro Sunagawa

Based on detailed investigations on morphology (evolution and variation in external forms), surface microtopography of crystal faces (spirals and etch figures), internal morphology (growth sectors, growth banding and associated impurity partitioning) and perfection (dislocations and other lattice defects) in single crystals, we can deduce how and by what mechanism the crystal grew and experienced fluctuation in growth parameters through its growth and post-growth history under unconstrained condition. The information is useful not only in finding appropriate way to growing highly perfect and homogeneous single crystals, but also in deciphering letters sent from the depth of the Earth and the Space. It is also useful in discriminating synthetic from natural gemstones. In this chapter, available methods to obtain molecular information are briefly summarized, and actual examples to demonstrate the importance of this type of investigations are selected from both natural minerals (diamond, quartz, hematite, corundum, beryl, phlogopite) and synthetic crystals (SiC, diamond, corundum, beryl).

Part A | 5

The morphology, perfection, and homogeneity of single crystals and the textures of polycrystalline aggregates vary depending on their growth conditions and are determined by the atomic process and mechanism of crystal growth, which occurs exclusively at the solid–liquid interface. Understanding how crystals grow at the atomic level is the key to understanding how and why single crystals can have various morphologies and degrees of perfection and homogeneity depending on their growth parameters. This will provide essential information for both the fundamentals and the applications of crystal growth. Based on such understanding, we may obtain useful hints to develop appropriate methods for the growth of single crystals with desired perfection and homogeneity. We may also decipher how and under what conditions crystals of terrestrial and planetary minerals, as well as biominerals, grew and what sort of changes they experienced, based on this understanding. The purpose of this chapter is to summarize presently available understanding of why and how the morphology, perfection, and homogeneity of single crystals are determined by their growth process. Since crystals growing in dilute multicomponent solution phases or by the chemical vapor transport (CVT) process show much wider variations in these properties than crystals grown in simpler monocomponent melt phase or by the physical vapor transport (PVT) process, actual examples to demonstrate how morphology, perfection, and homogeneity are controlled during growth processes will be selected from the former crystals, grown under unconstrained conditions, such as natural minerals or synthetic crystals whose growth process cannot be traced in situ.

Growth of crystals takes place uniquely at the solid–liquid interface, and the growth mechanism and growth rate versus driving force relations are determined by the structure of this interface, i. e., whether the interface is smooth or rough, and the roughening transition of smooth interfaces depending on the growth parameters. The process and mechanism of crystal growth at the atomic level are recorded in the surface microtopography of crystal faces. A surface microtopograph of as-grown faces provides information on the atomic processes of crystal growth. By examining this surface, the external morphology of crystals, i. e., the evolution of the bulk morphology of three-dimensional (3-D) crystals from polyhedral, to hopper, to dendritic forms (habitus and tracht) is determined. Also the origin of malformed morphologies of two-dimensional (2-D, epitaxial) and one-dimensional (1-D, whisker) crystals may be understood. Similarly, through growth process, various internal morphologies such as growth sectors, intrasectorial sectors, growth banding, imperfections (dislocation distribution), and inhomogeneities (element partitioning) can be determined and recorded as internal morphologies in nearly perfect single crystals. By applying sensitive methods to visualize or measure perfection and heterogeneity in single crystals, we may obtain useful information about how a single crystal grows and how its perfection and homogeneity are determined, and the whole growth history may be analyzed based on this information.

5.1 Background

Single crystals of various compounds have been synthesized in various sizes and forms, ranging from millimeter to meter scale in size, and from 3-D bulk, to 2-D thin film, to 1-D whisker in form, to meet corresponding requirements for scientific or industrial purposes. Bulk single crystals of large sizes are synthesized to obtain slices for semiconductor, optoelectronic, magnetic, telecommunication, and sensor devices, or to obtain synthetic gemstones, whereas millimeter size is sufficient for protein crystals for crystal structure analysis. Recent interest in nanotechnology has expanded the size ranges down to nanometer scale, and form to include hierarchical structures.

Since the size, form, perfection, and homogeneity of single crystals are essential factors for their use, various growth techniques have been designed and developed to obtain single crystals of appropriate size yet with desired form, perfection, and homogeneity, including:

1. Selection of appropriate ambient phase (melt, solution, vapor), i. e., selection of condensed or diluted phase, and of suitable solvent components
2. Homogenization of ambient phase by internal stirring or external rotation, or by applying magnetic or low gravity field
3. Using seed or substrate crystal to suppress unnecessary nucleation and control crystallographic orientation
4. Controlling temperature or concentration to realize a steady-state yet high rate of growth
5. Minimizing dislocation generation by selection and treatment of seed or by changing boule shapes during growth, or by other techniques.

In the synthesis of Si single-crystal boules larger than a few tens of centimeters across and longer than 1 m, techniques to control dislocation densities have been successfully realized to obtain so-called dislocation-free Si crystals. In the syntheses of compound semiconductors, ammonium dihydrogen phosphate $NH_4H_2PO_4$ (ADP), potassium dihydrogen phosphate KH_2PO_4 (KDP) or quartz single crystals, further efforts are still required to reach the same level as in Si crystals for the realization of *dislocation-free* crystals, although large-scale mass-production techniques and techniques to grow meter-size crystals are already well established. Polymer and protein single crystals are still in the state of characterizing dislocations and understanding how they are generated, whereas in nanometer-sized ultrafine crystals, control of morphology and hierarchical structure is the present main interest.

Compared with previous efforts to understand the nature and origin of crystals and control dislocations, less effort has been invested in understanding and controlling the chemical homogeneity of single crystals, i. e., element partitioning in relation to growth kinetics. The Burton–Prim–Slichter (BPS) formula [5.1] describing the relation between element partitioning and growth kinetics, proposed half a century ago, still plays an important role in this respect. This problem is still fresh in terms of the synthesis of perfectly homogeneous single crystals in solution systems more complicated than for the simpler melt growth of Si crystal.

In contrast to synthetic crystals, single crystals of natural minerals grow under given geological environments and unconstrained conditions. Their growth and postgrowth histories vary depending on geological environments and processes. In the growth and postgrowth histories, conditions may fluctuate or abruptly change, which modifies the morphology, perfection, and homogeneity of the crystal. Such variations and fluctuations are recorded within nearly perfect single crystals in the form of macromorphology (external forms), micromorphology (surface microtopography of crystal faces), and internal morphology (growth sectors, intrasectorial sectors, growth banding, the spatial distribution of lattice defects, and associated element partitioning). The morphological features, perfection, and homogeneity of single crystals of natural minerals, both terrestrial and extraterrestrial in origin, therefore provide the key to decipher how mineral crystals grew [5.2,3], as their perfection and homogeneity were determined during growth in ambient phases with complicated compositions under unconstrained conditions, provided that these can be visualized and characterized at the nm to μm scale. We have various sophisticated methods to characterize perfection and homogeneity within a nearly perfect single crystal. Nanometer-level information relating to growth mechanisms can be secured through surface microtopographic observations of crystal faces by means of various sensitive optical microscopy and interferometry techniques such as phase-contrast microscopy (PCM), phase-shifting microscopy (PSM), differential interference contrast microscopy (DICM), laser-beam scanning (LBS) microscopy, phase-shifting interferometry (PSI), and multiple-beam interferometry (MBI), as well as atomic force microscopy (AFM) and scanning tunneling microscopy (STM) [5.3, 4]. X-ray topography (XRT) is a powerful method to visualize the three-dimensional distribution of line and planar lattice defects, and micro-area x-ray fluorescence (MXRF), cathode-ray luminescence (CL), and laser-beam tomography (LBT) are useful characterization techniques to visualize the distribution of point defects and impurities in nearly perfect single crystals [5.3].

Information obtained through characterization of single crystals grown under unconstrained conditions provides useful background knowledge for the design of new techniques to control the perfection and homogeneity of synthetic single crystals, and also to distinguishing natural from synthetic crystals, since these crystals provide the full range of information relating to how imperfections and heterogeneities are generated and induced in single crystals during growth and their postgrowth history. To point out just one example demonstrating the importance of knowledge about the growth process of natural minerals in developing a new method for industrial purposes, we may mention the concept of *epitaxy*. Growth of guest phase on host phase with a definite crystallographic relation is known from many examples of natural minerals, and later this relation was used to grow thin films of single crystals to prepare electronic devices, by either vapor- or liquid-phase growth.

Since the perfection and homogeneity of single crystals are both related to the mechanism, kinetics, and process of growth, which are reflected in the morphological features of the crystals, it is essential to understand how and why morphological variations occur on growing crystals. To this end, single crystals grown from solution phase and under unconstrained conditions provide more useful information than those formed in simpler melt or vapor phases (physical vapor transport, PVT, or physical vapor deposition, PVD) under constrained conditions. It is the purpose of this chapter to summarize the presently available knowledge on the morphology, perfection, and homogeneity of single crystals of mainly natural minerals, but also synthetic crystals whose growth process cannot be traced in situ, such as synthetic crystals grown in high-temperature solution, high-pressure high-temperature solution, hydrothermal solution, and by chemical vapor transport (CVT) method. We pay most attention to single crystals grown from the solution phase and by CVT method, since they represent crystals grown in multicomponent and more complicated systems, and show all phenomena relating to crystal growth. Although large single crystals can be synthesized by methods using condensed melt phase, e.g., Czochralski, Verneuil, and Bridgman–Stockbarger methods, the solid–liquid interfaces in these synthesis are generally rougher than in the case of diluted solution or vapor growth, and thus morphological changes are less enhanced. In diluted solution and vapor growth, the solid–liquid interface is generally smoother than in condensed melt growth, leading to more variable morphological variations [5.3]. As compared with melt and PVT growth, solute–solvent interaction is involved in solution and CVT growth, thus providing additional information. We therefore purposely select crystals grown from solution and CVT, both natural and synthetic, to investigate the associated problems, although they are more complicated than those grown from pure melt or by PVT. Most crystal growth in nature, including crystallization in magma, pegmatite, and hydrothermal veins, as well as in regional and contact metamorphic rocks, can be regard as solution or CVT growth [5.2, 3, 5]. There is no pure melt or PVT growth in natural crystallization.

5.2 Smooth and Rough Interfaces: Growth Mechanism and Morphology

Crystal growth takes place uniquely at solid–liquid interfaces. Depending on the atomic structure of the solid–liquid interface, the growth mechanisms and normal growth rate R versus driving force $\Delta\mu/(k_B T)$ relations are different, resulting in different morphology. The larger the R, the earlier the face disappears from the growing crystal. To nucleate and grow a crystal, a driving force must overcome the respective energy barrier. Throughout this chapter, the driving force is expressed in a general form by the chemical potential difference between solid and liquid phases $\Delta\mu/(k_B T)$. The phase with the larger chemical potential will have to diminish while the phase with the smaller chemical potential will grow, until equilibrium is reached and $\Delta\mu = 0$. If we consider the growth of a solid phase from a vapor or solution at temperature T_B and solute concentration C, the chemical potential difference is obtained by $\Delta\mu = k_B T_B \ln S$, where $S = C/C_\infty$ is the supersaturation ratio defined with respect to the equilibrium concentration C_∞.

It has been well established [5.3] that:

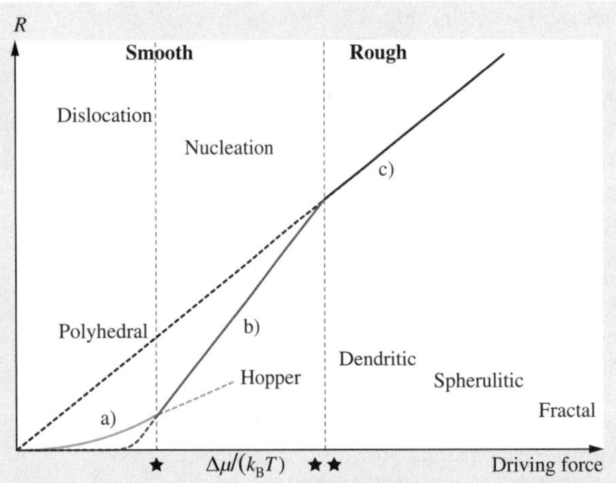

Fig. 5.1 Schematic illustration of the morphological evolution of a crystal on a normal growth rate R versus driving force $\Delta\mu/(k_B T)$ diagram. *Curve* (a) corresponds to spiral growth mechanism, *curve* (b) to 2DNG, and *curve* (c) to adhesive-type growth mechanism. Two critical driving forces $\Delta\mu/(k_B T^*)$ and $\Delta\mu/(k_B T^{**})$ are shown for one crystallographic face, and the regions where polyhedral, skeletal, dendritic, spherulitic, and fractal morphologies are expected are indicated (after [5.3, 5])

1. An adhesive-type growth mechanism operates on an atomically rough interface, whereas either two-dimensional nucleation growth (2-DNG) or a spiral growth mechanism operates on an atomically smooth interface.
2. Relations between normal growth rates R and driving forces $\Delta\mu/(k_B T)$ are different depending on the growth mechanism.
3. A smooth interface may transform to a rough interface with increasing growth temperature or driving force.

Taking all these factors into consideration, we may expect that the morphology of a single crystal will change with increasing driving force $\Delta\mu/(k_B T)$ (which can be correlated to the supersaturation σ or supercooling ΔT) from polyhedral, via hopper (skeletal), to dendritic forms (all being morphologies of single crystals). On further increases of the driving force, semispherulite to spherulite or more complicated morphologies of polycrystalline aggregate are expected to appear, as the nucleation rate increases. Two critical driving force conditions, $\Delta\mu/(k_B T^*)$ and $\Delta\mu/(k_B T^{**})$ are present with increasing driving force, judging from the growth rate versus driving force relations expected for different growth mechanisms [5.2, 3, 5]. Under driving force conditions above $\Delta\mu/(k_B T^{**})$, the interface is expected to be rough, while below this the interface will be smooth. Above $\Delta\mu/(k_B T^{**})$, the interface is rough and the principal growth mechanism is of adhesive type. The morphological stability of such an interface is lower, and more easily violated on rough than smooth interfaces. Morphological instability of a rough interface leads to cellular growth and the appearance of dendritic morphology. Below $\Delta\mu/(k_B T^{**})$, the interface is smooth and the principal growth mechanism is either 2DNG or spiral growth. Due to the presence of the Berg effect [5.6], i. e., a higher driving force along the edges or corners of a face on a polyhedral crystal than at the center of the face, growth layers originate near the edges and corners and advance inwards to the center of the face. Therefore, a crystal takes a hopper or skeletal form under the driving force condition between $\Delta\mu/(k_B T^*)$ and $\Delta\mu/(k_B T^{**})$. Below $\Delta\mu/(k_B T^*)$, spiral growth layers originating from the center of a face advance toward its periphery, leading to the formation of a polyhedral crystal bounded by flat low-index faces. In Fig. 5.1, the morphological evolution of a crystal is indicated schematically in relation to the driving force, interface roughness, growth mechanisms, and normal growth rate R versus driving force $\Delta\mu/(k_B T)$ relations [5.2, 3, 5]. The positions of $\Delta\mu/(k_B T^*)$ and $\Delta\mu/(k_B T^{**})$ on the driving force axis are different depending on the crystal face. The smoother the interface, the higher these values.

The morphological evolution from polyhedral, via hopper, to dendritic form with increasing driving force is expected theoretically and observed experimentally for many crystals grown in vapor and solution phases, such as snow crystals grown in vapor phase and silicate crystals formed in magmatic solution at elevated temperature. The reader may refer to [5.2, 3, 5] for details.

Under a driving force lower than the critical value $\Delta\mu/(k_B T^*)$ the morphology is polyhedral bounded by low-index crystal faces with slow normal growth rates. If growth continues for enough time and the crystal reaches equilibrium with the surrounding phase under a given thermodynamic condition and no further change is expected to occur, the crystal will be bounded only by the faces with the slowest, or second and third slowest, normal growth rates. The crystal morphology expected in the equilibrium state can be regarded as the *equilibrium form* under the given thermodynamic conditions. The sum of the surface free energy times the surface area will be minimum owing to the thermodynamic requirement of anisotropic material. In contrast, the equilibrium form of an amorphous material, such as a liquid, is a sphere. The equilibrium form was theoretically analyzed by *Curie* [5.7], *Wulff* [5.8], *Gibbs* [5.9], and more recently by *Kern* [5.10]. Through these analyses, a few important concepts relating to growth and morphology of crystals emerged, as summarized below:

1. The normal growth rates R of crystal faces are proportional to the surface energy γ. The smaller the surface energy, the slower the normal growth rate. This is strictly speaking valid only for molecular crystals [5.11, 12], but may be qualitatively applicable when the relative morphological importance of different faces is compared.
2. A polar diagram of *raspberry* shape, called a Wulff plot, can be drawn by connecting plots proportional to the surface energies.
3. The equilibrium form is obtained by connecting inscribed lines drawn on the cusps of the Wulff plot.
4. The equilibrium form corresponds to the form with minimum total surface free energy under a given thermodynamic condition, and not the growth forms, neither structure form.

In cubic crystals with equal axial lengths, polyhedral crystal will be bounded by only one crystallographically equivalent faces with the minimum surface energy. In anisotropic crystals, the equilibrium form of a polyhedral crystal is bounded by two or more crystallographic faces with the smallest, and second and third smallest, surface energies. Depending on the crystallographic system and axial relations, different characteristic growth forms appear, generally called the crystal habit, or more precisely habitus. Habitus is the characteristic form exhibited by a crystal species grown in an isotropic environment. Crystals belonging to the cubic system, or those with other systems but having nearly equal axial lengths, take equant habit. Crystals belonging to lower-symmetry systems, or with markedly different axial lengths, exhibit prismatic or platy habit. Crystals with much shorter axial length in the c-axis than the a- and b-axes take prismatic habit, whereas those with longer axial length in the c-axis take platy habit perpendicular to the c-axis. The expected polyhedral form of a given crystal is thus correlated to the symmetry elements involved in the unit cell, the crystal system, the crystal group, the space group or the chemical bonding in the crystal structure. Such morphology deduced from crystal structure alone can be called *structural form*, and be deduced theoretically based on types and symmetry elements, or by analyzing strong bonds in the structure, entirely neglecting the effect of thermodynamic or kinetic parameters. Examples of such analyses are the oldest Bravais–Friedel (B–F) law [5.13], an analysis based on leticular density or leticular spacing of unit-cell geometry, and Donnay–Harker's (D–H) extension of the B–F law [5.14], in which symmetry elements involved in 230 space groups are taken into consideration. In these analyses, the order of morphological importance of the crystal faces can be deduced by calculating leticular densities or spacing. The higher the leticular density or the wider the leticular spacing, the higher the order of morphological importance. The order of morphological importance may be evaluated based on statistical observations of the frequency of occurrence and relative development of faces, or by comparing surface microtopographic characteristics of faces on a crystal. However, B–F or D–H analyses do not predict whether the face belongs to a smooth interface or rough interface, or the surface microtopography of crystal faces.

On the other hand, Hartman–Perdok's periodic bond chain (PBC) theory [5.15–17] considers strong bond chains involved in crystal faces, and the connected net model of *Bennema* and *van der Eerden* [5.11] also considers chemical bonding in the structure. These analyses can correlate the result to interface roughness, i. e., whether a face is rough or smooth, and quantitatively evaluate surface energy terms by calculating the attachment energies of slices E_{att} [5.12]. In the H–P theory, crystal faces containing more than two PBCs are called flat (F) faces, which correspond to smooth interfaces, whereas those containing no PBCs are called kinked (K) faces corresponding to rough interfaces. Stepped (S) faces containing only one PBC correspond to an intermediate between F and K faces, and appear by piling up of edges of growth layers developing on adjacent F faces. No growth layers will develop on S or K faces. When a crystal has more than one F face, their order of morphological importance is determined by the strength of the PBCs involved in the respective faces, and follows the order of attachment energies [5.11, 12]. These analyses can also predict the morphology of growth spirals on different faces.

All these models predict the morphologies of polyhedral crystals, based purely on the structural factor, entirely neglected the effect of external factors. These morphologies are thus considered the *structural form*. They can be different from actually observed morphology, i. e., *growth forms*, but can be used as a criterion for analyzing growth forms or the origin of their variation.

The morphology of polyhedral crystals is determined by the internal structural factor and external factors, and thus is not necessary the same as the equilibrium or structural form. We have to take the following into consideration as possible factors affecting interface structures: ambient phases (condensed or diluted phases, involving solute–solvent interaction or not, and the strength of solute–solvent interaction), growth conditions (temperature, pressure, supercooling, supersaturation, diffusion, and convection), solvent and impurity components which modify surface or edge energy terms, and dislocations. Depending on how and under what conditions and by what processes polyhedral crystals are formed, different faces appear or disappear. Even crystallographically equivalent faces may develop with different sizes. Such morphology is called *growth forms* (plural), and offers information about how and under what conditions or conditional changes the crystal grew. Growth forms of polyhedral crystals may be bounded by more faces than the equilibrium or structural form, or by crystallographically equivalent faces of different sizes. In the same crystal habit (habitus) category, various forms may appear through different combinations and development of crystal faces. *Tracht* is a term that denotes variation within the same category of habitus, due to the combination and development of different faces. Malformation of polyhedral forms from the ideal habitus occurs for various reasons, such as anisotropy in the crystal itself (spatial distribution and density of active dislocations for growth) or in the ambient phase (solution flow, convection, and the presence or absence of obstacles such as substrate surface or inclusions), and the presence or absence of seed. Information obtainable through variation in habitus and tracht tells us how the crystal grew and how the condition fluctuated throughout its growth history. Such information is very useful in analyzing conditional changes in the growth history of crystals whose growth process cannot be traced in situ (natural minerals, or crystals synthesized in enclosed systems), and to design appropriate methods to control the growth parameters to obtain single crystals of desired perfection and homogeneity.

It is to be noted that even ultrafine crystals of nanometer size show polyhedral morphology bounded by flat crystallographic faces. Ultrafine particles of not only metals but also various compounds show polyhedral morphology; for examples see [5.18] and many other papers. Only crystals smaller than 1 nm show forms without crystal faces.

Another morphology to be mentioned is whiskers. Depending on the anisotropy involved in growth sites, extremely anisotropic morphology appears. Such crystals are called whiskers, and have attracted both scientific and application interests, due to curiosity about the origin of their unusual morphology and their much higher perfection than 3-D bulk crystals. Most whiskers are straight, but topological whiskers, such as helical, coil, ribbon, and rope forms, are also known. A variety of mechanisms to account for whisker growth have been proposed, but only a few models have been established. The vapor–liquid–solid (VLS) mechanism [5.19] put forward as a possible mechanism for Si whisker formation is such a case. Detailed discussion on the mechanism of whisker formation will be given in Sect. 5.4. There are other well-established mechanism for whisker formations, and a discussion of possible mechanism is given in Sect. 5.4 and can also be found in [5.3].

We shall analyze in more detail in Sect. 5.4 possible reasons for the formation of various growth forms. However, before dealing with the macromorphology of crystals, it is necessary to understand at the atomic level how crystal growth proceeds. This is recorded in the form of the surface microtopography of crystal faces.

5.3 Surface Microtopography

Since growth (and dissolution) takes place exclusively at the solid–liquid interface, growth (and dissolution) process are recorded in the form of the surface microtopography of the crystal faces. Well-developed as-grown low-index faces corresponding to F faces show surface microtopography characterized by the development of step patterns resembling contour lines of a geographic map. They consist of flat terraces and sharp steps, with elemental height of nanometer order originating from screw dislocations, as well as bunched macrosteps with thicker step height. On terraced portion, which is usually atomically flat, islandlike layer formed by two-dimensional nucleation may occasionally be seen, where condition is changed to terminate the growth, leading to shoot up of the driving force. Islandlike layer formed by 2DNG mechanism is not commonly seen on inorganic crystal grown freely, but is more frequently encountered on protein crystals, probably due to the higher driving force conditions necessary to grow protein crystals and the larger size of the growth units involved in their growth. At least based on the author's experience of surface microtopographic observations of a wide variety of inorganic crystals, features conclusively demonstrating 2DNG have never been observed. The author is always careful to draw a conclusion about 2DNG based on islandlike patterns.

The surface microtopography of S faces is characterized by straight striations parallel to edges with adjacent F faces, whereas those of rough K faces show rugged, hummocky, rough surfaces. On both S and K faces, neither spiral step pattern nor two-dimensional island are observed. Surface microtopographs characteristic of smooth F, intermediate S, and rough K faces are illustrated schematically in Fig. 5.2a–c, respectively.

When F faces receive dissolution (etching), various etch figures appear, which include point-bottomed etch pits (P-type) formed at outcrops of dislocations, flat-bottomed etch pits (F-type) formed at outcrops of

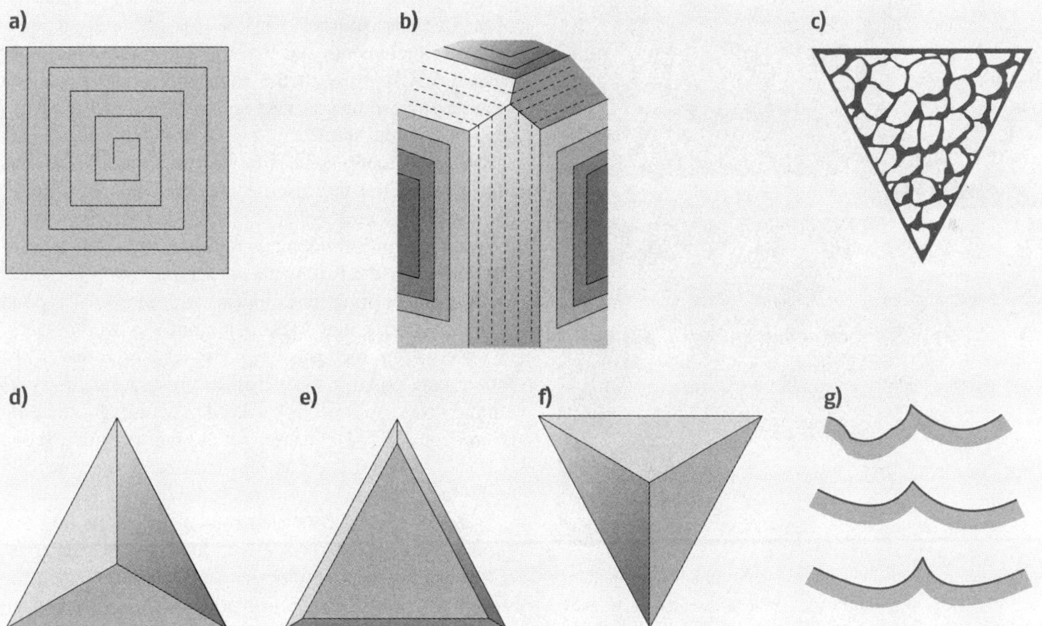

Fig. 5.2a–g Schematic illustrations of surface microtopographs expected to be observed on F (**a**), S (**b**), and K faces (**c**) formed by growth, and P-type (**d**), F-type (**e**) etch pits and etch hillock (**f**), and concave steps (**g**) due to two-dimensional recession of smooth growth steps on an F face. Microtopographs (**a–c**) are due to growth, whereas microtopographs (**d–g**) are due to dissolution (etching). Microtopographs (**d–f**) are illustrated by oblique illumination, whereas in (**g**), higher sides of steps are *shadowed*

point defects or impurity atoms, etch hillocks formed around obstacles such as dislocations decorated by impurities, and concave and rugged steps formed by two-dimensional recession of smooth growth steps. These are illustrated schematically in Fig. 5.2d–g.

If sophisticated methods to visualize and measure nanoscale steps are used, a spiral step pattern originating from isolated screw dislocations with elemental Burgers vector, and constant step separation can be seen on well-developed crystal faces. An elemental spiral originating from an independent screw dislocation has a profile consisting of atomically flat terraces and steep steps with height equal to the component of the Burgers vector perpendicular to the face, and with constant step separation. Depending on the growth conditions and ambient phases, the ratio of step separation λ to step height h varies. On crystals grown in very diluted vapor phase, the ratio is on the order of $1\times10^3-1\times10^5$, whereas on the same face of the same crystal grown

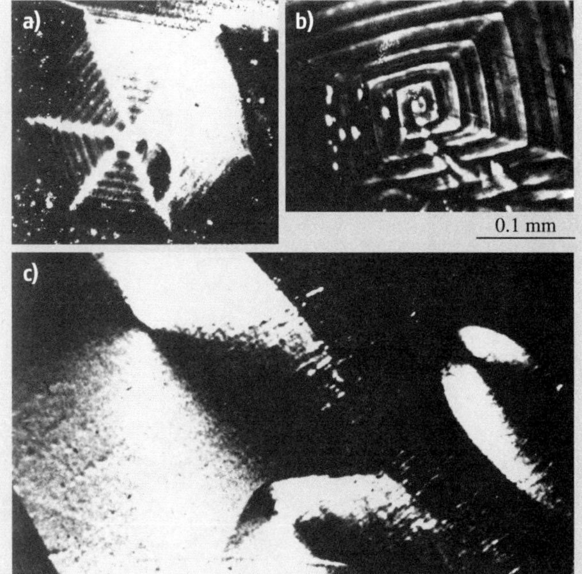

Fig. 5.3a–c Variation in the roundness of steps of growth spirals observed on different crystal faces of a hydrothermally grown synthetic beryl crystal. Regular hexagonal spirals are seen on the morphologically most important (0001) face (a), slightly rounded rectangular spirals are observed on the second important $(10\bar{1}0)$ face (b), and rounded growth hillocks are observed on the far less important $(21\bar{3}1)$ face (c). Growth features become more rounded with decreasing morphological importance

in less diluted solution phase it is on the order of $1\times10^2-1\times10^3$ [5.20]. It is to be noted that the profile of an elemental growth spiral is such that, after walking on an extremely flat plateau for 1–10 km, one meets a sharp cliff 1 m in height. Phase-contrast (PCM), interference contrast (DICM), and phase shifting (PSM) microscopes can reveal such spiral steps with nanoscale height, and multiple-beam (MBI) and phase-shifting (PSI) interferometry can measure the height of such thin steps. The step height of an elemental spiral is equal to the unit cell height, or its fraction in the direction perpendicular to the face. The step separation of an elemental spiral corresponds to twice the radius of the critical two-dimensional nucleus under the given conditions. The step separation λ of an ideal elemental spiral is equal to $19r^*$, where r^* is the radius of the critical two-dimensional nucleus. r^* is determined by the edge free energy γ and the driving force $\Delta\mu = k_B T_B \ln S$ according to

$$r^* = 2\gamma \frac{\rho}{m k_B T_B \ln S}\,, \qquad (5.1)$$

where m is the mass of one molecule and ρ is the density of the nucleus volume.

With increasing $\Delta\mu/(k_B T)$, λ becomes narrower when crystals grow in the same ambient phase. γ is modified by solute–solvent interaction and impurities adsorbed on the steps.

When the step separation is wider than the resolution limit of the microscopy adopted, the ideal spiral pattern is discernible. with increasing driving force, the step separation becomes narrower than this resolution and individual steps become impossible to resolve. Such growth spirals will appear in the form of growth hillocks with pointed summit, bounded by vicinal faces. If observation methods with higher horizontal resolution and equally high vertical resolution, such as atomic force microscopy (AFM), scanning tunneling microscopy (STM), or laser beam scanning microscope (LBSM) are applied, spiral steps may become discernible on the summit of such growth hillocks.

Growth spirals with elemental step heights take circular or polygonal form depending on the roughness of the spiral steps. A circular spiral appears when the step is rough, whereas a polygonal form appears when the step is smooth. So the same concept applied to the roughening transition of a smooth interface is applicable in this case too. The circular form appears under higher driving force conditions, whereas the polygonal form is seen when the driving force is below the critical value for the two-dimensional roughening tran-

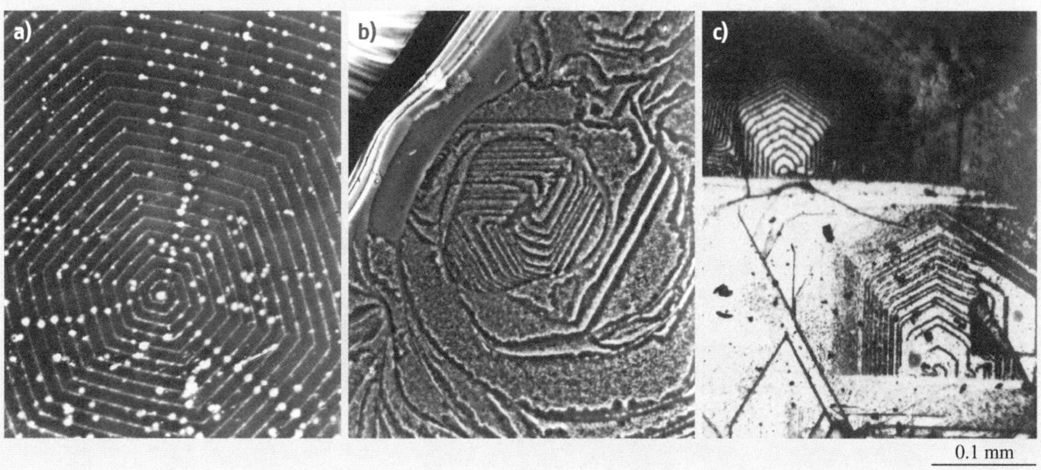

Fig. 5.4a–c Symmetry of polygonal growth spirals observed on crystals of different crystal systems: (**a**) hexagonal spiral observed on (0001) face of a hexagonal system, SiC, (**b**) triangular spiral observed on (0001) face of a trigonal system, hematite and (**c**) five-sided spiral containing only one mirror symmetry plane observed on (001) face of a monoclinic crystal, phlogopite. In (**b**), the crystal receives weak etching, forming rugged steps due to two-dimensional recession (dissolution) of smooth growth steps. In (**c**) spiral steps with height of 1 nm are decorated by selective nucleation of foreign crystals along the spiral steps

sition. The critical points for the roughening transition to take place may differ on the same crystal face depending on the growth parameters as well as on different faces on the crystal grown under the same conditions. On the same faces of crystals grown under different conditions, circular or polygonal spirals may be seen depending on the growth conditions. More polygonal spirals appear on crystals grown under lower driving force conditions than those grown under higher driving force conditions. Polygonal spirals may be observed on the morphologically most important face, whereas circular ones are seen on morphologically less important faces on the same crystal grown under the same conditions. Representative examples to demonstrate such variation is shown in Fig. 5.3. When spirals take the polygonal form, their symmetry follows that of the face, since the steps are determined by PBCs involved in the face. Figure 5.4 shows representative examples of various morphologies of growth spirals observed on the faces of crystals belonging to different crystal systems.

Growth spirals with elemental step height bunch together to form thicker macrosteps while advancing

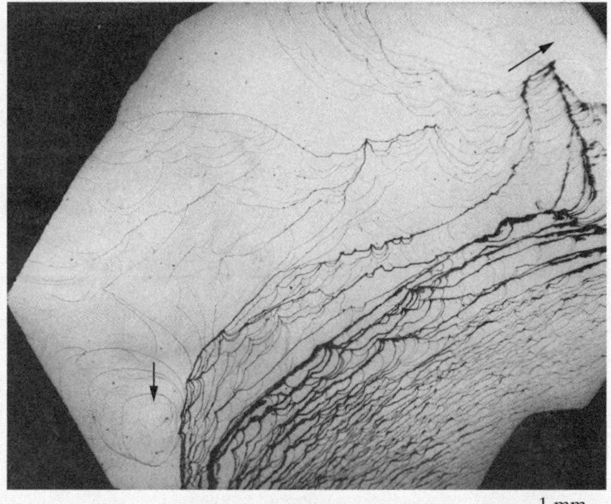

Fig. 5.5 An example of a step pattern observable on a whole face. More perturbed, bunched macrosteps appear closer to the edges of the face at the *bottom right*. All perturbed macrosteps seen at the lower right are bunched macrosteps originating from elemental spiral centers indicated by the *arrows*. Also note the wider step separation along the upper edges, which is considered as due to a higher driving force along the edge than at the center of the face, i. e., the Berg effect. Hematite (0001) ▶

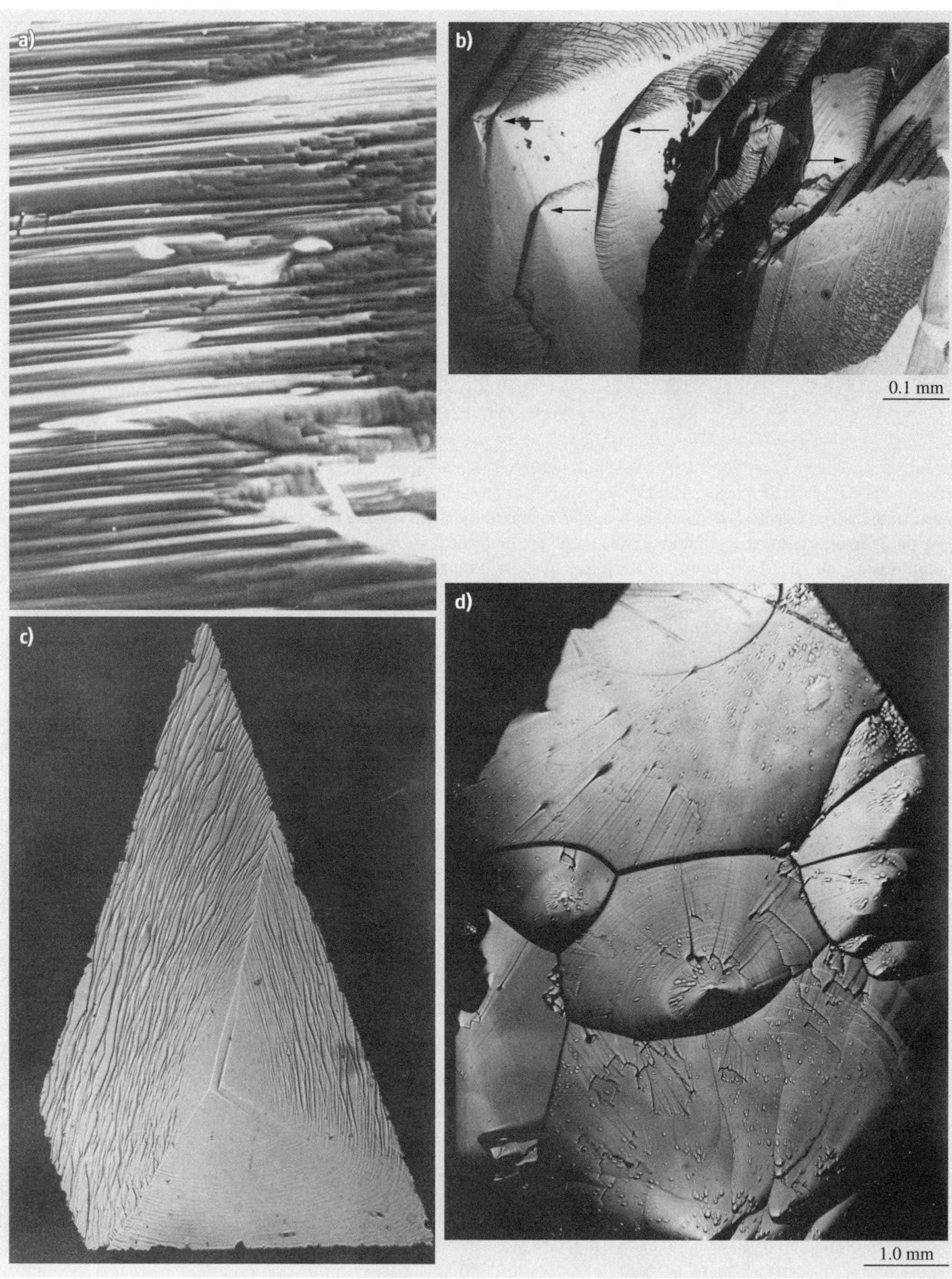
a)
b)
0.1 mm
c)
d)
1.0 mm

Fig. 5.6a–d Surface microtopographs of prism (**a**,**b**) and rhombohedral (**c**,**d**) faces of natural (**a**,**c**) and synthetic (**b**,**d**) quartz crystals. *Arrows* in (**b**) indicate spiral centers ◄

laterally, due to either impurity adsorption along the step or fluctuations in the driving force over the surface. Since the advancing rate of a bunched macrostep is slower than that of elemental steps, the step pattern is perturbed more as it advances laterally. Since the driving force near the edges and corners of a crystal face is higher than at the center of a face due to the Berg effect [5.6], the step separation becomes narrower and the step pattern becomes more perturbed when advancing close to edges. Figure 5.5 shows such an example.

On real crystals, it is exceptional for a whole crystal face to be entirely covered by one spiral step pattern, with a constant step separation, originating from an outcrop of an independent screw dislocation. Several or a large number of growth spirals appear on one face, originating from several dislocations. So, step patterns observable on the same crystallographic face can differ from crystal to crystal, or among crystallographically equivalent faces, even if the crystal grew in similar ambient phases under similar growth conditions. However, overall surface microtopographs show characteristics corresponding to the crystal growth environment and conditions. The same hematite crystals occurring at different localities, but formed in similar ambient phases, i. e., natural CVT due to postvolcanic action, show surface microtopographic characteristics according to which different localities can be identified. Prism and rhombohedral faces of natural and synthetic quartz crystals show markedly different surface microtopographies according to which natural and synthetic crystals can be easily discriminated, even if both show the same hexagonal prismatic habit. In Fig. 5.6, surface microtopographs of prism (Fig. 5.6a,b) and rhombohedral (Fig. 5.6c,d) faces of natural (Fig. 5.6a,c) and synthetic (Fig. 5.6b,d) quartz are compared. The marked differences between natural and synthetic quartz crystals are due to the differences in growth rates (natural crystals grew slower under lower driving force conditions than synthetic crystals) and solution chemistry (natural quartz grows in neutral hydrothermal solution, whereas synthetic quartz in alkaline solution).

5.4 Growth Forms of Polyhedral Crystals

Polyhedral crystals bounded by flat crystallographic faces are formed by the spiral growth mechanism under conditions lower than $\Delta\mu/(k_B T^*)$. If a crystal grows in an isotropic environment, i. e., in an ambient phase with a concentric diffusion gradient toward the growing crystal, the growth form is determined simply by the relative normal growth rates of the faces present on the crystal surface. The morphology eventually reaches a form corresponding to a structural or equilibrium form. Crystallographically equivalent faces develop equally in size. If additional factors that may induce anisotropy in the mass transfer in the ambient phase or in the distribution of dislocation outcrops active as growth centers are involved, polyhedral growth forms deviate or are malformed from their ideal form. Even crystallographically equivalent faces will develop differently. As possible factors to induce such anisotropy, we may mention the following:

1. Convection or directional flow in the ambient phase, which induces anisotropy in mass transfer. This may result in not only modification of growth rates but also in the distribution of inclusions from which dislocations are newly generated.
2. The presence of seed or substrate surface. Seed modifies the growth form, and the surface of seed or substrate affects normal growth rates by newly generating dislocations on the interface.
3. The presence of reentrant corners or concentration of dislocations in twin junctions due to twinning. These provide sites for preferential growth.
4. Anisotropic distribution of active growth centers, i. e., outcrops of screw-type dislocations, for various reasons, such as new generation of dislocations from inclusions.
5. Impurity elements selectively adsorbed along growth steps or foreign compounds that selectively cover the growing surface due to epitaxial relation, or selective adsorption that suppresses the normal growth rate.

Since an ideally isotropic environment is not expected in real systems, actually observed polyhedral crystals show various forms which deviate or are malformed from ideally expected forms, i. e., the structural or equilibrium form. Even a simple octahedral crystal bounded by crystallographically equivalent {111} faces may only take polyhedral forms such as tetrahedron, triangular or hexagonal plate, or even elongated rodlike form, due to anisotropic development of crystallographically equivalent faces. In Fig. 5.7, various polyhedral forms appearing due to anisotropic development of crystallographically equivalent {111} faces are shown. We shall in the following first analyze growth forms expected in an isotropic environment, and then proceed to the analysis of possible reasons for such deviations.

The normal growth rate R of a face growing by the spiral growth mechanism is determined by the height h, the separation between successive steps λ, and the advancing rate v of the spiral step $R = hv/\lambda$. Since h corresponds to the Burgers vector of dislocation, it is different on different faces. Since λ is equal to $19r1^*$, and

$$\Delta G(r^*) = 16\pi\gamma^3 \frac{[v]^2}{3\Delta\mu^2}\,, \tag{5.2}$$

where r^* is the radius of the critical two-dimensional nucleus, γ is the edge free energy of the two-dimensional nucleus, v is the molar volume of the bulk nucleated phase, and $\Delta\mu$ is the chemical potential difference.

Important parameters that affect R, and hence modify the habitus and tracht of polyhedral crystals when they grow in an isotropic environment, are those that modify the edge free energy and driving force. The following are considered to be the major factors:

1. Driving force, i. e., supercooling or supersaturation.
2. Ambient phase. Crystals of the same species growing from different ambient phases, i. e., those grown from melt, solution or vapor phase, will show different growth forms.
3. Solute–solvent interaction energy, i. e., crystals of the same species grown from solutions with different solvent compositions will show different growth forms.
4. Impurity elements, which modify the edge free energy by adsorbing on spiral growth steps. Foreign compounds which can selectively adsorb or epitaxially grow on particular faces. These may suppress the normal growth rate of rough or less smooth interfaces which grow with high growth rate in the pure system, resulting in marked morphological changes.

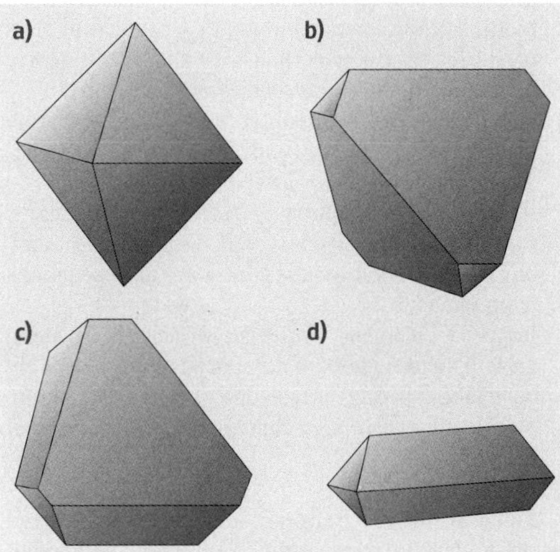

Fig. 5.7a–d A simple octahedral crystal bounded by {111} only **(a)** may take various malformed forms: tetrahedral **(b)**, triangular platy **(c)** or elongated rodlike **(d)** due to anisotropy in either environmental factors or in the crystal itself

Examples demonstrating the effect of these factors are well known and can be found in older works by *Wells* [5.21] and *Buckley* [5.22]. Wells demonstrated various examples of variation of growth forms due to effects 1, 2, and 3, and Buckley summarized his observations on effect 4. Observations on the variation of habitus and tracht of mineral crystals in relation to their modes of occurrences and the analysis of their origins may be found in [5.2, 3, 5].

In real systems, polyhedral crystals grow in environmental phases that deviate from the ideal isotropic situation. Mass (or heat) transfer will be anisotropic due to convection or directional solution flow. It was noted as early as the 17th century by *Hook* [5.23] that alum crystals formed on the bottom of a beaker take hexagonal platy habit bounded by only {111} faces, as compared with the nearly simple octahedral habit of crystals formed on a string immersed in the solution, i. e., those growing in a nearly isotropic environment. Around a crystal growing on the bottom of a beaker, mass transfer to the face adhering to the bottom is strongly suppressed as compared with the faces directly facing to the solution, resulting in remarkable malformation from the ideal octahedral morphology.

Reentrant corners provided by twinning also play a similar role, resulting in triangular or hexagonal platy habit, since preferential growth occurs at a reentrant corner due either to the geometry or the concentration of dislocations in the twin junction plane.

Convection or directional flow of solution is universally involved in mineral formation in hydrothermal veins in nature and in hydrothermal synthesis in the laboratory. When crystals grow in an impure system, precipitation and incorporation of solid grains of foreign minerals or mother liquid phase as inclusions in the growing crystal is also anisotropic, which will enhance growth rate anisotropy. Since such inclusions act as sites generating new dislocations, the density of outcropping dislocations on crystallographically equivalent faces become different, resulting in different growth rates. Depending on the conditions of solution flow, laminar or turbulent, the rate of mass transfer and the precipitation of foreign particles may be different. In turbulent flow, the growth rate is enhanced and more inclusions are included on the rear side of the solution flow than on the front side. Variation of normal growth rates or the precipitation and inclusion of foreign particles are recorded in the forms and sizes of growth sectors, the width of growth banding, and the distribution of centers of dislocation generation detectable in single crystals. Such internal morphology of single crystals offers important information for the analysis of growth history and conditional variations during growth. These will be described in more detail in Sect. 5.5.

When aiming to synthesize large single crystals of high perfection at high growth rate in the laboratory, seed crystals are generally used to suppress unnecessary spontaneous nucleation. Other techniques, such as the application of an alternating temperature to dissolve unnecessary nuclei, agitation and stirring of the solution, rotation of the crucible, application of a magnetic field, and growth under microgravity conditions, are applied to homogenize the ambient phase.

In preparing the seed, it is common to adopt a crystallographic orientation bounded by rough or less smooth interfaces to secure a higher growth rate and obtain forms appropriate for industrial use. The crystallographic orientations and surfaces of seed are generally chosen as different from those observed on freely grown forms. *Y*-bar seed for the hydrothermal synthesis of quartz and a seed plate parallel to $[21\bar{3}1]$, often used in the hydrothermal synthesis of emerald, are such examples. The seed surface is often treated by chemical etching before the onset of new growth. This results in markedly different as-grown morphology of syn-

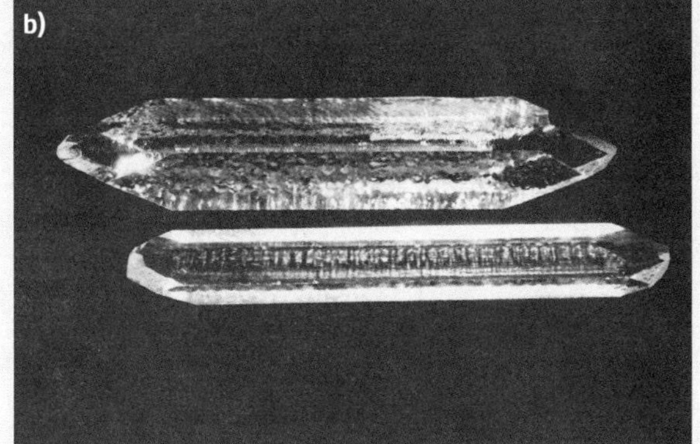

Fig. 5.8a,b Natural quartz crystal grown without seed (**a**) and synthetic crystals grown on seed plate (Y-bar) (**b**). The *c*-axes are set vertically in both photographs ▶

thetic crystals from those grown freely. Seed surfaces are usually those not seen on freely grown crystals, and belong to the category of rough interfaces. They often develop to a large size on synthetic crystal. On natural quartz crystals, $\{0001\}$ faces never appear unless a special effect is involved, whereas on synthetic quartz crystals these faces become large. In Fig. 5.8a,b natural and synthetic quartz crystals are compared with the same crystallographic orientation. Natural quartz crystals show hexagonal prismatic form elongated in the c-axis, and bounded by prism $\{10\bar{1}0\}$, two types of rhombohedral $\{10\bar{1}1\}$ and $\{01\bar{1}1\}$ faces, whereas synthetic crystals show prismatic form elongated along the Y-axis and perpendicular to the c-axis and bounded by basal $\{0001\}$, prism $\{10\bar{1}0\}$, $\{11\bar{2}0\}$, rhombohedral $\{10\bar{1}1\}$, $(01\bar{1}1)$ and trigonal pyramid $\{11\bar{2}2\}$ faces. $\{0001\}$, $\{11\bar{2}0\}$, and $\{11\bar{2}2\}$ faces are not commonly observed on freely grown quartz crystals. The difference is simply because the growth of synthetic quartz is forced to terminate before the crystal attains its final form. If growth proceeds further, synthetic quartz crystals will eventually show similar hexagonal prismatic form to natural quartz crystals. Spontaneously nucleated quartz crystals without seeds on the holder in the same autoclave take trigonal or hexagonal prismatic form, similar to natural crystals, except with more exaggerated trigonal symmetry than natural quartz.

Although not common, we may occasionally identify seed crystals in single crystals of natural minerals as well. The seed is formed elsewhere in a different geological environment and later incorporated into a new environment where further growth takes place later. Growth morphology may be different between the seed and the newly grown portion, and dislocations may be newly generated on the seed surface.

In epitaxial growth, the morphology of the guest crystal is different from that of freely grown crystal, since an additional interface energy and steps are introduced between the host crystal surface and the guest crystal due to their epitaxial relation. Epitaxially grown guest crystal usually takes more flattened or elongated forms. Many examples are known among mineral crystals.

Whiskers are single crystals with highly anisotropic forms. In most cases, they take straight forms, elongated in a certain crystallographic direction and bounded by low-index crystal faces [5.24]. Kinked whiskers are also sometimes observed. Recently, topological whiskers showing twisted, curved, helical, screw, Möbius ring, and other topological forms have been reported [5.25]. To account for the origin of such highly anisotropic morphology of single crystals, we have to assume the presence of a preferential unique growth site, where growth can occur while growth on other sites is suppressed. In the case of the vapor–liquid–solid (VLS) mechanism for Si whiskers [5.19], Si is continually supplied in vapor phase (V), which dissolves in Au particles to form a eutectic Au-Si liquid droplet (L), in which only Si nucleates and grows as whiskers (S). Since growth occurs only at the site of the nucleus, Si crystal grows as a whisker, with a cap of Au-Si eutectic liquid droplet at the tip, and Si is continually supplied from the vapor. In the eutectic solution (L) phase, for thermodynamic reasons (i. e., the lower melting point of the solution than of the pure solute and solvent phases and the phase to nucleate, and that growth in the solution is determined by the composition of the liquid phase), only Si is nucleated in the solution phase, providing a unique growth site, leading to whisker growth. Growth exclusively occurs at the root of whiskers in the eutectic liquid droplets. The VLS mechanism has been well established in many other example systems, including mineral crystals.

Another well-established mechanism for whisker formation was demonstrated in the case of KCl or NaCl crystals [5.26]. When KCl or NaCl aqueous solution in a wineskin is kept in the dark, whiskers of these crystals grow from the outer surface of the wineskin. These whiskers are hollow along their length, indicating that crystal growth occurs as soon as the solution is transported through capillaries in the skin and exposed on the skin surface or the tip of the hollow whisker, since supersaturation sharply increases. Growth uniquely occurs at the tip and, as long as the capillary is present, growth continues, leading to hollow whiskers.

5.5 Internal Morphology

Polyhedral crystals are bounded by several crystallographically equivalent or different faces that grow at different normal growth rates R. Crystal faces with smaller R become larger, whereas those with larger R diminish in size or terminate as growth proceeds. So rough interfaces will disappear very soon from the external form, and only smooth faces will become large. Among smooth interfaces, the morphologically

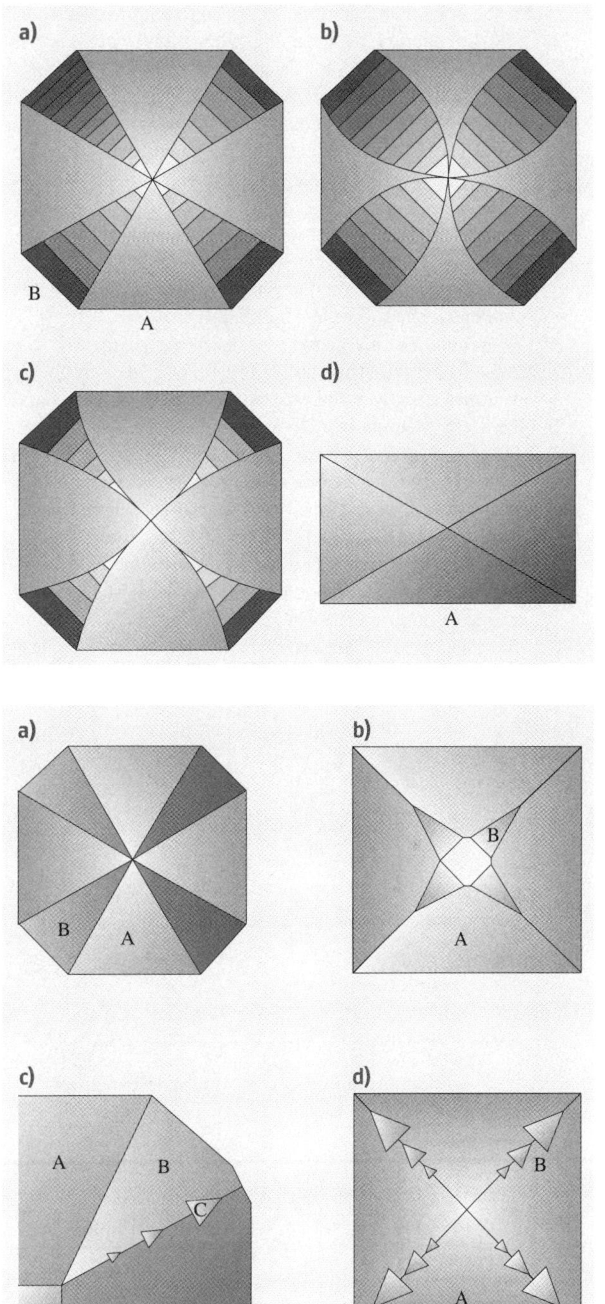

Fig. 5.9a–d Schematic illustration of growth sector boundaries. **(a)** Straight growth sector boundaries appear when R_A and R_B are constant; convex **(b)** or concave boundaries **(c)** appear when R_A or R_B is decreasing. Growth sector boundaries in a crystal bounded by crystallographically equivalent A faces, but with different growth rates are indicated in **(d)** ▶

most important face develops as the largest face while morphologically less important faces develop smaller. Unless some factors suppress or enhance the normal growth rates, the final morphology is the same as that structurally expected. However growth forms are different from ideal forms. Even crystallographically equivalent faces grow at different R when a crystal grows in a real environmental phase in which flow or convection currents or substrate surface are present. The densities of active growth centers (outcrops of dislocations) will not be the same in a real system on growing surfaces of crystallographically equivalent faces, leading to different growth rates. Growth forms of polyhedral crystals may vary as they grow, due to variations and differences in the normal growth rates in different directions as well as changes in the growth conditions.

Through the growth process, pyramidal portions are formed in a single crystal, with their summits at the initiation of the face and the base at their terminations either within the crystal or on the final as-grown surface. These pyramidal portions observable in single crystals are called growth sectors, and can be visualized even in perfectly clean crystals by the naked eye when investigated by applying appropriate methods such as polarization microscopy, CL and laser-beam tomography, x-ray topography, and etching. In the Russian literature, they are called growth pyramids.

Depending on the growth history, various forms of growth sectors appear. If a face grows steadily throughout the whole growth history, the growth sector takes a regular pyramidal form with the summit at the center

Fig. 5.10a–d Schematic illustrating various types of growth sectors. **(a)** Two faces, A and B, grew at a constant growth rate. **(b)** The growth rate of sector B increases as growth proceeds and the sector terminates within the crystal. **(c)** Face C appears due to conditional change, and the corresponding growth sector soon terminates due to the original rapid growth rate. **(d)** Intermittent growth sectors, indicating repeated changes in growth conditions ▶

of the crystal and the base at the final as-grown surface. The boundaries between the adjacent growth sectors detected when a crystal is bisected through the center may be straight, convex or concave, depending on the relative growth rates of the faces, as illustrated schematically in Fig. 5.9. Growth sector boundaries may be detected even between neighboring crystallographically equivalent faces. Since sector boundaries are places where advancing growth steps on the adjacent faces meet, where strain concentrates, they can be detected by appropriate methods. They are often seen by the concentration of mother liquid inclusions.

When one face grows at a much higher growth rate than the neighboring faces, the former growth sector becomes narrower and tapered as growth proceeds and the sector boundaries may terminate within the crystal. This forms a center cross pattern, with tapering growth sectors. If the face appears intermittently, zigzag sector boundaries appear. Figure 5.10a–d shows schematically various types of growth sectors. Clearly, morphologically important faces that behave as smooth interfaces under the growth condition form growth sectors that persist throughout the whole growth history, while morphologically less important faces that behave as less smooth or rough interfaces form tapering or intermittent growth sectors. In these sectors, the summits and base are just opposite to ordinary observable growth sectors. The base of a pyramid appears first, and the boundaries are tapered as growth proceeds. This indicates that the face appeared when conditions changed to suppress the normal growth rate of the face, which otherwise grows with a higher growth rate. The presence of intermittent growth sectors indicates that an abrupt conditional change took place to suppress the normal growth rate of the face, leading to the appearance of the face, which soon disappears due to the rapid growth rate under the given conditions. Tapered growth sectors may represent either the coexistence of two smooth interfaces with different smoothness, or smooth and rough interfaces. In the case of the coexistence of smooth and rough interfaces, we have to assume an effect of external factors to suppress the normal growth rate of the rough interface. Otherwise, growth sectors corresponding to the rough interface will very quickly disappear from the crystal.

Even if the final morphology is a polyhedron bounded by flat faces, this does not automatically guarantee that the crystal took polyhedral form throughout its growth history, i. e., was formed exclusively under conditions lower than $\Delta\mu/(k_B T^*)$ throughout its whole growth history. They may start as spherulite or den-

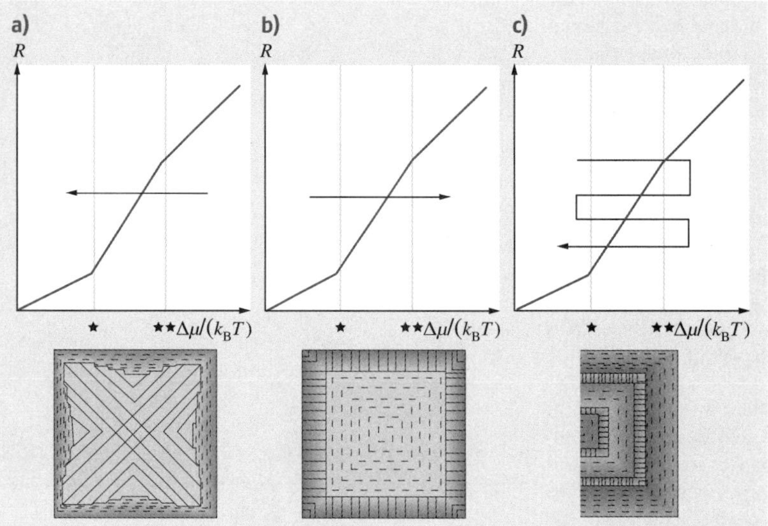

Fig. 5.11a–c Schematic illustrating the internal morphologies of single crystals formed first by dendritic growth followed by 2DNG and spiral growth (**a**). Fibrous (dendritic) overgrowth on earlier formed polyhedral crystal grown by 2DNG or spiral growth, due to a sharp increase of the driving force by conditional change at the latest stage, is indicated in (**b**). A case of repeated conditional change is indicated in (**c**). Corresponding conditional changes are indicated by *arrowed lines* on the respective growth rate versus driving force diagrams shown above

dritic forms under conditions higher than $\Delta\mu/(k_BT^{**})$, and 2DNG or spiral growth may take place later when the driving force drops below $\Delta\mu/(k_BT^{*})$. The earlier formed dendritic or spherulitic forms may be found at the center of polyhedral crystals formed through such process, as illustrated schematically in Fig. 5.11. The skeleton and the present size of a polyhedral crystal may have been constructed during the earlier stage of dendrite formation.

So, through the analysis of internal morphology, we may assess how the as-grown morphology of a crystal has changed and which faces persisted and were morphologically important throughout its growth history, and how and why morphologically less important faces appeared or disappeared during the whole growth history of a crystal.

Although less distinct, growth sectors of smaller size can be detected within one growth sector when sensitive observation methods such as polarization microscopy or cathode luminescence tomography are used to detect chemical heterogeneity. Such growth sectors are called intrasectorial growth sectors, or vicinal sectoriality. They appear due to different advancing rates v of spiral growth layers or impurity incorporations between different vicinal faces of a spiral growth hillock developing on an F face. On the surface of an F face growing by spiral growth mechanism, a spiral growth hillock with polygonal form appears, being composed of a few vicinal faces and following the symmetry of the face. The vicinal faces appear by piling up of spiral steps, since the advancing rates of spiral growth layers are different in different directions. Due to the difference in the advancing rate, impurity concentrations can be different between different vicinal faces. This forms vicinal sectoriality or intrasectorial growth sectors. A good example of vicinal sectoriality was reported in [5.27], as revealed by CL on polygonal growth spirals developing on $\{10\bar{1}0\}$ face of synthetic quartz crystal. Brighter CL images corresponding to higher concentration of impurity Al are seen on only two among six vicinal faces forming a polygonal growth spiral. The higher concentration of Al in these vicinal faces results in the appearance and development of s$\{11\bar{2}1\}$, x$\{51\bar{6}1\}$ faces on natural quartz crystals and further to the appearance of S$\{11\bar{2}2\}$ faces on synthetic quartz crystals grown on seed [5.27]. Vicinal sectoriality is also reported on rhombohedral faces of calcite crystal [5.28]. The appearance of vicinal sectoriality due to impurity adsorption indicates that element partitioning is controlled by the growth kinetics.

Partitioning or the distribution of impurity elements is controlled by both thermodynamic factors and growth kinetics. Assuming growth under a constant thermodynamic condition, element partitioning is principally controlled by growth kinetics. The ratio of element partitioning between the ambient phase and crystal is defined by the effective distribution coefficient, K_{eff}. Elements with $K_{eff} < 1$ are incorporated less in the growing crystal and accumulate more in front of the growing crystal in the ambient phase, while those with $K_{eff} > 1$ are incorporated more into the crystal as growth proceeds. *Burton* et al. [5.1] indicated theoretically that K_{eff} depends on the normal growth rate R, the diffusion constant D, and the thickness of the diffusion boundary layer δ as

$$k_{eff} = \frac{k_0}{k_0 + (1 - k_0)\exp(-R\delta/D)} . \quad (5.3)$$

Since the normal growth rates differ depending on the crystallographic orientation and interface structures, element partitioning will be different between smooth and rough interfaces and among different crystallographic faces.

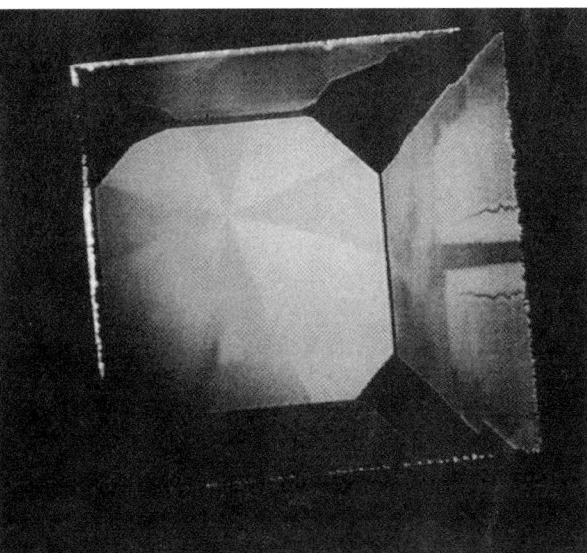

Fig. 5.12 Cathodoluminescence tomograph of cubo-octahedral crystal of HPHT synthetic diamond. CL intensity in {111} growth sectors is high, whereas no or weaker CL intensity is seen in {100} growth sectors, indicating selective and higher partitioning of nitrogen in morphologically more important {111} growth sectors than in less important {100} faces (courtesy of GAAJ)

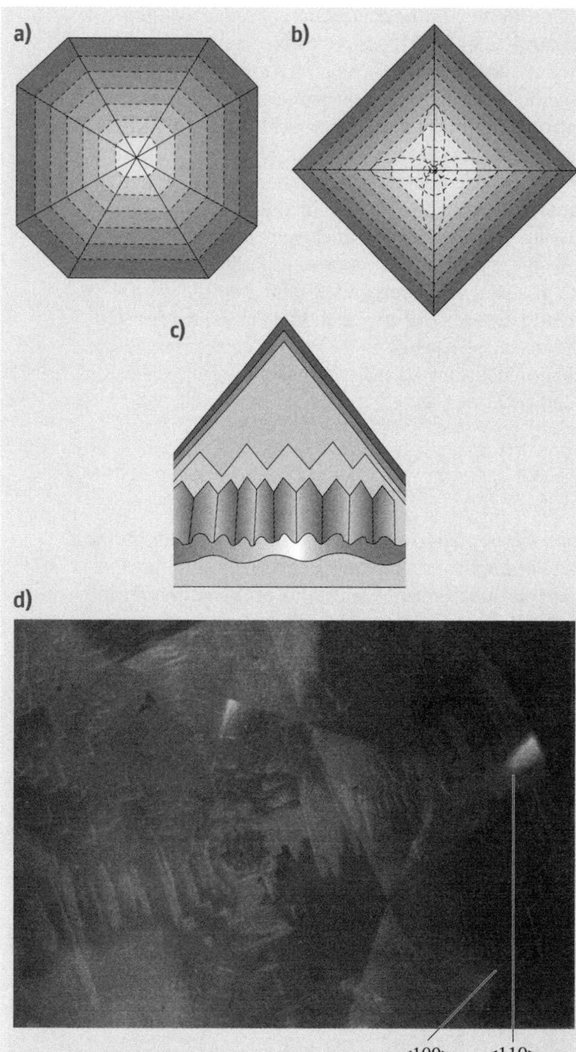

Fig. 5.13a–d Straight growth banding observable in the growth sectors corresponding to (**a**) two smooth interfaces and (**b**) one smooth interface and another rough interface, in which growth banding is not straight but hummocky. Transition of smooth to rough interface, followed by morphological instability of the interface to form cellular growth and the development of microfacets in the recovery process from rough to smooth interface, is indicated in (**c**). (**d**) CL tomography of a round, brilliant cut diamond, indicating the development of micro {111} facets during the recovery process from seed cuboid bounded by rough interfaces

The normal growth rate R of a rough interface is higher than that of a smooth interface. However, the lateral advancing rate v of growth layers on a smooth interface is much higher than the normal growth rate on a rough interface. The higher the growth rate or the step advancing rate, the higher the probability of incorporating impurity elements at the growth front. Therefore, it is anticipated that impurity elements with $K_{eff} < 1$ will be more concentrated in growth sectors of smooth interface than those of rough interface. This was initially observed in Si single crystals grown from the melt phase by Czochralski method. In melt growth of Si by CZ method, the solid–liquid interface follows an isothermal profile, the interface is mainly rough, and an adhesive-type growth mechanism operates principally. However, at the central portion of a growing single-crystalline boule, a facet of smooth {111} interface appears, forming a central, narrow growth sector formed by smooth interface surrounded by those formed by rough interface growth. It was found that Bi distributes more in the central growth sector of faceted growth on {111} smooth interface than in the major growth sectors of the rough interfaces. This anisotropic distribution of Bi was understood as due to the much higher advancing rate v of growth layers on a smooth interface than the normal growth rate R of a rough interface.

From the observation on Si crystal grown from the melt phase, it is anticipated that impurity elements with $K_{eff} < 1$ will be more concentrated in the growth sector of smooth face with the slowest normal growth rate than in growth sectors of both rough interface and smooth interface with lower morphological importance. Figure 5.12 shows an example demonstrating this: a CL tomograph of high-pressure high-temperature synthetic diamond crystal bounded by {111} and {100} faces showing brighter zones corresponding to higher concentration of nitrogen in {111} growth sectors than in {100} growth sectors, which show no CL intensity. In the growth of high-pressure high-temperature (HPHT synthetic) diamond, both {111} and {100} behave as smooth interfaces, but {111} is morphologically more important than {100} [5.29]. Natural diamonds also indicate higher concentration of nitrogen in smooth {111} growth sector than in rough {100} growth sector. In natural diamond growth, {111} behaves as a smooth interface, whereas {100} exclusively behaves as a rough interface. In synthetic quartz, Al impurity was found to be selectively adsorbed on growth steps of certain vicinal faces of growth spirals developing on the morphologically most important $\{10\bar{1}0\}$ faces [5.27].

Impurity elements with $K_{\text{eff}} < 1$ accumulate more as growth proceeds, in the ambient phase, to form a concentration gradient in the diffusion boundary layer surrounding the growing crystal. This modifies the chemistry of the ambient phase in the diffusion boundary layer, affecting the normal growth rate, the critical driving force for 2DNG, diffusion rates, etc., and leads to the formation of a band with a gradient of impurity concentration parallel to the growing surface. Due to coupling of the accumulation of impurity on the growing interface, their diffusion, and the resulting change in the normal growth rate, an alternating succession of bands with varying concentration of impurity (and also point defects) appears parallel to the growing surface, even if the crystal grew under nearly constant growth conditions. Such bands may be less distinct and have nearly uniform and narrow spacing. When growth parameters change abruptly, a more distinct growth band is formed. So, within one growth sector, distinct but not uniformly spaced bands and less distinct but uniformly and narrowly spaced bands appear. Since these bands appear through growth and represent the morphology of the crystal at successive stages, this banding is called growth banding. Analysis of growth banding provides important information about how the crystal grew and the morphology changes throughout its growth process. Growth bands in the growth sectors of smooth interfaces exhibit straight banding, whereas in those of rough interface growth, hummocky growth banding appears (Fig. 5.13a,b). If a smooth interface transforms to a rough interface during growth due to a conditional change, and the rough interface changes to a smooth interface in further growth, the appearance of a wavy and hummocky band and its transformation to a cellular pattern, followed by transformation from a rough to smooth interface with the appearance of microfacets in the recovery process, may be traced by observation of growth banding (Fig. 5.13c,d). The evolution of the crystal morphology in the formation of single crystal may also be traced. Dissolution of earlier formed crystal followed by regrowth may be identified through the observation of the banding pattern in single crystal. All sorts of event may be discernible by the analysis of heterogeneities in single crystals. Figure 5.13a–c shows schematically a few cases to be expected, and Fig. 5.13d shows an actual example of the appearance of microfacets during the recovery process. Figure 5.14 shows schematically the internal morphology of emerald and ruby crystals, indicating earlier formation of core portion by 2DNG or spiral growth, on which dendritic growth took place to form the arms and branches, followed by 2DNG or spiral growth filling the interstices of the dendritic arms. These are called trapiche emerald or ruby. Various terms are used to express the corresponding portions, as indicated in the figure [5.30, 31], but the growth mechanism and history are the same in both minerals. The skeleton and the size of the present crystal are determined by this growth history.

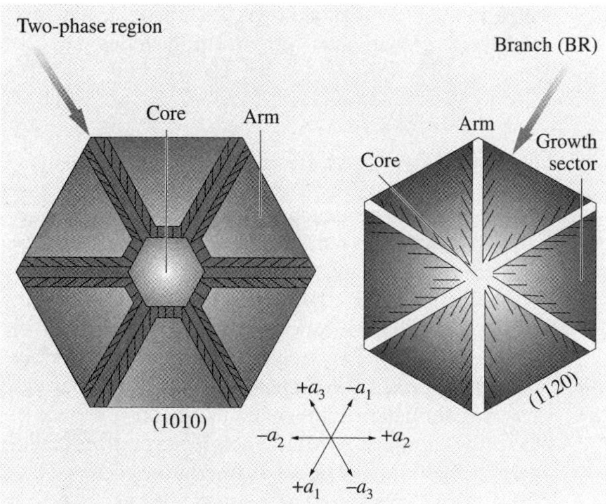

Fig. 5.14 Trapiche emerald (*left*) and ruby (*right*), formed by dendritic growth forming the skeleton of the crystal, followed by growth on smooth interface filling the interstices of dendritic arms (after [5.30, 31])

If a polyhedral crystal receives weak dissolution (etching), corners, edges, and outcrops of point and line defects are preferential sites to be attacked. The polyhedral crystal will be rounded off with etch pits on the surfaces. Straight growth banding observable in growth sectors is cut by the rounded external surface. All natural diamond crystals show rounded morphology and etch pits, indicating that their characteristic rounded forms are due to dissolution experienced during the ascent period from depth to the Earth's surface, during which they experienced pressure–temperature conditions labile for diamond. When dissolution occurs to form rounded crystal, on which regrowth later took place, earlier straight growth banding is intersected by rounded discontinuity, followed later by the appearance of microfacets and straight growth banding. A wide variety of internal morphology is encountered in polyhedral single crystals of natural minerals, and their growth and postgrowth histories can be analyzed if their internal morphology is properly analyzed.

5.6 Perfection of Single Crystals

Dislocations are generated where lattice planes advancing from different sources meet with mismatches or displacements during growth. There can be various origins of lattice mismatches during the growth process of a crystal. In the nucleation stage the nucleus may be bounded by rough interfaces, but soon smooth interface starts to appear. Lattice mismatch may occur through this transformation from rough to smooth interface at the earliest stage of growth. When dendritic arms conjugate, or when solid or liquid inclusions are enclosed into the growing crystal, lattice mismatch occurs at places where they are enclosed. When bunched macrosteps advancing on smooth interface meet, lattice mismatch may also occurs. In these cases, dislocations with large Burgers vector are often generated on the growing surface. Since such dislocations are energetically unfavorable, they dissociate into many dislocations with elemental Burgers vector. When seed crystal is used, its surface provides preferential sites to generate new dislocations, since inclusions are likely trapped on etched seed surface, from where dislocations are newly generated, together with dislocations inherited from the seed. A variety of spatial distributions of dislocations can therefore be encountered in single crystals grown freely under unconstrained conditions and with seed crystal. When crystals grow in the interstices of solid crystals of other minerals, such as in the formation of metamorphic rocks, numerous dislocation centers will be

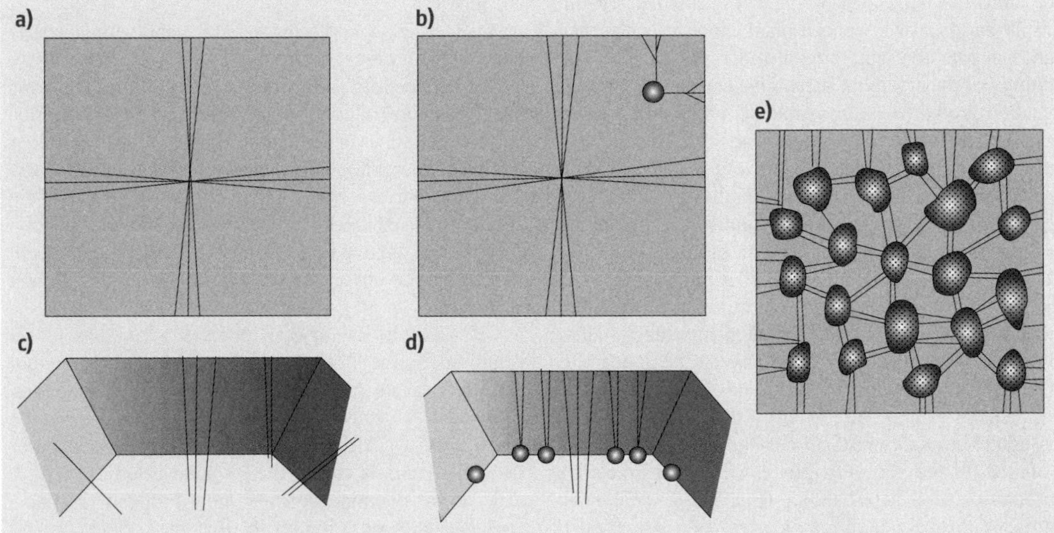

Fig. 5.15a–e Schematic illustrations of the spatial distribution of dislocations in single crystals. (**a**) Dislocations originating at the center of a crystal and radiating in a bundle nearly perpendicularly to the growing faces for energetic reasons. This is a representative distribution of dislocations observed in freely grown crystal in diluted vapor or solution phase under unconstrained conditions. (**b**) Dislocations are also generated from a point where an inclusion is enclosed. Dislocations with a large Burgers vector dissociate into many dislocations with elemental Burgers vector. (**c**,**d**) Comparison of the inheritance of dislocations in the seed to the newly grown portion (**c**) and the generation of new dislocations from inclusions formed on the seed surface (**d**). (**c**) As-grown crystal is used as seed, whereas (**d**) etched crystal is used. (**e**) When growth occurs in the interstices of solid grains, such as expected in metamorphism, a single crystal may show a large number of points from where dislocations are generated

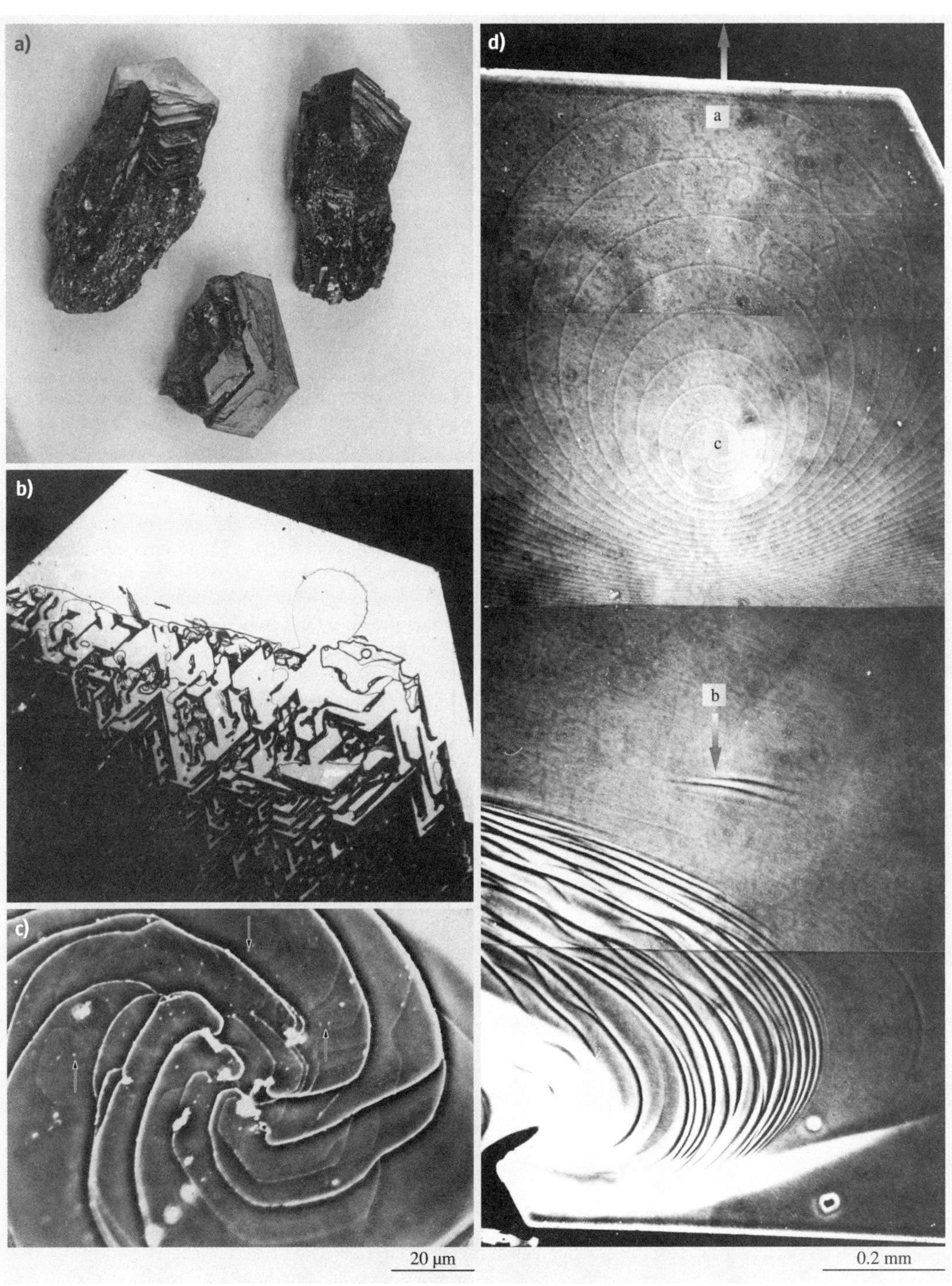
a)
b)
c)
d)
a
c
b
20 μm
0.2 mm

Fig. 5.16a–d How SiC crystals grow by the Acheson method (impure CVT) (**a–c**), and (**d**) by the Lely method (pure CVT or close to PVT). (**a**) Macro photograph of SiC crystals grown by the Acheson method, showing earlier dendritic growth followed by the appearance of {0001} face. (**b**) Ordinary reflection photomicrograph showing appearance of flat {0001} face through conjugation of microfacets of {0001} at the tips of dendrite arms, and (**c**) positive phase-contrast photomicrograph of growth spirals with larger and elemental (*arrows*) step heights, observable on flat {0001} surface such as shown in (**b**). Spiral steps originating from dislocations with larger Burgers vector show much brighter contrast than those originating from elemental (*arrows*) Burgers vector. Coexistence of growth spirals with larger and elemental Burgers vectors is due to the dissociation of dislocations with larger Burgers vector. (**d**) A positive phase-contrast photomicrograph showing an example of growth spirals commonly observed on SiC crystals synthesized by the Lely method. *Arrows* a and b indicate the direction to the center and the wall of the crucible, respectively, and c the spiral center. Step separation is eccentric from c to a and from c to b, due to the flow of source vapor (*gradient*) over the surface [5.32]. Bunching of spiral steps and perturbation of step morphology is seen in direction b ◄

observed in the crystals. These are illustrated schematically in Fig. 5.15, and a few representative examples will be described in the following.

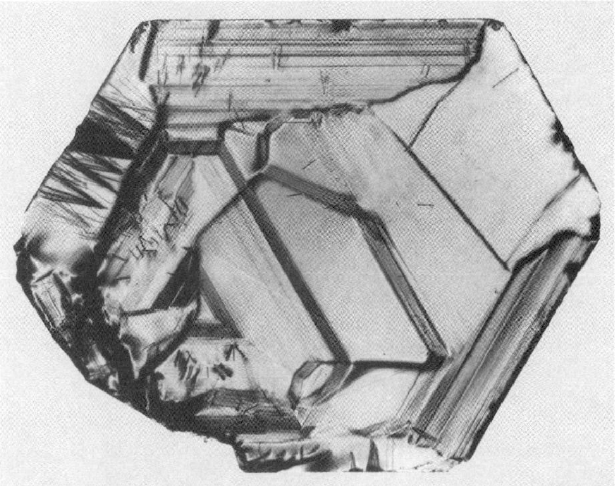

Fig. 5.17 X-ray topograph of a slice cut perpendicular to the c-axis of a prismatic quartz crystal from pegmatite (x-ray topography, courtesy of T. Yasuda)

A wide variety of spatial distributions of dislocations are observed in freely grown single crystals, such as natural minerals or crystals grown from solution phase. Since the spatial distribution of dislocations in single crystals provides important information relating how the crystal grew and what sort of conditional changes took place during its growth history, we may analyze these based on their investigations, coupled with observations of its surface microtopography and internal morphology. Three examples are now described.

SiC crystals synthesized by the Acheson method grow as dendrite at the initial stage on the wall of the reaction crucible. The dendritic arms conjugate together and the basal {0001} face starts to appear, inclined with respect to the length of the dendrites [5.33]. This can be seen clearly in Fig. 5.16a,b. Dislocations generated at the points where dendritic arms are close usually have large Burgers vector and dissociate into many dislocations with elemental Burgers vector. As a result, on the {0001} surface, coexistence of growth spirals with step heights corresponding to large and elemental Burgers vector is often observed. Figure 5.16c shows an example of growth spirals often observed on SiC crystals grown by the Acheson method. In the case of SiC crystals synthesized by the Lely method, more ideal growth spiral originating from independent dislocation are generally observed. However, these spirals often show eccentric step separation (Fig. 5.16d) due to the surface supersaturation gradient [5.32].

Figure 5.17 shows an x-ray topograph of a slice cut perpendicular to the c-axis of a prismatic quartz crystal occurring in a granitic pegmatite. Growth sectors corresponding to $\{10\bar{1}1\}$, $\{01\bar{1}1\}$, and $\{10\bar{1}0\}$ faces and sector boundaries, growth banding, and dislocation bundles generated from inclusions are observed in contrast images. Distinct anisotropy in the development of growth sectors, in the spacing of growth banding, and in the distribution and density of inclusions and dislocation bundles can be noticed among crystallographically equivalent growth sectors. The observed anisotropies imply that the quartz crystal grew under the effect of solution flow. On the growing $\{10\bar{1}1\}$ and $\{10\bar{1}0\}$ faces, facing the solution flow, more inclusions are precipitated, from where more dislocations are newly generated than on the surface growing on the opposite side.

Figure 5.18a,b shows x-ray and CL tomographs of a pear-shaped brilliant cut diamond, respectively. The squarelike pattern at the center in Fig. 5.18a is an outline of a tiny cuboid crystal whose overall form is close to cubic, although not bounded by flat {100} faces but by rough, near-{100} faces. The cuboid was formed elsewhere and transported into a different growth environment, where it acted as a seed for further growth under the new conditions. In Fig. 5.18a, it is noted that dislocations are newly generated on the surfaces of the seed cuboid and radiate in the form of bundles running parallel to ⟨100⟩, although there are also a few dislocations inherited from the seed. The Burgers vector of these dislocations is 100, which is different from those generally observed in most gem-quality diamonds, i. e., 110. In Fig. 5.18b, it is seen that growth of diamond on the seed transforms from rough {100} to smooth {111} morphology through the appearance of many {111} microfacets, indicating morphological evolution in the recovery process from rough to smooth interfaces [5.34]. The most likely place where the seed cuboid was formed is considered to be in ultrahigh-pressure high-temperature (UHPHT) metamorphic rocks formed by plate subduction. In these metamorphic rocks, minute diamond crystals are formed in porphyroblastic (crystals developed much larger than the coexisting ones) silicate minerals, such as garnet and zircon [5.35]. It is argued that these diamond crystals were formed in silicate–carbon liquid droplets formed by partial melting of porphyroblastic silicate mineral containing unmelted carbon [5.34]. The carbon source is assumed to be of organic origin, subducted from oceanic sediments. This explains the much higher concentration of diamond, attaining up to 2% in UHPHT metamorphic rock, as compared with the very low content of diamond (on the order of ppm) in mantle-originated ultramafic rocks. When these UHPHT metamorphic rocks are subducted deeper and digested in magma in the mantle, this diamond acts as a seed on which further diamond grows under much lower driving force conditions [5.34]. This example demonstrates that a large-scale geological movement or cycle is recorded in the form of the internal morphology, perfection, and homogeneity within a small crystal, provided that this information can be properly deciphered.

Natural diamond crystals experience a severe post-growth history. They grew in the depth of the Earth, under diamond-stable high-pressure high-temperature conditions, in silicate or carbonate solution phase, and were uplifted at great speed by volcanic action, to be quenched as a metastable phase through volcanic eruption. In the rapid ascent process, diamond crystals suffer partial dissolution, resulting in rounded forms and the formation of etch pits. They also experience plastic deformation, forming dislocation tangles and exsolution (precipitation) of impurity nitrogen. Natural diamonds are classified into type I and II, which differ in various physical properties related to their different nitrogen content. Type I contains higher nitrogen content and thus corresponds to C-N alloy, whereas type II contains

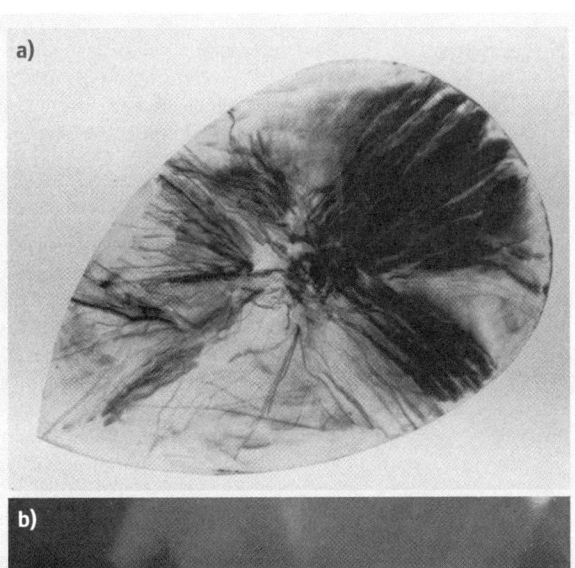

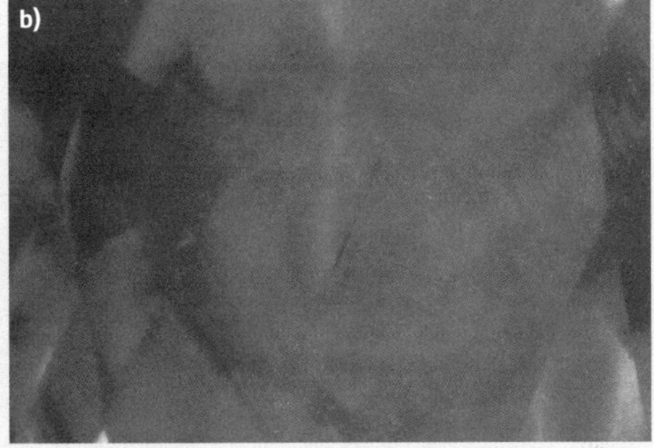

Fig. 5.18a,b X-ray topograph (**a**) and cathodoluminescence tomography (**b**) of a pear-shaped brilliant cut diamond. (**a**) demonstrates that most of this crystal was formed by growth under new conditions on a seed cuboid, seen at the center, which was formed elsewhere under different conditions and transported to the new conditions. Most dislocations with Burgers vector 100 are generated on the seed surface. (**b**) demonstrates that the morphology transformed from rough cuboid to octahedral bounded by smooth {111} faces via the appearance of a series of {111} microfacets, which is discernible from the distribution of brighter CL contrast corresponding to selective partitioning of impurity nitrogen to smooth {111} faces

far less nitrogen and corresponds to pure C. Due to this difference, their plastic deformation behavior is different, type I being plastically stronger than type II. As a result, when both types experience the same stress history, type II suffers greater plastic deformation. Among natural diamond crystals, type II crystals exclusively show irregular forms without crystallographic faces, whereas type I crystals exhibit polyhedral form with rounded corners and edges. Type I crystals exhibit dislocation bundles radiating from the center of a crystal on x-ray topographs, whereas type II crystals show bending with portions of slightly different orientations, indicating that type II crystals are heavily deformed compared with type I crystals, and sometimes even broken into pieces [5.36].

Observations on the perfection, homogeneity, and internal morphology of natural diamond crystals demonstrate that the whole growth and postgrowth histories are recorded in small crystals of diamond. We can see in a tiny diamond crystal the whole large-scale geological movements experienced by the crystal, provided that this record can be properly deciphered. Understanding the crystal growth mechanism, morphology, perfection, and homogeneity form the basis for properly reading this message sent from the depth of the Earth. It also indicates that natural and synthetic gemstones, including diamond, can be discriminated by these investigations, since both grow in solution phases but with different solvents (natural diamonds from silicate or carbonate, synthetic diamonds from metallic) and have different growth histories, although both are the same crystal species [5.29]. It is also possible to fingerprint two brilliant cut diamonds obtained and fashioned from one rough stone [5.34].

References

5.1 J.A. Burton, R.C. Prim, W.P. Slichter: The distribution of solute in crystals grown from the melt, Part 1, Theory, J. Chem. Phys. **21**, 1987–1991 (1953)

5.2 I. Kostov, R.I. Kostov: *Crystal Habits of Minerals* (Pensoft, Sofia 1999)

5.3 I. Sunagawa: *Crystals – Growth, Morphology and Perfection* (Cambridge Univ. Press, Cambridge 2005)

5.4 I. Sunagawa: Surface microtopography of crystal faces. In: *Morphology of Crystals, Part A*, ed. by I. Sunagawa (Reidel, Dordrecht 1987) pp. 321–365

5.5 I. Sunagawa: Growth of crystals in nature. In: *Materials Science of the Earth's Interior*, ed. by I. Sunagawa (Reidel, Dordrecht 1984) pp. 63–105

5.6 W.F. Berg: Crystal growth from solutions, Proc. R. Soc. Lond. Ser. A **164**, 79–95 (1938)

5.7 P. Curie: On the formation of crystals and on the capillary constants of their different faces, J. Chem. Educ. **47**, 636–637 (1970), translation of Bull. Soc. Franc. Min. Cryst. **8**, 145-150 (1885)

5.8 G. Wulff: Zur Frage der Geschwindigkeit des Wachstums und der Auflösung der Kristallflächen, Z. Krist. **34**, 449–530 (1901), in German

5.9 J.W. Gibbs: On the equilibrium of heterogeneous substances. In: *The Scientific Papers of J. W. Gibbs*, Vol. 1 (Longman Green, London 1906)

5.10 R. Kern: The equilibrium form of a crystal. In: *Morphology of Crystals, Part A*, ed. by I. Sunagawa (Reidel, Dordrecht 1987) pp. 77–206

5.11 P. Bennema, J.P. van der Eerden: Crystal graphs, connected nets, roughening transition and the morphology of crystals. In: *Morphology of Crystals, Part A*, ed. by I. Sunagawa (Reidel, Dordrecht 1987) pp. 1–75

5.12 P. Hartman: Modern PBC. In: *Morphology of Crystals, Part A*, ed. by I. Sunagawa (Reidel, Dordrecht 1987) pp. 269–319

5.13 A. Bravais: Les systemes formes par des pointes distributes regulierement sur un plan ou dans l'espace, J. Ecol. Polytech. **XIX**, 1–128 (1850), in French

5.14 J.D.H. Donnay, D. Harker: A new law of crystal morphology extending the law of Bravais, Am. Mineral. **22**, 446–467 (1937)

5.15 P. Hartman, W.G. Perdok: On the relations between structure and morphology of crystals. I, Acta Cryst. **8**, 49–52 (1955)

5.16 P. Hartman, W.G. Perdok: On the relations between structure and morphology of crystals. II, Acta Cryst. **8**, 521–524 (1955)

5.17 P. Hartman, W.G. Perdok: On the relations between structure and morphology of crystals. III, Acta Cryst. **8**, 525–529 (1955)

5.18 R. Uyeda: Crystallography of metal smoke particles. In: *Morphology of Crystals, Part B*, ed. by I. Sunagawa (Reidel, Dordrecht 1987) pp. 367–508

5.19 R.S. Wagner, W.C. Ellis: Vapor-liquid-solid mechanism of single crystal growth, Appl. Phys. Lett. **4**, 89–90 (1964)

5.20 I. Sunagawa: Vapor growth and epitaxy of minerals and synthetic crystals, J. Cryst. Growth **43**, 3–12 (1978)

5.21 A. Wells: Crystal habit and internal structure I + II, Philos. Mag. Ser. 7 **37**, 184–236 (1946)

5.22 H.E. Buckley: *Crystal Growth* (Wiley, New York 1951)

5.23 R. Hook: *Micrographia* (Royal Society, London 1665)

5.24 E.I. Givargizov: *Highly Anisotropic Crystals* (Reidel, Dordrecht 1986)

5.25 I. Sunagawa, Y. Takahashi, H. Imai, S. Yamada: Topological whisker bundles of amphibole and frost column of quartz, J. Cryst. Growth **276**, 663–673 (2005)

5.26 Y. Aoki: Growth of KCl whiskers on KCl crystals including the mother liquids, J. Cryst. Growth **15**, 163–166 (1972)

5.27 M. Kawasaki: Growth-induced inhomogeneities in synthetic quartz crystals revealed by the cathodoluminescence method, J. Cryst. Growth **247**, 185–191 (2003)

5.28 R.J. Reeder, J.C. Grams: Sector zoning in calcite cement: Implication for trace element distributions in carbonates, Geochim. Cosmochim. Acta **51**, 187–194 (1987)

5.29 I. Sunagawa: The distinction of natural from synthetic diamonds, J. Gemmol. **24**, 489–499 (1995)

5.30 K. Nassau, K.A. Jackson: Trapiche emeralds from Chivor and Muzo, Colombia, Am. Mineral. **55**, 416–427 (1970)

5.31 I. Sunagawa, H.-J. Berhardt, K. Schmetzer: Texture formation and element partitioning in trapiche ruby, J. Cryst. Growth **206**, 322–330 (1999)

5.32 I. Sunagawa, I. Narita, P. Bennema, B. van der Hoek: Observation and interpretation of eccentric growth spirals, J. Cryst. Growth **42**, 121–126 (1977)

5.33 I. Sunagawa: Surface micro-topography of silicon carbide, Sci. Rep. Tohoku Univ. Ser. III **12**, 239–275 (1974)

5.34 I. Sunagawa, T. Yasuda, H. Fukushima: Fingerprinting of two diamonds cut from the same rough, Gems Gemol. **Winter**, 270–280 (1998)

5.35 N.V. Sobolev, V.S. Shetsky: Diamond inclusions in garnets from metamorphic rocks: A new environment for diamond formation, Nature **343**, 742–746 (1990)

5.36 I. Sunagawa: A discussion on the origin of irregular shapes of type II diamonds, J. Gemmol. **27**, 417–425 (2001)

6. Defect Formation During Crystal Growth from the Melt

Peter Rudolph

This chapter gives an overview of the important defect types and their origins during bulk crystal growth from the melt. The main thermodynamic and kinetic principles are considered as driving forces of defect generation and incorporation, respectively. Results of modeling and practical in situ control are presented. Strong emphasis is given to semiconductor crystal growth since it is from this class of materials that most has been first learned, the resulting knowledge then having been applied to other classes of material.

The treatment starts with zero-dimensional defect types, i.e., native and extrinsic point defects. Their generation and incorporation mechanisms are discussed. Micro- and macrosegregation phenomena – striations and the effect of constitutional supercooling – are added. The control of dopants by using the nonconservative growth principle is considered. One-dimensional structural disturbances – dislocations and their patterning – are discussed next. The role of high-temperature dislocation dynamics for collective interactions, such as cell structuring and bunching, is shown. In a further section second-phase precipitation and inclusion trapping are discussed. The importance of in situ stoichiometry control is underlined. Finally two special defect types are treated – faceting and twinning. First the interplay between facets and inhomogeneous dopant incorporation, then main factors of twinning including melt structure are outlined.

6.1 Overview

The quality of single crystals and devices made therefrom are very sensitively influenced by structural and atomistic deficiencies generated during the crystal growth process. It is the chief task of the crystal grower to determine the conditions for their control, minimization or even prevention. Crystalline imperfections include point defects, impurity and dopant inhomogeneities, dislocations, grain boundaries, second-phase and foreign particles, twins, and so on. Some defect types, such as point defects, are in thermodynamic equilibrium and are therefore always present. This is due to the thermal excitations and entropic disordering forces at temperatures $T > 0$. Further, each crystal is bounded by surfaces with interface characteristics deviating from volume perfection. These facts prevent the growth of ideal, perfect crystals. Hence, in practice only optimal crystals are achievable.

Over more than a half-century of development of melt growth, most of the important defect-forming mechanisms have become well understood. Historical aspects of this progress were recently summarized by *Hurle* and *Rudolph* [6.2, 3]. Today, there exists an enormous knowledge about defect genesis in as-grown crystals supported by demanding theoretical fundamentals and computational modeling. As a result, the present state of technology makes it possible to produce crystals of remarkably high quality with tailored parameters fitting the demands of the device industry quite well. However, that is not to say that all problems are already solved. Thus, the present chapter will also cover still open questions and help to find optimal measures of defect engineering.

In this section first the defect types will be categorized in the classical manner of zero-, one-, two- and three-dimensional defects. Then some effects of defects on device properties will be covered.

6.1.1 Defect Classification

The international standard crystal lattice defects (defects in short), sketched in Fig. 6.1, are usually classified according to their dimension as follows [6.1].

Zero-dimensional defects are *point defects*, often referred to by the unpopular name *atomic-size defects*, which include the intrinsic defect types *vacancies*, *interstitials*, and in compounds, *antisites*. If extrinsic atoms are invoked unintentionally (as residual *impurities*) or intentionally (as *dopants*) they occupy interstitial or substitutional (lattice) positions. At growth temperatures, point defects are isolated and usually electrically charged. The charge state of point defects can lead to their interaction with electrically active dopants, creating *point defect complexes*.

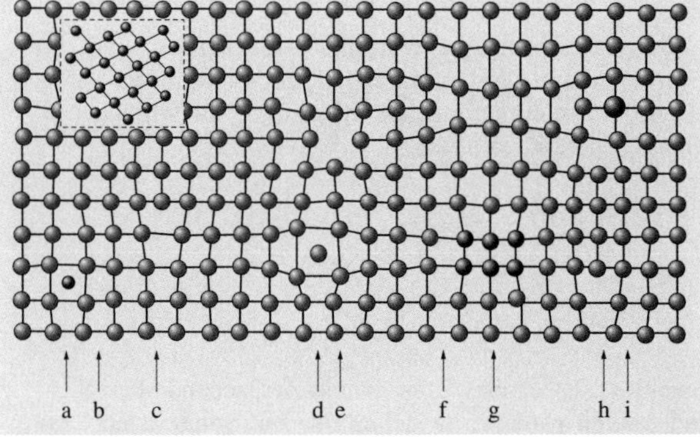

Fig. 6.1 Scheme of real crystal lattice with defects (after *Föll* [6.1]). a – interstitial impurity atom, b – incongruous inclusion, c – edge dislocation, d – self-interstitial atom, e – vacancy, f – vacancy-type dislocation loop, g – precipitate of impurity atoms, h – interstitial-type dislocation loop, i – substitutional impurity atom

One-dimensional defects include all kinds of *dislocations*, i. e., perfect screw and edge dislocations, mixed dislocations, partial dislocations (always in connection with a stacking fault), and dislocation loops. The propagation and interaction of dislocations over mesoscopic distances is the subject of *dislocation dynamics*. The collective screening behavior of dislocations contributes to their rearrangement in dipoles, walls, and networks, as well as under simultaneous stress in cell patterns and bundles. Whether the cell patterning is driven energetically or by a self-organizing process in the framework of equilibrium or nonequilibrium thermodynamics, respectively, is still the subject of research.

Two-dimensional defects are grain boundaries, stacking faults, phase boundaries, facets, and twins. A *low-angle grain boundary* structure is formed by the mechanism of dynamic polygonization and still belongs conventionally to a single-crystalline state. In contrast, *large-angle grain boundary* structures are formed by polycrystalline growth due to spontaneous or foreign nucleation processes. *Facets* are formed along atomically smooth planes, indicating the tendency of high-quality crystals to form polyhedra. They cause extrinsic point-defect and temperature-field inhomogeneities due to their fast lateral growth kinetics and enhanced radiation reflectivity, respectively. Grown-in *twins* are originated by a false stacking sequence, especially when the two-dimensional nucleus on a growing facet is disoriented.

Three-dimensional defects include second-phase particles (*precipitates*), intrinsic vacancy conglomerates (*microvoids*), and foreign particles (*inclusions*). It is important to differentiate between precipitates and inclusions, which are mostly confused in the literature. Whereas precipitates and microvoids are formed by supersaturation-driven condensation of intrinsic point defects, i. e., interstitials and vacancies, respectively, inclusions are melt–solution droplets, gas bubbles, and foreign microparticles incorporated at the growing melt–solid interface, especially when the melt composition deviates from the congruent melting point or contaminations are present. The two types of defects usually differ in their size, being 10–100 nm for precipitates and 1–10 μm for inclusions.

6.1.2 Consequences of Crystal Defects for Devices

Defects have deleterious effects on the performance, reliability, and degradation behavior of devices. Following the early classification of *Pick* [6.4] defects influence the:

- Structural properties (vacancies and interstitials may change the lattice constant; grains affect the single crystallinity)
- Chemical properties (defects participate in chemical reactions; their redistribution causes composition inhomogeneities)
- Electronic properties (defects occupy a specific state in the band structure)
- Scattering properties, i. e., the defect interacts with particles (phonons, photons of any energy, electrons, positrons, etc.).

As can be seen, the interaction processes are manysided and require wide interdisciplinary research with direct correlation to advanced technical progress. In fact, a large part of the worldwide technology progress depends on the control and manipulation of defects in crystals, above all in the semiconductor and optical industries, but also in biotechnology and many others. There are an enormous number of monographs and publications dealing with this topic. In the following, however, only selected examples of correlation between device characteristics and defects will be touched upon. An instructive review about defects in semiconductors and their electronic properties is given by *Mahajan* [6.5].

Point defects determine the basic properties of the materials used in devices. Parameters such as the specific resistance of semiconductors, conductance in ionic crystals, or diffusion properties in general, which may appear to be intrinsic properties of a material, are in fact defect dominated. In optical devices the transmittance, birefringence, and refractive index are influenced by the density and distribution of intrinsic and extrinsic point defects very sensitively. For instance, high-quality electrooptical and nonlinear optical devices of $LiNbO_3$ require an extremely accurate constant Li/Nb ratio (congruent composition) to ensure birefringence homogeneity of $(5-7)\times10^{-5}$ [6.6]. Intrinsic point defects can influence the conduction type in semiconductors. For instance, vacancies and interstitials in silicon can have distinct acceptor and donor energy levels within the bandgap, respectively [6.5]. As_{Ga} antisites in GaAs are deep-level donors (EL2). Their density, which depends on the deviation from stoichiometry, determines the compensation doping level by a shallow acceptor (carbon) in order to ensure semi-insulating property in high-frequency circuits [6.7]. Point defects and their diffusion have a strong impact on the noise characteristics through an interaction of charge carriers with the fluctuating local ionic surrounding. In (Hg,Cd)Te infrared photodiodes there is a linear dependence between the $1/f$ noise power and the fraction of ionized Hg vacancies providing p-type conductivity [6.8]. In some cases, however, a redistribution of extrinsic defects is even desirable. For instance, in micro laser waveguides for integrated optics (e.g., $LiNbO_3$ fibers) the accumulation of certain dopants (Mg) near the surface helps to concentrate the laser beam in the center by the effect of refractive-index cladding [6.9].

As is well known, *dislocations* are defects influencing the quality of nearly all types of devices insofar as they act as getters for point defects and dopants so that they can contribute to electrical and chemical inhomogeneity. They decrease transmission in lenses, affect light intensity in laser rods, and influence the mechanical stability in piezo- and acoustoelectric transducers. Dislocations in substrates are transformed as threading dislocations into the epitaxial layers grown on them. When an overcritical misfit between the lattice parameters of a substrate and epilayer exists, misfit dislocations are generated at their interface. Both threading and misfit dislocations play an important role in the rapid degradation of (Ga,Al)As/GaAs lasers. The degradation of light-emitting semiconductor diodes and lasers follows from the fact that dislocations cause nonradiative recombination and decrease luminescence efficiency. They reduce the minority-carrier lifetime and, when the spacing between them is comparable to the diffusion length, luminescence efficiency breaks down [6.10]. Dislocations can contribute to electronic behavior, especially, in diamond and zincblende structures where the cores of glide and shuffle set dislocations are associated with dangling bonds [6.5]. Recently it was shown that small screw dislocations and threading edge dislocations are the most common defects in 4H-SiC homoepitaxial devices produced by chemical vapor deposition on SiC substrates. As their densities increase, the breakdown voltage of Schottky devices is decreased [6.11]. The central parameter of field-effect transistors (FETs) is the turn-on threshold voltage (V_{th}), the fluctuations of which across a wafer

Table 6.1 Selected examples of defects in melt-grown crystals demonstrating their adverse effect on device quality (BPT – bipolar transistor, PD – photodiode, IR – infrared, KTN – $K(Ta_xNb_{1-x})O_3$, ME – microelectronics, MC – multicrystalline, LED – light-emitting diode, LD – laser diode, PVE – photovoltaic efficiency, UV – ultraviolet, YAG – $Y_3Al_5O_{12}$, MOS – metal–oxide–semiconductor, HBT – heterostructure bipolar transistor, MMIC – monolithic microwave integrated circuit, NLO – nonlinear optic, AO – acoustooptic)

Material	Defect type	Device version	Adverse effect
	Zero-dimensional		
Si	Interstitial Si_i	BPT, PD	Ionized donor level, carrier traps
CdTe	Vacancy V_{Cd}	Radiation detector	Shallow donor reducing electrical resistivity
(Hg,Cd)Te	Vacancy V_{Hg}	IR photodiodes	Increased $1/f$ noise power
GaAs	Antisite As_{Ga}	Radiation detector	Deep level trap reducing carrier lifetime
KTN	Ta-rich striations	Optical modulator	Optical inhomogeneity, refractive index change
	One-dimensional		
Si	Dislocation loops	ME circuits	Swirl formation, shorts
Si	Dislocations	MC-Si solar cell	Affecting miniority carrier lifetime
GaAs	Dislocations	LED, LD	Nonradiative recombination
(Cd,Zn)Te	Cell structure	Radiation detector	Impediment of electron transport
	Two-dimensional		
Si	Grain boundaries	MC-Si solar cell	Decreasing PVE by impurity (Fe) gettering
InP	Twins	ME circuits	Decreasing usable crystal gain
CaF_2	Grain boundaries	UV lenses	Light scattering and radiation damage
YAG	Facets	Solid state laser	Refractive index variation, optical loss
	Three-dimensional		
Si	V_{Si} clusters (voids)	MOS circuits	Gate oxide degradation by local thinning
GaAs	As precipitates	HBT, MMIC	Impairment of wafer polishing and epitaxy
$LiNbO_3$	Eutectic inclusions	NLO modulators	Light scattering, birefringence impairment
$PbMoO_3$	Pb-rich inclusions	AO transmitter	Light scattering, reduced transmission

must be minimal to ensure high device yield. In ion-implanted GaAs FETs a shift of V_{th} around dislocations was observed which has been explained not only by the dislocation presence but also by the enhanced GaAs antisites and As interstitials on dislocations [6.12]. Thus, in order to remove the decorating defects from dislocations today each as-grown GaAs crystal is postannealed before it is applied for device technology. Note that dislocations themselves are practically immune to post-growth thermal treatment.

Cellular structures of dislocations and *grain boundaries* are two-dimensional defects responsible for harmful optical and electrical inhomogeneities. For instance, across semi-insulating {100} GaAs wafers a mesoscopic resistivity variation is observed due to the accumulation of As_{Ga} antisite defects (EL2) within the cell walls [6.13]. Subgrain boundaries also impede the electron transport, as in $Cd_{1-x}Zn_xTe$ radiation detectors [6.14]. In general, since grain boundaries are defects in the crystal structure, they tend to decrease the electrical and thermal conductivity of the material. Additionally, the high interfacial energy and relatively weak bonding in grain boundaries makes them preferred sites for the onset of corrosion and the precipitation of new phases from the solid. The presence of a small-angle grain boundary structure in melt-grown LiF crystals prevents their use in monochromators and x-ray analyzers due to the high light and x-ray diffraction scatterings, respectively [6.15]. Increasing quality is demanded for CaF_2 lenses, which are used in deep-ultraviolet (UV) semiconductor microlithography. The stepwise reduction of the exposing wavelength down to

157 nm, correlating with the smallest circuit structure size, requires a dramatic improvement of the growth and annealing conditions of CaF_2 crystals. The highest transmission and lowest radiation damage can be only achieved when the crystals are free of grain boundaries [6.16].

It is noteworthy that there is also a certain interest in crystals with mosaic structure. For instance, diffraction lenses for nuclear astrophysics show an improved reflection power when crystals with mosaicity of 20–50 arcs (e.g., $Ge_{1-x}Si_x$) are used [6.17]. Further, in nanocrystalline materials, controlled reduction of grain size to nanometer scale leads to many interesting new properties including a great increase in strength [6.18]. Therefore, the further development of knowledge about collective dislocation interactions in growing crystals is of general practical relevance for both targets, i.e., both suppressing and promoting cellularity.

Precipitates and *inclusions* are second-phase particles of autonomous crystallographic structure and chemical composition forming an interfacial boundary with the matrix material. They induce local parameter and stress fluctuations and, therefore, mostly misfit dislocations [6.19,20]. In liquid encapsulation Czochralski (LEC) GaAs substrates arsenic precipitates affect the device properties of epitaxial-type metal-semiconductor field effect transistors (MESFETs). They also cause the formation of small surface oval defects on molecular beam epitaxy (MBE) layers [6.5]. Principally, inclusions impair the surface quality of wafers during the polishing process. In lenses and backside radiation detectors they reduce the transmission quality by light scattering.

Table 6.1 summarizes selected correlations between device quality impairments and the responsible defect types.

6.2 Point Defects

At all temperatures above absolute zero, equilibrium concentrations of vacancies, self-interstitials, and in the case of compound semiconductors, antisite defects will exist. This is because point defects increase the configurational entropy, leading to a decrease in free energy. Thus, such native point defects are always presented in as-grown crystals. However, their concentration can be influenced by the growth conditions very sensitively. In silicon a nearly defect-free situation due to vacancy–interstitial annihilation can be achieved by selection of a certain relation between the temperature gradient and crystallization rate. In compound crystals in situ control of stoichiometric growth conditions can minimize the intrinsic defect density.

Extrinsic point defects are incorporated arbitrarily or deliberately as impurities or dopants, respectively. Today, the purification techniques of the starting charge materials are of such a high standard that total residual impurity concentrations fall below the frozen-in contents of native point defects. In elemental crystals (e.g., silicon) purity levels less than $10^{14}\,cm^{-3}$ and in compound crystals (e.g., GaAs) values below $10^{15}\,cm^{-3}$ can be achieved. Therefore, their influence on the crystal lattice parameter and electrical parameters is of secondary significance. Of course, due to a contaminated growth atmosphere or because of strong chemical affinities it can happen that certain arbitrary elements are introduced in enhanced concentrations, such as oxygen in CaF_2 or carbon in silicon, for example. In contrast, for light-emitting devices the required dopant concentrations markedly exceed those of impurities and usually exceed $10^{18}\,cm^{-3}$ or amount even to $10^{21}\,cm^{-3}$, as for Cr^{3+} in ruby laser crystals. In these cases homogeneous incorporation is of essential technological importance but is complicated by the natural segregation effect that may lead to macro- and microdistributions. A characteristic structural impairment during growth from doped or incongruent melts can arise from constitutional supercooling – the interplay between rejected dopants or excess atoms, diffusion, and heat transfer at the growing melt–solid interface. Growing-in dopants interact with native point defects, which are isolated and mostly ionized at growth temperatures. As a result, the physical efficiency of dopants can be reduced by compensation and complex formation.

It is the aim of the crystal grower to understand these interactions on the atomic scale and determine their correlations to the growth conditions in order to master chemical and electronic homogeneities as much as possible.

6.2.1 Native Point Defect Generation

Thermodynamics

All thermodynamic processes strive to minimize the free energy. Applied to the crystallization process this means that the single-crystalline state is a normal one because the free thermodynamic potential G (free potential of Gibbs) is minimal if the *crystal growth units* (atoms, molecules) are perfectly packed in a three-dimensional ordered crystal structure, i. e., the atomic bonds are saturated regularly. Because the sum of the atomic bonds yields de facto the potential part H, i. e., the enthalpy part of the internal crystal energy $U = H - PV$ (where P is pressure, and V is volume), the process of ordering responsible for adjustment of the *crystal periodicity* is characterized by the minimization of enthalpy ($H \rightarrow$ min).

On the other hand, however, an ideally ordered crystalline state would imply an impossible minimal entropy S. Thus, the opposite process of increasing entropy, i. e., disordering ($S \rightarrow$ max) gains relevance with increasing temperature T. This is expressed by the basic equation of the thermodynamic potential of Gibbs

$$G = U + PV - TS = H(\downarrow) - TS(\uparrow) \rightarrow \min . \quad (6.1)$$

Hence, crystallization is composed of two opposite processes:

i) Regular
ii) Defective arrangement of the *growth units*.

Considering this dialectics of ordering and disordering forces at all temperatures above absolute zero it is not possible to grow an absolutely perfect crystal. In reality *no ideal* but only an *optimal* crystalline state can be obtained. In other words, in thermodynamic equilibrium the crystal perfection is limited by incorporation of a given concentration of native point defects n.

Neglecting any effects of volume change, defect type, and defect interplay, at constant pressure the equilibrium defect concentration n can be determined from the change of thermodynamic potential by introducing the defect as

$$\Delta G = \Delta H_d - \Delta S_d T \rightarrow \min , \quad (6.2)$$

with $H_d = nE_d$, the change of internal energy due to the incorporation of n defects, depending on the total defect formation energy E_d, and $S_d = k_B \ln\{(N!)/[n!(N-n)!]\}$ the accompanying change of entropy (configurational entropy), where k_B is the Boltzmann constant and N the total number of possible sites. After substitution and application of Stirling's approximation for multiparticle systems such as a crystal ($\ln N! \approx N \ln N$, $\ln n! \approx n \ln n$, $\ln(N-n)! \approx (N-n)\ln(N-n)$) (6.2) becomes

$$\Delta G = nE_d - k_B T \times [N \ln N - n \ln n - (N-n)\ln(N-n)] . \quad (6.3)$$

Setting the first derivative of (6.3) as $\partial \Delta G / \partial n = 0$ to yield the energetically minimum defect concentration n_{min}, and considering $N \gg n$, the *perfection limit* of a crystal is

$$n_{min} = N \exp\left(\frac{-E_d}{k_B T}\right) , \quad (6.4)$$

which is exponentially increasing with temperature. Setting $N = 5 \times 10^{22}$ atoms cm^{-3} and $E_d = 1$ eV (the vacancy formation energy in metals) the minimum defect concentrations n_{min} at 1000 and 300 K are about 5×10^{17} and 10^6 cm^{-3}, respectively. Note that, in the case of formation of vacancy–interstitial complexes (Frenkel defects), the value of $n_{min}^{(F)}$ is somewhat modified and yields $\sqrt{N_{is}N} \exp(-E_d^{(F)}/k_B T)$, where N_{is} is the total number of interstitial positions depending on the given crystal structure, and $E_d^{(F)}$ is the energy of formation of a Frenkel defect. More fundamental details are given in *Kröger*'s compendium [6.21], which remains even today one of the basic guides for the crystal grower.

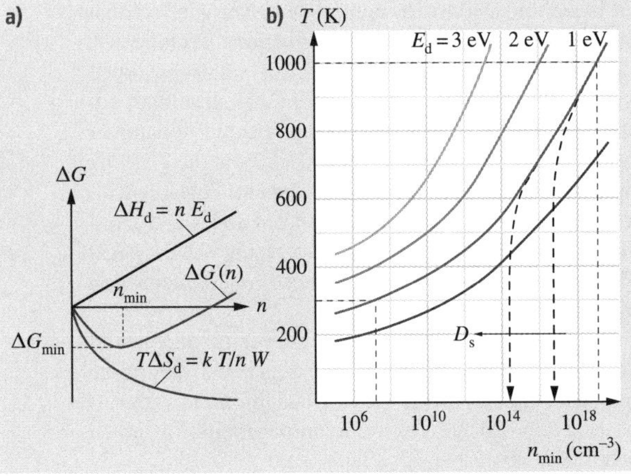

Fig. 6.2 (a) Schematic illustration of the equilibrium defect concentration (*perfection limit* n_{min}) obtained by superposition of defect enthalpy H_d and entropy S_d using (6.2–6.4). (b) Minimum defect concentration versus temperature at various defect energies E_d according to (6.4). *Dashed lines* show the *freezing-in* courses of high-temperature defects for different migration coefficients D_s

Figure 6.2a,b shows the functions $\Delta G(n)$ and $n_{min}(T)$ schematically. Due to the limitation of diffusion rate, a certain fraction of high-temperature defects freeze in during the cooling-down process of as-grown crystals (broken lines) and exceed the equilibrium concentration at room temperature markedly (Fig. 6.2b). In other words, in practical cases the intrinsic point-defect concentration is still far from thermodynamic equilibrium.

In principle, the total defect formation energy consists of various factors: $E_d = E_{form} + U_{vib} + S_{vib}T$ where E_{form} is the relevant defect formation energy, and U_{vib} and S_{vib} are the energy and entropy terms for the vibration contributions to the free energy. Of course, each contribution in its turn depends on temperature. Their exact experimental determination requires high-purity crystals and extremely precise analytical techniques, for instance, measurements of the thermal dependence of dilatation combined with precision x-ray analysis of the lattice constant. The native point defect concentration is proportional to the difference between the relative increase of length and the change of the lattice constant as $n \sim (\Delta L/L - \Delta a/a)$, whereas a positive or negative amount identifies the prevailing presence of vacancies or interstitials, respectively. No dilatation effect appears when the number of vacancies equals the number of interstitials. In principle, such measurements and most others begin to fail at defect concentrations below $\approx 10^{17}\,cm^{-3}$. Therefore, theoretical treatments come to the fore, e.g., quasichemical, molecular dynamics (MD), ab initio, and first-principle calculations. Table 6.2 shows the formation energies and entropies of native point defects near the melting point in some important materials. The activation energies of defect migration are also given.

Note that these values for compound semiconductors differ in the literature markedly and Table 6.2 is only of approximate character. Additionally, one has to consider the multiple charge stages of native defects showing various energies depending on the position of the Fermi level [6.24].

It is clear that a given native point-defect content corresponds to each crystalline phase, like a *solute component* in a system with ideal mixing. Compound materials show a phase extent, termed the existence or homogeneity region, that deviates from the stoichiometric composition by a certain value. Assuming a conservative compound system AB with defect balance, the maximum deviation δ can be expressed in terms of the concentrations of the native point defects in each sublattice as [6.30]

$$\delta = \delta A - \delta B = \{[A_i] - [V_A] + 2[A_B] - 2[B_A]\} - \{[B_i] - [V_B] + 2[B_A] - 2[A_B]\}\,, \quad (6.5)$$

where $[A_i]$, $[B_i]$, $[V_A]$, $[V_B]$, $[A_B]$, and $[B_A]$ are the interstitial, vacancy, and antisite contents of A and B, respectively. In simple terms, by using (6.4) for each defect type in (6.5) the maximum equilibrium defect solubility in a given AB crystal at each temperature becomes identical to the solidus curve of the existence region in the phase diagram. Table 6.3 shows the maximum phase extents of some compound materials. As can be concluded from (6.4), in a cooling crystal the solidus curves take a retrograde course leading under realistic cooling rates to freeze in of high-temperature nonequilibrium defects (Fig. 6.2b). In standard-grown

Table 6.2 Selected formation energies, entropies, and activation energies of migration in selected materials (A, B_i – interstitial, $V_{A,B}$ – vacancy, A_B – antisite, k_B – Boltzmann constant, $^{\pm i}$ – charge stage and E_f in the mid-gap)

Material	Defect	Formation energy E_f (eV)	Formation entropy S_f (k)	Activation energy of migration E_m (eV)	Reference
Cu	Cu_i	1.1			[6.22]
	V_{cu}	0.78–1.2	1.5	0.52–0.62	[6.22]
Si	Si_i	1.1, 3.46	1.4	0.937	[6.5,23]
	V_{Si}	2.3, 2.48	−3.7	0.457	[6.5,23]
GaAs	As_i	(5.5^{2+})			([6.24])
	V_{Ga}	2.59 (3.1^{2-})	32.9	1.7	[6.10] ([6.24]) [6.25]
	V_{As}	2.59 (3.7^{1+})	1.1		[6.10] ([6.24])
	As_{Ga}	3.21 (2.0^{2+})			[6.10] ([6.24])
CdTe	Te_i	1.97 (3.67)			[6.26] ([6.27])
	Cd_i	0.96	11.1	2.47–2.67	[6.28,29]
	V_{Cd}	3.55, 3.84 (4.7)	−5.6		[6.29] ([6.27])
	Te_{Cd}	0.81 (2.29)			[6.26] ([6.27])

Table 6.3 Maximum widths of phase extent δ_{max} of selected compounds [6.6, 19, 31]

Material	InP	GaAs	CdTe	CdSe	PbTe	SnTe	$LiNbO_3$
δ_{max} (mole fraction)	5×10^{-5}	2×10^{-4} (2×10^{-3})	1×10^{-4} (3×10^{-4})	5×10^{-4}	1×10^{-3}	1×10^{-2}	≈ 5
Side of maximum deviation and congruent melting point	In-rich	As-rich	Te-rich	Cd-rich	Te-rich	Te-rich	Nb_2O_5-rich

compound crystals the real intrinsic point-defect concentration at room temperature is between 10^{15} and $10^{17}\,cm^{-3}$. This is about 7–9 orders of magnitude higher than the calculated values in thermodynamic equilibrium (if E_d is assumed to be 1–2 eV).

For silicon crystals near to the melting point it is accepted that vacancies and interstitials are simultaneously present in concentrations of about $10^{14}-10^{15}\,cm^{-3}$ and that they can recombine very rapidly [6.32]. In the end, the ratio between them within the cooled crystal is determined by the relation of the atomistic transport processes (e.g., diffusion and thermodiffusion), the recombination rate, and the applied growth parameters (e.g., the temperature gradient and pulling velocity). The situation in compound materials, e.g., semiconductors such as $A^{III}B^{V}$ and $A^{II}B^{VI}$, is more complicated and less well studied. The equilibrium point defect concentrations at the melting point tend to be much higher than in Si. *Hurle*'s calculations [6.30, 33] in GaAs of the concentrations of the principal native point defects at the melting temperature yield values about $10^{17}-10^{19}\,cm^{-3}$ for $[V_{Ga}]$, $[Ga_i]$, and $[As_i]$, $[V_{As}]$, respectively. This is comparable to or even greater than the intrinsic carrier concentrations. Due to their isolated, and usually electrically charged, states they can influence the position of the Fermi level. This results in a complex interaction between electrically active dopants and the native point defects, which can exist in more than one charge state (Sect. 6.2.2).

Generation and Incorporation Kinetics

Principally, increasing the temperature raises the probability of thermal (vibrational) and entropy-driven (configurational) generation of Frenkel defects. Depending on the temperature gradients acting along the crystal and the cooling rate the initially coincident concentrations of vacancies and interstitials can differ markedly due to differing migration rates (Table 6.2). As a result, one defect type can move to the crystal surface, leaving defects of Schottky type. Further, during the crystallization process, point defects can be incorporated from the melt at the propagating melt–solid interface. Hence, one has to distinguish between:

i) Intrinsic defect generation
ii) Incorporation from outside.

In any case, the principles of both energy minimization and electrical neutrality act as *regulators* of concentration and charge balancing as far as equilibrium conditions occur. Lowering the energy enhances the recombination probability of interstitials and vacancies during the cooling-down process of the growing crystal. Heat and mass flows, however, can lead to deviations from thermodynamic equilibrium.

During crystal pulling from the melt or unidirectional solidification, native point defects undergo various types of transport kinetics, i.e., capture at the interface due to the crystal translation or phase boundary propagation (often designated *convection*), Fickian diffusion by jumping via interstitial sites, kick-out or vacancy occupation mechanisms, and temperature-gradient-driven thermal diffusion (the Soret effect). Until now silicon is the material best studied. The prevailing transport mode at each temperature and in each crystal region has been estimated by dimensionless numbers of point defect dynamics [6.34]. The numbers of Péclet $Pe_{I,V} = vL/D_{I,V}(T)$, Damköhler $Da_{I,V} = k_r C^{eq}_{V,I}(T)L^2/D_{I,V}(T)$, and Soret $ST_{(I,V)} = D_T/D_{I,V}(T) \approx v k_B T^2/H^F_{I,V} G_T D_{I,V}$ compare the convection, recombination, and thermodiffusion, respectively, with Fickian diffusion, where v is the crystal growth velocity, L is the characteristic length, k_r is the recombination reaction constant, $C^{eq}_{V,I}$ is the equilibrium concentration, D_T is the coefficient of thermodiffusion, $D_{I,V}(T)$ is the coefficient of Fickian diffusion of interstitials and vacancies, k_B is the Boltzmann constant, T is absolute temperature, $H^F_{I,V}$ is the formation enthalpy, and G_T is the temperature gradient at the melt–solid interface. Setting the parameters for silicon [6.23, 34] one can see that convective and Fickian diffusion flows are comparably strong and compete with each other in the hot crystal section, that recombination proceeds very fast at high temperatures, but that thermodiffusion contributes only marginally. Thus, variations of the pulling rate can shift point-defect transport between convection- and diffusion-dominated regimes.

As was proved experimentally [6.35], at high pulling rates the flux of vacancies dominates over that of self-interstitials. On the other hand, at low pulling rates or with high temperature gradients, interstitials are in excess. This fact is of high significance for in situ control of the native point defect type and content during growth (see later).

The situation in semiconductor and oxide compound crystals is not nearly as well understood as in silicon. There are still no detailed studies demonstrating the correlation between temperature gradient at the growing interface, crystallization rate, and content of the given intrinsic defect types. Of course, one reason is the presence of dislocations acting as effective getters (Sect. 6.3) that make it difficult to study the point defect dynamics in the pure form. In principle, as demonstrated in (6.5), the situation in multicomponent materials is additionally complicated due to the presence of equivalent defect types in each sublattice and the formation probability of antisites. It is established that the As_{Ga} antisite in GaAs crystals, known as the neutral EL2 defect with charged states $EL2^{+}$ and $EL2^{2+}$, does not appear at the growth temperature but forms during the cooling-down process. In material grown from a near-congruent composition, i. e., slightly As-rich melt, the As_i content removes the $[V_{Ga}]$ by producing As_{Ga} as soon as the lowering temperature evokes supersaturation [6.33]. A nearly identical mechanism takes place in CdTe. In cooling Te-saturated samples the antisite Te_{Cd} does form and may become important as singly and doubly ionized mid-gap donors $Te_{Cd}^{1+,2+}$ responsible for compensation of the shallow native acceptor $V_{Cd}^{1-,2-}$ [6.36].

For understanding the defect trapping kinetics at a propagating melt–solid interface, first of all one has to clarify whether the front moves by atomically rough or smooth morphology. This can be estimated by the Jackson factor [6.37] $\alpha = (w/u)\Delta H/k_B T$, where ΔH is the enthalpy of crystallization, w is the number of nearest neighbors of an atom on the face, u is the number of nearest neighbors of an atom in the crystal, k_B is the Boltzmann constant, and T is absolute temperature (see also Sect. 6.5). Whereas metals show α values of about 1, i. e., a rough interface, dielectrics crystallize mostly with atomically smooth interfaces due to typical α values larger than 2. In between stand the semiconductor materials with their covalent bondings and, hence, α factors around 2. In this case the crystallographic binding structure along the crystallizing plane (w/u) determines whether the interface is microscopically smooth or rough. Semiconductors with diamond, zincblende, and wurtzite structures grow from the melt along most directions by the atomically rough mode. They tend to form atomically smooth interfaces only on their most close-packed (i. e., {111}) planes (see also Sect. 6.5).

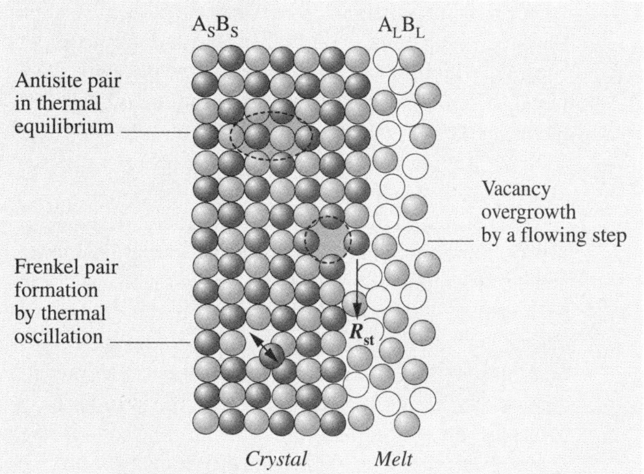

Fig. 6.3 Sketch of intrinsic point defect kinetics of an AB compound at an atomically flat melt–solid interface propagating by step-growth mechanism (R_{st} – lateral step-growth rate)

On atomically rough interfaces atoms can be added singly without the need for nucleation, i. e., at very low chemical potential difference between solid and liquid phases. As a result, possible defect sites are added to the crystal under quasi-equilibrium conditions. On the contrary, for atomically smooth planes much higher supercooling is required in order to initiate two-dimensional (2-D) nucleation followed by layer-by-layer growth. In such a case vacancies, interstitials and even foreign atoms, possibly delivered by the adjacent liquid boundary layer, can be overgrown very rapidly and, hence, incorporated in metastable states if their diffusion rate back to the melt is not high enough. The sketch in Fig. 6.3 demonstrates this situation. According to *Chernov* [6.38] equilibrium incorporation takes place only when the lateral step rate R_{st} fulfils the inequality

$$R_{st} < \frac{D_{IF}}{h} , \quad (6.6)$$

where D_{IF} is the interface diffusion coefficient (10^{-4}–10^{-5} cm^2/s) and h is the step height. The actual step growth velocity can be estimated by the relation

$$R_{st} = \beta_i \Delta T , \quad (6.7)$$

where β_i is the kinetic step coefficient [6.38, 39], which for silicon is 50 cm/(s K) [6.40], and ΔT is the supercooling, which has been determined at growing {111} facets of silicon crystals to be a maximum of 7 K [6.41]. Hence, at very high step flow rates, i. e., relatively high supercooling, nonequilibrium (metastable) point-defect incorporation can result.

It seems that the fluid phase should deliver sufficient imperfections due to its structural instability and disordered character. However, as recent MD simulations have demonstrated, between the liquid and crystal surface a characteristic transition region of several atomic layers exists, within which the thermal motion of the atoms decreases and the structure approaches the crystalline one [6.39]. In fact, as was ascertained by high-resolution in situ transmission electron microscopy, the {111} solid–liquid interface of Si has already a well-ordered transition layer on the atomic scale, which is compatible with defect-free 1×1 Si-{111} surface [6.42]. It can be assumed that the presence of such an ordered transition region should stabilize the growth kinetics against incorporation of an excess of deficiencies. According to *Motooka*'s MD simulations [6.43], point defects are formed directly at the interface due to the density misfit between the liquid and solid phases.

Point Defect Engineering

In order to ensure specified electrical and optical qualities of single crystals and devices made therefrom the control of native point defects plays an important technological role, especially if harmful secondary reactions are evoked by point imperfections, such as precipitation (Sect. 6.4.1) and microvoid generation. From an experimental point of view, there are two possible ways of defect mastering:

i) In situ control during the crystal growth process
ii) Postannealing of the crystal bulk or pieces cut therefrom.

The first method is of increasing interest due to its cost benefits.

In the case of low-dislocation or even dislocation-free crystal growth, the complete native point defect dynamics come into play. The absence of dislocations leads to a quasihomogeneous defect interplay without catalytic effects. As a result the relatively high supersaturation promotes vacancy and/or interstitial condensations within the crystal volume, escalating to unfavorable micro and mesoconglomerations. For instance, in silicon at high pulling rates, vacancies are incorporated in excess and condense during cooling down to form octahedral voids of ≈ 100 nm. On the other hand, at low pulling rates interstitials are in excess, forming a network of dislocation loops. In between, a defect-free region is obtained which is bounded by unwanted oxidation-induced stacking faults (OSF). The balance between the number of vacancies and interstitials is the controlling factor. Based on these experimental observations three ways to achieve microdefect-free silicon have been worked out [6.35]:

i) Keeping the growth conditions within the defect-free regime, which is approximately $\pm 10\%$ around the critical ratio $v/G_T = 1.34 \times 10^{-3}\ \text{cm}^2/(\text{K min})$, where v is the growth rate and G_T is the temperature gradient at the interface. However, such a small tolerance permits only very low pulling velocities of about 0.5 mm/min. *Falster* and *Voronkov* [6.44] described the in situ outdiffusion of interstitials. In this case, the crystals are pulled under interstitial-rich conditions and maintained at high temperatures for extended times, thus utilizing the very high migration speed of interstitials. However, extended cooling times with very low cooling rates are required.
ii) Keeping a maximum pulling rate with fast cooling followed by a wafer annealing process to reduce the grown-in defect sizes.
iii) Using a cost-optimized approach, a so-called *flash wafer* step [6.35], where only a thin Si layer of 0.5 μm is deposited onto the wafer surfaces. This combines maximum pull rate and fast cooling with low-cost treatment.

The in situ control of native point defects in compound crystals is coupled with the feasibility of accurate composition control during the growth process and, therefore, with exact knowledge of the phase diagrams. Taking the T–x-phase projections (where x is the mole fraction) one can find numerous AB compounds with a region of homogeneity containing the strict stoichiometric composition and extending to both sides of stoichiometry, such as CdTe and PbTe, for example [6.19]. However, there are also compounds with asymmetric phase extents shifted to anion or cation excess, even not including the stoichiometric composition, such as GeTe and SnTe, for example [6.45]. Also in $LiNbO_3$ the prevailing part of the existence region is located on the Nb_2O_3-rich side [6.46]. Unfortunately, for a lot of compounds the exact shape and width of the homogeneity region has not yet been clarified, especially when very small phase extents exist, as in InP, InAs,

Part A | 6.2

Fig. 6.4a–c Crystal growth arrangements for in situ stoichiometry control by partial vapor pressure of the volatile element. (**a**) Horizontal Bridgman (HB) method. (**b**) Vertical gradient freeze (VGF) technique. (**c**) Vapor-pressure-controlled Czochralski (VCZ) method ▶

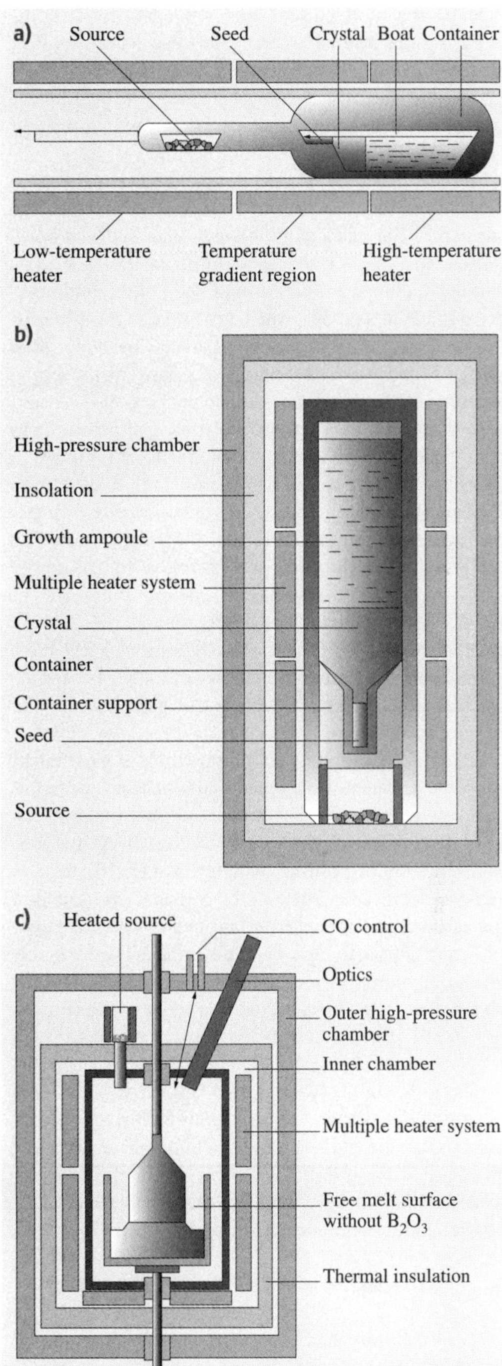

SiC, and $PbMoO_4$, or if phase relation analysis is difficult, as for ZnO and GaN. Even for GaAs only recently have thermochemical calculations [6.47, 48] confirmed the earlier and newer growth observations [6.49] that the phase extent is deviated exclusively towards the As-rich side. In principle, for materials with phase extents containing or touching the stoichiometric composition, it is obvious from theoretical calculations and annealing studies that stoichiometric growth conditions guarantee a minimum total native point defect concentration.

A method that has been well known for a long time is the in situ control of near-stoichiometric growth by applying a fixed-temperature vapor source of the volatile component in the horizontal Bridgman (HB) method without covering the melt [6.50]. In this technique there is direct contact of the vapor phase with the crystallizing melt–solid phase region, which guarantees near phase equilibrium conditions during the whole growth run (Fig. 6.4a). However, the technical limitation of the HB method for scaled-up production of crystals with diameters larger than 3 inches is commonly known. The vertical Bridgman (VB) and vertical gradient freeze (VGF) techniques have also been introduced, using an extra source for control of the vapor phase [6.51] (Fig. 6.4b). A certain transport transient of the vapor species has to be considered due to the complete covering of the crystallization front by the melt column. As a result the equilibrium temperatures for the vapor pressure source can somewhat differ between the techniques. Vapor-pressure-controlled HB, VB, and VGF have been successfully used to grow near-stoichiometric compound semiconductors, such as CdTe crystals [6.28]. If highly purified material is used even high-resistivity crystals can be obtained [6.52], due to intrinsic defect annihilation and compensation. However, ensuring a highly reproducible, stable intrinsic semi-insulating state over the whole crystal length during the whole growth run is very complicated due to native point defect segregation. A more promising method appears to be compensation doping (Sect. 6.2.2). In general, for VB and VGF, being obviously the most favorable melt growth techniques for semiconductor compounds, stoichiometric processing by in situ vapor pressure control is not yet possible on the industrial scale. The continuously decreasing height of the melt column and segregation phenomena during the normal freezing process would require a well-controlled source temperature program that is well fitted to the growth rate. As a result, the stoichiometry is still tuned in cut wafers by postgrowth annealing if required for a given application [6.53].

Much less widely practiced is in situ control of stoichiometry in the Czochralski growth of semiconductor compounds. For this a modified technique without boric oxide encapsulant is required to influence the melt composition by the partial pressure of the volatile element. During the 1980s the hot-wall Czochralski (HWC) technique with As (P) source was advanced by *Nishizawa* et al. [6.54, 55] to analyze the variations of the physical parameters and dislocation density in GaAs and GaP crystals as functions of the melt composition. In recent years in situ control of the stoichiometry in GaAs by vapor-pressure-controlled Czochralski (VCz) technique [6.56] without boric oxide encapsulation has been under investigation [6.49, 57] (Fig. 6.4c). The mole fraction of the melt is controlled in the range of $0.45 \leq x_L \leq 0.50$ by partial arsenic pressures of 0.02–2.1 MPa, adjusted via the temperature of the As source from 540 to 650 °C. The authors [6.57] have demonstrated that near-stoichiometric growth conditions with a Ga-rich melt reduces the As_{Ga} antisite and V_{Ga} concentrations.

To date, for dielectric compounds, Czochralski growth experiments with precise in situ vapor pressure control are still rare. This is first of all due to the much higher growth temperatures and chemical aggressiveness of oxygen or fluorine, which make it difficult and expensive to insert a chemically resistant and gastight inner growth chamber with an extra source. However, some activities have been described. For instance, high-quality composition-controlled $Bi_{12}GeO_{20}$ (BGO) crystals were grown in an inner chamber made of platinum [6.58]. References [6.59, 60] describe a growth arrangement to control the stoichiometry of $PbMoO_4$ crystals with markedly improved optical transmission by using a MoO_3 evaporation source within a bell covering around the pulling crystal. *Baumann* et al. [6.6] demonstrated the importance of precise melt composition control for the growth of $LiNbO_3$ to obtain crystals with high axial homogeneity of birefringence.

6.2.2 Extrinsic Point Defect Incorporation

Thermodynamics

In growing crystals with dopant concentrations below the solubility limits, the matrix is regarded as contributing one component in a phase diagram and the solute another. Thus, in growing such crystals from the melt, the system can be considered a binary one. The equilibrium between the chemical potentials of added species (i. e., solvent) i in the liquid and solid phases $\mu_{il}(x,T) = \mu_{is}(x,T)$ yields

$$\mu_{is}^0 + k_B T \ln x_{is}\gamma_{is} = \mu_{il}^0 + k_B T \ln x_{il}\gamma_{il} , \qquad (6.8)$$

where μ^0 is the standard potential and γ is the interaction activity between i and atoms or molecules of the matrix. Setting $\mu_{il}^0 - \mu_{is}^0 = \Delta\mu_i^0 = \Delta h_i^0 - \Delta s_i^0 T$ and $s_i^0 = h_i^0/T_{mi}$, where h_i^0 and s_B^0 are the intensive standard enthalpy and entropy, respectively, T_{mi} is the melting point of the dopant, and $\Delta h_{Mis,l}^0 = k_B T \ln \gamma_{is,l}$ is the enthalpy of mixing, the transformed equation (6.8) becomes [6.31, 45]

$$\frac{x_{is}}{x_{il}} = k_0 = \exp\left[-\frac{\Delta h_i^0}{k}\left(\frac{1}{T} - \frac{1}{T_{mi}}\right) + \frac{\Delta h_{Mil} - \Delta h_{Mis}}{k_B T}\right] , \qquad (6.9)$$

where $k_{0i} = x_{is}/x_{il}$ is the (thermodynamic) *equilibrium distribution coefficient*, which can be assumed to be a constant for residual impurity or low dopant concentrations if the solidus and liquidus curves admit linearization. A more exact treatment, however, is provided by the description of the real solidus and liquidus courses by use of regression functions [6.61]. In most phase diagrams the solidus and liquidus courses can be approximated by polynomials. For instance, a polynomial of second order describes the temperature dependence of the solidus and liquidus as $T_{is,l} = p_{is,l}x_{is,l}^2 + q_{is,l}x_{is,l} + T_m$ where p and q are the regression coefficients and T_m is the melting point of the crystal matrix. The concentration dependence of the distribution coefficient then becomes $k_{0i}(x) = (p_{is}x_{is} + q_{is})/(p_{il}x_{il} + q_{il})$. For the Al–Si system within the temperature region 660 °C (T_{mAl})–577 °C the coefficients are $p_{Sis} = 15.6065$, $q_{Sis} = -77.1694$, $p_{Sil} = -0.0371$, and $q_{Sil} = -6.2958$, for example [6.61].

Much more complicated is the situation of nonstoichiometric compound growth. Usually, the boundaries of the existence regions, being equivalent to the solidus curves of the excess component, show strong nonlinear T,x-behavior. Here a definition of the k_0 value for the given excess component becomes quite difficult, all the more so due to uncertainties in the shape of the phase extent. Such conditions take place during growth of semiconductor compounds. In [6.28, Fig. 4.5] an attempt was made to estimate the segregation coefficients for Te and Cd in CdTe from the compound solidus courses on the Te- and Cd-rich sides. On the Te-rich side between the liquidus temperatures 1092 and 1080 °C, i. e., from the congruent melting point composition $x_{\text{cmp}} = 0.5 + 5\times 10^{-6}$ up to a tellurium excess of $\delta x_{\text{Te}} = 5\times 10^{-5}$, the k_0 value for Te changes between 0.8 and 0.1, respectively.

Interaction with Intrinsic Point Defects

Dopants can become incorporated on lattice sites or interstitial positions. In compounds, depending on their position in the periodic table, dopants may occupy either one or another or even both sublattice sites. In general, their incorporation efficiency is affected by the point defects that are present, which at growth temperatures are isolated and usually electrically charged [6.30]. The charge state of point defects depends on the position of the Fermi level, leading to their complex interaction with electrically active dopants. As a result the dopant solubility can be markedly influenced, and controversially, doping can affect the native point defect solubility.

The incorporation reaction of a dopant from the melt X_l with a neutral vacancy of the B sublattice V_B^0 in a AB semiconductor compound is assumed to be $X_l + V_B^0 = X_B^+ + e$, where X_B^+ is the substituting donor and e is a free electron. If the material is uncompensated the carrier concentration is $n = [X_B^+]$ and the segregation coefficient is

$$k_0 = \frac{[X_B^+]}{[X_l]}\,, \tag{6.10}$$

being constant as long as the dopant concentration is much lower than the intrinsic content of electron–hole pairs $n_i = [e_i] = [h_i]$ produced by the ionized native point defects. This follows from an effective value of the segregation coefficient k which was derived by *Chernov* [6.38] considering mass balance, electroneutrality, and k_0 from (6.10) as

$$k = \frac{k_0}{\sqrt{1+\frac{[X_B^+]}{n_i}}} = \frac{k_0}{\sqrt{1+\frac{[X_B^+]}{n_i}}}\,. \tag{6.11}$$

As can be seen, when $[X_B^+] \ll n_i$, the value of k equals the equilibrium distribution coefficient k_0. However, once $[X_B^+]$ rises to $\approx n_i$ the effective segregation coefficient k begins to fall. In other words the dopant efficiency decreases. Physically, this is due to the growing charge state of the intrinsic point defect ($V_B^0 \to V_B^+ + e$) in order to preserve the electroneutrality of the crystal $n = (n_i^2/n) + [X_B^+] + [V_B^+]$. This phenomenon has been well known for a long time from numerous melt-grown semiconductor compounds, such as GaAs doped with Te and Si [6.33] and CdTe doped with Ag and Cu [6.62].

The influence of the electron or hole concentration generated by the ionized dopants on the charge state of the native point defects is named the Fermi-level effect [6.63], whereby the degree of ionization depends on the position of the Fermi level [6.24]. For instance, the charge state of the Ga vacancy in GaAs changes from neutral V_{Ga}^0 in p-type material through double negatively charged V_{Ga}^{2-} in the midgap to a triple negatively charged V_{Ga}^{3-} in n-type material, which affects the compensation level and enhances the *complex formation* probability too. An example of a complex reaction in the common form is

$$\left(V_A^{3-} + 3h\right) + \left(X_B^+ + e\right) \leftrightarrow \left(V_A\,X_B\right)^{2-} + 2h\,, \tag{6.12}$$

as found in the form of $(Te_{As}V_{Ga})^{2-}$ or $(Si_{Ga}V_{Ga})^{-}$ in GaAs crystals.

Another factor influencing the incorporation efficiency of impurities and dopants is the nonstoichiometry of compounds. The greater the deviation from the stoichiometric composition during growth, the higher the possibility of vacancy generation within the compositionally impoverished sublattice. As a result the incorporation density of atoms at the growing interface occupying these vacant sites increases with nonstoichiometry. This was demonstrated in [6.62], where CdTe crystals were grown by vapor-pressure-controlled Bridgman method (Sect. 6.2.1) from different melt compositions containing Ag, Cu, and P dopants. Whereas Ag and Cu atoms occupy the Cd vacancies when growing proceeds from Te-rich melt, P takes up the Te vacancies on growth from Cd-rich melt. As has been demonstrated, the concentration of the neutral substitutions $[Ag_{Cd}^0]$, $[Cu_{Cd}^0]$, and $[P_{Te}^0]$, determined in as-grown crystal samples by absorption-calibrated photoluminescence [6.64], increased with Te or Cd enrichment of the melt, respectively. Figure 6.5 shows the variation of the substitutional concentration $[Ag_{Cd}^0]$ as a function of the starting silver concentration in the melt $[Ag_l]$ for various tellurium excess. At stoichiometric growth conditions the concentration of Ag_{Cd}^0 is drastically reduced down to $2\times 10^{13}\,\text{cm}^{-3}$. Compared with that, a Te excess of $5\times 10^{17}\,\text{cm}^{-3}$ leads to incorporation of Ag atoms on V_{Cd} sites at about one order of magnitude higher concentration (note the prevailing part of Ag atoms, is incorporated according k_0 in interstitial positions, probably acting as deep donors). The results underline the

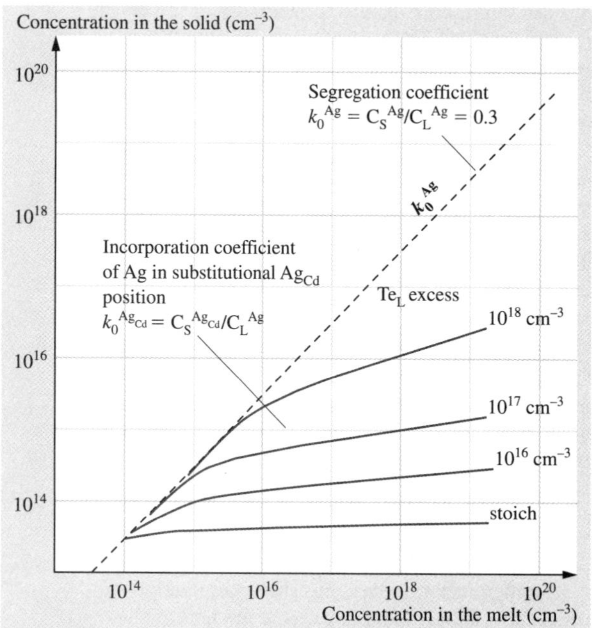

Fig. 6.5 Incorporation of silver atoms into substitutional Ag_{Cd} positions in CdTe crystals versus the degree of deviation from stoichiometry (Te excess) of the melt (after *Rudolph* et al. [6.62])

importance of in situ stoichiometry control in growth of both compound semiconductor crystals and dielectric crystals.

Segregation Phenomena

The redistribution of extrinsic atoms, arbitrarily or deliberately incorporated into a growing crystal, as well as the rearrangement of intrinsic point defects prove to be an essential challenge for the crystal grower. Such phenomena are due to the segregation effect and temperature gradients within the cooling crystal producing chemical and physical crystal inhomogeneities.

The deviation of the equilibrium segregation coefficient (6.9) from unity causes segregation phenomena during melt growth which can be treated in terms of an effective segregation coefficient $k_{\text{eff}} = x_{is}/x_{il}{}^{(\infty)}$, where $x_{il}^{(\infty)}$ is the mole fraction of the dopant in the melt far away from the interface. During an actual freezing process the solute is rejected ($k_0 < 1$) or preferentially absorbed ($k_0 > 1$) by the propagating solid–liquid interface, forming an enriched or depleted solute boundary layer in front of it.

The width δ_S of this boundary layer is determined by the growth rate R and by the diffusive and convective species transport in the melt, which is very often difficult to predict. A very popular model which is commonly used in melt growth was introduced by *Burton*, *Prim*, and *Slichter* (BPS) [6.65] for steady-state segregation

$$k_{\text{eff}} = \frac{k_0}{k_0 + (1-k_0)\exp\left(\frac{-R\delta_S}{D}\right)} . \tag{6.13}$$

The advantage of this model is the ease of its use for plotting experimental data by fitting δ_S. By applying the Cochran flow solution at the surface of an infinite rotating disk, *Levich* [6.66] obtained the expression $\delta_S \approx 1.6D^{1/3}\nu^{1/6}\omega^{-1/2}$ (where D is the diffusion coefficient in the melt, ν is the kinematic viscosity of melt, and ω is the rotation frequency). Later, *Ostrogorski* and *Müller* [6.67] quantified δ_S more physically for situations where natural convection rather than rotating disc flow dominated by considering its dependence on lateral convection velocity and the length of the interface. They defined $k_{\text{eff}} = (1+\Lambda)/(1+\Lambda/k_0)$ where $\Lambda = (\nu_D\delta_S)/(7.2\nu L)$, ν_D is the convective velocity at the edge of the boundary layer, ν is the growth velocity, and L is the length of the interface. The value of k_{eff} is of central importance to explain macro- and microsegregation phenomena.

Macrodistribution

The macrodistribution describes the courses of a given solute concentration over its radius and along the as-grown crystal axis. Whereas the radial rearrangement is first of all affected by the shape of the melt–solid interface, the axial distribution depends on the growth velocity and level of convection in the melt. The degree of segregation depends on the extent to which the solute segregation coefficient (k_0) differs from unity. For instance, an enrichment of the dopants around the crystal center occurs at a concave- or convex-shaped interface and $k_0 < 1$ or > 1, respectively. Against this, the solute is concentrated near the crystal periphery in the case of a convex interface and $k_0 > 1$. Therefore, only a nearly flat crystallization front can guarantee a homogeneous radial composition distribution.

The axial segregation function for unidirectional solidification in a completely mixed melt is described by the well-known Scheil equation (6.14), being valid for a stable planar crystal–melt interface and conservative mass balance, i. e., no solute evaporation from the melt or recharging into the melt takes place during the whole crystallization process. When the growth process is started with uniform melt concentration x_{i0}, the concentration in the crystal x_{is} at a distance z from the initial growth face is

$$x_{is} = x_{i0} k_{eff} \left(\frac{1-z}{L_0} \right)^{k_{eff}-1} , \qquad (6.14)$$

where L_0 is the length of the charge and the process considered is one in which the whole charge is initially molten and then frozen. From (6.13) and (6.14) it follows that k_{eff} equals unity at high growth rates or/and motionless melts, i. e., high values of δ_S in (6.13). In such cases high macroscopic distribution uniformity within the crystal can be reached. However, v is limited by the onset of morphological instability, as shown below. On the other hand, if the melt is stirred, the boundary layer is removed by the melt flow ($\delta_S \to 0$) and k_{eff} equals k_0. In this case the axial distribution of the solute is typically inhomogeneous.

In order to obtain high axial chemical homogeneity in modern crystal growth processes often a nonconservative (open) system [6.68] with continuous dopant recharging (or extraction) is applied. This is practiced in the growth of semi-insulating GaAs crystals by applying carbon as a shallow acceptor for compensation of the deep intrinsic donor defects EL2 (As_{Ga}). As follows from (6.14), in the conservative case with complete melt mixing the C distribution, and hence the degree of compensation, which determines the electrical resistivity, becomes decreasing along the crystal axis due to $k^C_{eff} \approx k^C_0 \approx 2$. By contrast, in a nonconservative system the axial distribution can be homogenized by proper in situ control of the CO fugacity within the growth chamber atmosphere delivering the C species for the melt via CO decomposition at the interface between boric oxide encapsulant and Ga–As melt. This is arranged by a controllable CO content within the working gas [6.69], a process well matured on the industrial scale [6.70]. Figure 6.6 compares the axial carbon distribution along GaAs crystals grown under nonconservative doping conditions [6.71]. For such case the Scheil equation (6.14) has to be modified as [6.72]

$$x_{is} = \frac{\eta k_{eff} x_{iv}}{k_{eff} + \eta - 1} \times \left(k_{eff} x_{i0} - \frac{\eta k_0 x_{iv}}{k_{eff} + \eta - 1} \right) \left(\frac{1-z}{L_0} \right)^{k_{eff}+\eta-1} , \qquad (6.15)$$

with

$$\eta = \frac{D_{i(B_2O_3)}}{h_{B_2O_3} v} \left(\frac{r^2_{cruc}}{r^2_{crys}} - 1 \right) . \qquad (6.16)$$

x_{iv} is the concentration of dopant species in the gas phase at the upper interface of boric oxide, $D_{i(B_2O_3)}$ is the transport coefficient of the dopant species in boron oxide ($\approx 5 \times 10^{-8}$ cm^2/s for carbon [6.72]), $h_{B_2O_3}$ is the height of the boric oxide encapsulant, and r_{cruc} and r_{crys} are the radius of the crucible and crystal, respectively.

As is well known, a higher macroscopic homogeneity can be achieved by applying single-pass zone-melt techniques. A relative short first-to-freeze transient is followed by a compositionally uniform level region, the concentration of which equals that of the starting (polycrystalline) source rod, i. e., $x_{is} = x_{i0}$. The axial distribution function is

$$x_s = x_0 \left[1 - (1 - k_{eff}) \exp\left(\frac{-k_{eff} z}{L_z} \right) \right] , \qquad (6.17)$$

with zone length L_z and distance $z < L_0 - L_z$. The final part of solidification when $z > L_0 - L_z$ represents a normal freezing process and the concentration profile is given by (6.14). Single-pass zone growth is successfully used not only for production of extremely highly purified silicon by floating-zone melting [6.73] but also to obtain homogeneous mixed crystals of the

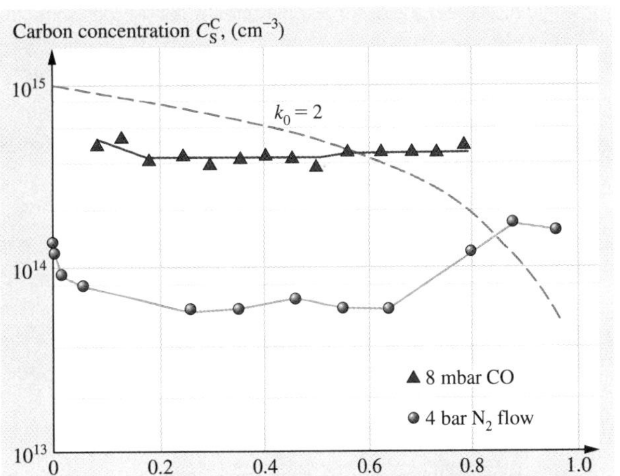

Fig. 6.6 Axial carbon distribution in LEC (▲) and VCZ (●) GaAs crystals grown under nonconservative conditions by controlled CO gas concentration (8 mbar) and rinsing nitrogen working gas ($\approx$ 4 bar) [6.56], respectively. The *dashed curve* shows the theoretical Scheil distribution with $k_0 = 2$ (the equilibrium segregation coefficient of carbon) in case of conservative growth regime with complete melt mixing ($k_0 = k_{eff}$) and starting concentration $x_{CO} = 5 \times 10^{14}$ cm^{-3}

types $A_{1-x}B_x$ or $AB_{1-x}C_x$, even with widely spread liquidus and solidus curves. High-quality $Hg_{1-x}Cd_xTe$ and $Cd_{1-x}Zn_xTe$ crystals with axial uniformity of the constituents can be grown by the traveling-heater method (THM) from a Te-rich zone [6.74, 75]. Uniform oxide mixed crystals have been grown by a special heater-immersed zone melting technique [6.76]. Another use is microzone melting for the growth of various oxide fiber crystals by laser-heated pedestal growth (LHPG) [6.77]. Today this technique is applied for the production of nearly preparation-free microlasers for wavelength conservation by higher-harmonic generation [6.78].

Microinhomogeneities (Striations)

Microinhomogeneities are short-range composition fluctuations with characteristic spacing ranging from 1 μm to 1 mm, usually modulating the macrodistribution as *fine structure*. Such oscillations are found in nearly all crystals and are visible under light microscopy as *striations* on crystal cuts, especially after etching (Fig. 6.7a,b). They are one of the most investigated crystal defects [6.2, 3], and can seriously limit the device application of the given crystal (Sect. 6.1). Nonuniform segregation of solutes on the microscale is due to a variety of mechanisms, as recently summarized by *Scheel* [6.80]. Principally, one has to differentiate between kinetically and thermally induced striations.

Kinetic striations appear in the discontinuous (step-by-step) growth mode of atomically flat interfaces or facets (Sect. 6.5) when macrosteps are formed by bunching. Their repeated rapid lateral growth tends to trap the adjacent solute in alternating nonequilibrium concentration. At misoriented growth planes the propagation direction of the macrosteps, and hence striations, are inclined to the macroscopic course of the interface. This phenomenon is well known from liquid phase epitaxy (LPE) experiments [6.81].

Thermally induced striations are generated by nonsteady growth velocities $\Delta v/v$, leading to variation of k_{eff} (6.13). Oscillating interface rates are caused by temperature fluctuations, which can be induced by:

i) Rotation of Czochralski crystals in thermally asymmetric melts [6.82]
ii) Convective instabilities [6.45, 67].

Vibrations and pressure fluctuations have also been identified as sources of growth rate variations. After *Hurle*'s treatment [6.83], for certain frequencies of temperature oscillation, resonant coupling should occur between thermal and solute fields to give large compositional amplitudes.

For low-frequency oscillations with fluctuation periods $\tau > 2\delta_S/D$ (longer than about 10 s) the concentration changes are associated with growth rate R changes

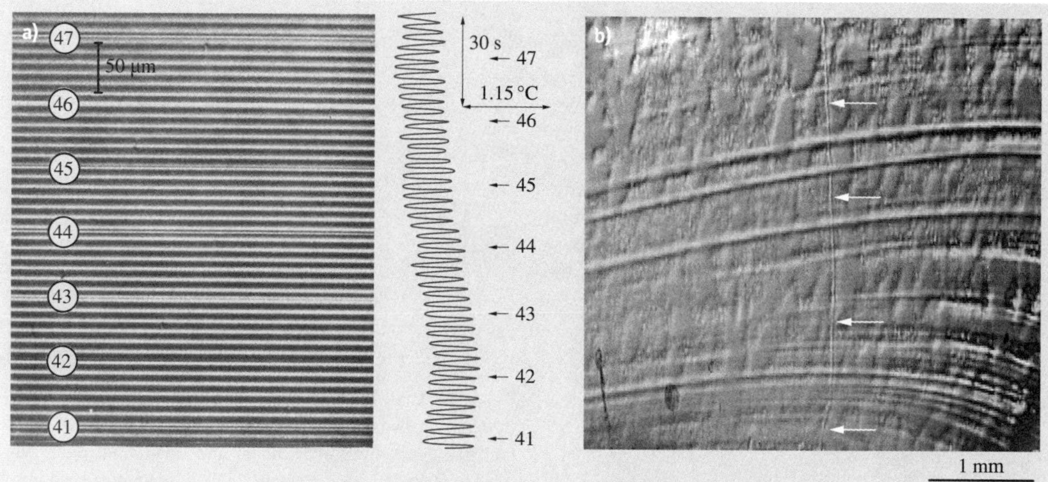

Fig. 6.7a,b Striations in semiconductor crystals. **(a)** Section of Te-doped InSb crystal grown in the presence of thermal oscillations in the melt [6.79] (after [6.45]; reproduced by permission of Springer). **(b)** Section of an undoped VCZ GaAs crystal grown from incongruent As-rich melt with compositional striations marking the interface shape. The *arrows* show the trace of a dislocation bundle propagating perpendicularly to the interface (courtesy of U. Juda from IKZ Berlin)

according to the steady-state BPS relation (6.13)

$$\frac{\Delta x_{is}}{x_{is}} = \frac{\Delta R}{k_{eff}} \frac{\partial k_{eff}}{\partial R} = \frac{\Delta R}{R} \frac{(1-k_0)\left(\frac{R\delta_S}{D}\right)\exp\left(\frac{-R\delta_S}{D}\right)}{k_0+(1-k_0)\exp\left(\frac{-R\delta_S}{D}\right)} . \tag{6.18}$$

The maximum composition amplitude $[(\Delta x_{is}/x_{is})/(\Delta R/R)]_{max} = (1-k_0)(R\delta_S/D)$ occurs under conditions where $R\delta_S/D \ll 1$. By contrast, for high-frequency growth rate oscillations ($\tau < 2\delta_S/D$), the reaction time of the mass diffusivity is no longer able to follow the thermal agility. Thus, the amplitudes of the compositional fluctuations are increasingly reduced with increasing perturbation frequency f as

$$\frac{\Delta x_{is}}{x_{is}} = \frac{\Delta R}{R}(1-k_0)\frac{\frac{R\delta_S}{D}}{\left(\frac{2f\delta_S^2}{D}\right)^{1/2}} . \tag{6.19}$$

At frequencies higher than 10 Hz the relative concentration fluctuations fall to less than 10%. Therefore, low-frequency fluctuations affect the crystal homogeneity much more than do high-frequency ones. In other words, a melt–solid interface acts as a *low-pass filter*. Unfortunately, in crystal growth melts convective frequencies in the range of $0.1{-}0.5\,s^{-1}$ are typical and the relation (6.18) must be used.

There are two general ways to damp temperature oscillations within the melt (when growth under microgravity is not considered):

i) Brake the buoyancy convection streams
ii) Minimize the temperature differences.

For the first method the application of various kinds of magnetic fields proves to be very effective for melts with electrical conductivity [6.84]. In the second case measures for temperature homogenization by using high-frequency melt mixing are required. From the above discussion, it follows that they are not dangerous for crystal homogeneity. Appropriate techniques are the accelerated crucible rotation technique (ACRT) [6.85] and ultrasonic vibration stirring [6.86].

Note that, for fundamental research and technology developments, the presence of striations proves to be of certain advantage because of their ability to mark the interface shape. Striation analysis along longitudinal crystal cuts reveals the time and course characteristics of the crystallization front during the whole growth process. Note that striations appear even in high-purity compound crystals grown from slightly off-congruent melting point composition. For instance, in standard semi-insulating GaAs crystals grown from slightly As-rich melts, striations of very small and harmless amplitudes can be resolved by use of diluted Sirtl with light (DSL) etching [6.87] (Fig. 6.7b). They reflect the alternating incorporation of As-related native point defects [6.88]. For some materials it is important to consider the relatively high diffusion coefficients in the solid, leading to a leveling effect [6.89]. For instance, in melt-grown (Hg,Cd)Te mixed crystals, striations are not revealable due to the extremely high D_S^{Hg}.

6.2.3 Constitutional Supercooling – Morphological Instability

Under certain conditions, especially if the melt is not mixed by convection or stirring (i. e., if the solution boundary layer δ_s is well developed), the interface can become morphologically unstable. Both an enriched ($k_0 < 1$) or depleted ($k_0 > 1$) solute layer δ_s, showing a typically exponential concentration course (increasing or decreasing, respectively) at the growing interface, give rise to constitutional instability, especially if the corresponding liquidus temperatures of the concentration course exceed the actual temperature course

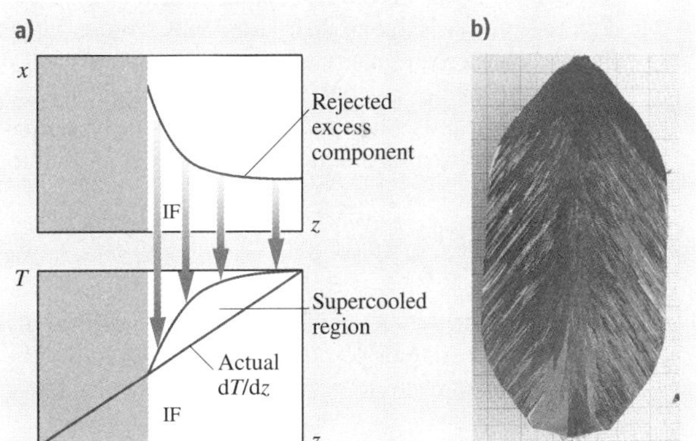

Fig. 6.8a,b Constitutional supercooling of the growing melt–solid interface. **(a)** Sketch demonstrating the supercooling effect at the interface due to the rejection of impurity, dopant or excess component (*above x–z* diagram) that leads to higher equilibrium melt temperature distribution compared with the actual temperature gradient (*below T–z* diagram; x – concentration, T – temperature, z – axial direction). **(b)** Undoped LEC InP crystal with features of morphological instability (cellular interface) grown from In-rich melt with too high a pulling rate (courtesy of M. Neubert and A. Kwasniewski from IKZ Berlin)

(Fig. 6.8a). Then random formation of a projection on the interface advances that portion of the interface into the region of increased supercooling, where it can grow more rapidly, causing lateral segregation of solute that suppresses growth in the neighboring region. As a result a close-packed array of such projections is formed on the length scale of the lateral diffusion distance D/v. Such cellular interface morphology produces harmful columnar grain boundary structures with redistributed concentration and dislocation densities leading, finally, to polycrystallinity (Fig. 6.8b). This is exactly what happens in most metal alloy systems [6.90].

For semiconductors, however, faceted interfaces can be obtained. Once the amplitude of the projections grows to the point that their interface with the melt becomes tangential to a {111} faceting direction (Sect. 6.5), microfacets form on the interface [6.2, 3]. This occurs even if the dopant is not the cause of the constitutional supercooling. A noncongruent melt produces rejection of the component in excess and this lowers the liquidus temperature in exactly the same way as a solute. Because of the development of microfacets, the morphology of the resulting cellular structure is orientation dependent (not to be confused with the polygonized cell structure described in Sect. 6.3.2). In principle, dielectric (oxide) crystals that grow under conditions of constitutional supercooling always show faceted interfaces due to their atomically smooth interface kinetics on all important crystallographic planes.

First, *Tiller* et al. [6.91] deduced theoretically the condition for prevention of constitutional supercooling (i. e., preservation of morphological stability of the interface) as

$$\frac{G}{R}=\frac{m x_{\mathrm{il}}(1-k_0)}{k_0 D}\,, \tag{6.20}$$

where G is the temperature gradient in the melt at the interface and m the slope of the liquidus from the T–x-phase diagram projection (all other parameters are introduced above). Of course, relation (6.20) is an approximation that ignores some stabilizing factors. In particular, a face growing by an atomically flat mechanism is more stable against disturbances [6.38]. Further, the convection can stabilize the interface morphology by effective removal of the solute layer. Indeed, the application of artificial melt mixing techniques, such as ACRT [6.85], ultrasonic vibration stirring [6.86] or time-dependent (i. e., rotating, alternating, traveling) magnetic fields [6.92–94], can help to ensure morphological stability very effectively.

A linear stability analysis predicting the exact conditions of onset of the morphological instability was later developed by *Mullins* and *Sekerka* [6.95], completing equation (6.20) by consideration of the heat diffusivity in the melt and solid. In subsequent years the theory was widely extended to include higher-order bifurcations and additional physical effects such as melt flow, atomic kinetics, Soret diffusion, applied electric fields, etc. All treatments showed that the Tiller criterion (6.20) can serve as a good approximation for the crystal grower. A detailed review is given in [6.96].

Constitutional supercooling can also appear during growth of undoped compound crystals growing from noncongruent melts. As shown above, in such cases the equilibrium segregation coefficient k_0 cannot be assumed to be constant. *Wenzl* et al. [6.97] modified the Tiller criterion (6.20), replacing the liquidus–solidus relation with the difference between the congruent melting composition x_{cmp} and the deviation from it in the melt x_{l}. When x_{cmp} is close to the stoichiometric composition, as can be assumed for most compounds, one can set $x_{\mathrm{cmp}} \approx 0.5$ and the undercritical growth velocity becomes [6.97]

$$R \le \frac{GD}{(0.5-x_{\mathrm{l}})}\left(\frac{\mathrm{d}T}{\mathrm{d}y}\right)^{-1}, \tag{6.21}$$

where D is here the diffusion coefficient of the excess component in the melt and $\mathrm{d}T/\mathrm{d}y$ is the slope of the liquidus at the given x_{l}. Obviously, R can be chosen relatively large close to the congruent melting point ($\mathrm{d}T/\mathrm{d}y \approx 0$). However, it falls drastically with increasing deviation and becomes for GaAs at $x_{\mathrm{l}} = 0.55$ (i. e., 5 at. % As excess, $D \approx 10^{-5}\,\mathrm{cm^2/s}$, $\mathrm{d}T/\mathrm{d}y \approx -200\,\mathrm{K}$ [6.97]) about 4 mm/h, being markedly lower than the growth rate of standard crystals growing from near-stoichiometric melt composition. Figure 6.8b shows an image from the longitudinal cut of an InP crystal grown with overcritical growth velocity from an In-rich melt. The polycrystalline columnar structure induced by morphological instability is quite perceptible.

6.3 Dislocations

Contrary to native point defects (Sect. 6.2), dislocations are not in thermodynamic equilibrium, and hence in principle are preventable. However, at present only silicon and germanium standard crystals can be grown dislocation-free. This is because in compound crystals the situation is complicated by much higher intrinsic point defect content and lower critical resolved shear stress τ_{CRSS} (e.g., in GaAs and CdTe near the melting point, ≈ 0.5 MPa instead of ≈ 10 MPa in Si) markedly increasing the dislocation mobility and multiplication probability.

Dislocations in a growing crystal which come from the seed are termed grown-in dislocations. In the course of crystal cooling within a given temperature field the development of the dislocation density correlates closely with the thermomechanical stress induced by temperature nonlinearities within the crystal volume. It is the aim of dislocation engineering to achieve uniaxial low thermal gradients in order to minimize the elastic stress, and hence the dislocation density. To achieve this computational modeling of nonstationary plastic deformation proceeding within the crystal is today absolutely essential.

The phenomenon of dislocations interacting with each other at high temperatures is driven by mutual screening of the individual energetic fields and also by dissipative ordering processes in the framework of nonequilibrium thermodynamics. These processes, described by the principles of dislocation dynamics (DD), lead to characteristic collective rearrangements into cell patterns, bundles, and lineages. It can be shown experimentally that the cell size correlates with the mean dislocation density and acting elastic stress. Depending on material parameters, stress value, and the ripening level of dynamic polygonization, the cells transform into low-angle grain boundaries. Dislocation cells and bundles are not observed under low linear temperature gradients, stoichiometric growth conditions, and nearly flat interfaces. See also Chap. 4 in the present Handbook.

6.3.1 Dislocation Types and Analysis

Dislocations are linear crystallographic defects, or irregularities, within a crystal structure. Such linear defects cannot start or end within the crystal. They enter or leave the crystal through its surfaces, with the exception of dislocation loops closed within the crystal volume. The presence of dislocations strongly influences many of the properties of materials (Sect. 6.1) because of lattice distortion, local stress field, electrical activity, and getter ability for point defects. As defined by the angle between the dislocation line and the Burgers vector there are two primary dislocation types: *edge dislocations* and *screw dislocations* with an angle of 90° and 0°, respectively. *Mixed dislocations* are intermediate combinations between these. In the diamond and zincblende structure 30° and 60° dislocations appear within the {111} planes with Burgers vector $\boldsymbol{b} = a_0/\sqrt{2}$ along the $\langle 110 \rangle$ directions (a_0 = lattice constant). In semiconductor A^{III}–B^{V} compounds 60° dislocations are the prevailing ones, amounting in GaAs to 30% of all dislocation types present [6.98].

It is experimentally and theoretically well established that, independent of the growth method, the density and distribution of dislocations in melt-grown crystals are due to a thermoplastic relaxation of thermally, and to a much lower extent constitutionally, induced stress during growth. Principally, for growing crystals a differentiation between the terms *generation* and *multiplication* of dislocations is recommended. The formation of new dislocations within an ideal dislocation-free crystal under normal growth conditions is nearly impossible. Stresses of the order of the material strength limit $\tau \approx 10^{-2}$–$10^{-1}G$ (where G is the shear modulus ≈ 10–70 GPa) would be required, being much higher than that usually obtained in standard growth processes (around 10 MPa at most). Such values are not even high enough to penetrate Shockley partials in zincblende structures from microscopically stepped crystal surfaces, which requires stresses of at least ≈ 100 MPa [6.99]. Dislocations can be generated in the form of Frank loops due to collapsing vacancy agglomerations, interstitial disks, and interface misfits between foreign-phase inclusions and matrix (Fig. 6.1). The possibility of such origins is enhanced in nonstoichiometric material. The majority of dislocations in as-grown crystals, however, originate from glide- and climb-assisted elongation (bowing out) and multiplication (dissociation, cross-glide) of primary existing faults grown-in from the seed crystal.

There exist numerous reliable methods to analyze dislocations such as:

i) High-resolution transmission electron microscopy (HRTEM)
ii) X-ray Lang topography, enabling Burgers vector analysis
iii) Laser scattering tomography (LST), ascertaining the spatial dislocation courses
iv) Fully automatized etch pit density (EPD) mapping.

Even high-temperature synchrotron x-ray and transmission electron imaging techniques make it possible to study the dislocation kinetics at the crystallization front of some materials in situ [6.100]. As a result, knowledge about dislocation types and their mobility in correlation with the growth conditions as well as their interaction with dopants is today well developed and a large number of related papers are available. For III–V compounds, a comprehensive summary was published by *Sumino* and *Yonenaga* [6.101], for example.

6.3.2 Dislocation Dynamics

Basic Considerations

Dislocations are the elementary carriers of plastic flow, relaxing the elastic stress, which affects the crystal structure. Plastic relaxation (i. e., deformation) of crystalline solids is related to the motion and multiplication of dislocations. This is a basic process within growing crystals by which dislocations are stored. The study of propagation and interaction of dislocations over mesoscopic distances is the subject of dislocation dynamics [6.102]. Because the collective behavior of dislocations is complicated, to date very little is known about the effects that individual dislocations have on each other when they come into close proximity, and more generally about the evolution of collections of strongly interacting dislocations near the melting point. In this chapter basic processes and some characteristic features of dislocation rearrangements into cells and bundles during crystal growth from the melt will be presented.

Dislocations can move within the glide plane by *glide* or *slip* in the direction of b (edge dislocations) or orthogonal to b (screw dislocation) or at a certain angle (30° and 60° dislocations). In addition, screw dislocation can *cross-slip* from one glide plane to another. *Climb* is the motion of dislocations perpendicular to the glide plane. Note that even climb and cross glide are responsible for spatial dislocation (see below).

The glide mobility $\boldsymbol{v}_{\mathrm{g}}$ is given by

$$\boldsymbol{v}_{\mathrm{g}} = \boldsymbol{v}_0 \tau_{\mathrm{eff}}^m \exp\left(\frac{E_{\mathrm{a}}}{k_{\mathrm{B}}T}\right) , \tag{6.22}$$

where $\boldsymbol{v}_0$ is the material constant, of the order of the magnitude of the Debye frequency, E_{a} is the activation energy (Peierls potential), $\tau_{\mathrm{eff}} = \tau - A\sqrt{\rho_0}$ is the effective shear stress on dislocations, m is the stress exponent, τ is the acting shear stress, $A = Gb/2\pi(1-\nu)$ is the strain hardening factor, G is the shear modulus, ν is Poisson's ratio, and ρ_0 is the mobile dislocation density.

Usually, at higher temperatures, the process of climb constitutes the dominant mode of dislocation motion. It is thermally activated and therefore dependent upon the diffusivity of vacancies or interstitials to the dislocation core. A phenomenological expression for dislocation climb velocity $\boldsymbol{v}_{\mathrm{cl}}$ is given by [6.103]

$$\boldsymbol{v}_{\mathrm{cl}} = B\left(\frac{D_{\mathrm{s}}}{b}\right)\left(\frac{G\Omega}{k_{\mathrm{B}}T}\right)C_{\mathrm{j}}\left(\frac{\gamma_{\mathrm{SF}}}{Gb}\right)^2\left(\frac{\tau}{G}\right) , \tag{6.23}$$

where B is a constant on the order of 10^3, Ω is the atomic volume, C_{j} is the concentration of jogs, and γ_{SF} is the stacking fault energy (all other parameters are introduced under relation (6.22)).

The cross-slip mechanism can proceed effectively only in case of relatively high stacking fault energy. Even in semiconductor compounds with zincblende structure containing characteristic partial dislocations (Shockley partials), cross-slip can be restrained due to a large equilibrium stacking fault distance between them, inversely related to the stacking fault energy γ_{SF} as $d_{\mathrm{Sh}} = Ga_0^2/(24\pi\gamma_{\mathrm{SF}})$. InP and CdTe crystals show the lowest stacking fault energies among the semiconductor compounds, and therefore reduced cross-slip probabilities (as a result, dislocation patterning is hindered).

As it is well known the stress force exerted by a dislocation on other dislocations is long range. Moving dislocations tend to minimize their individual stress field by mutual field screening that reduces the overall system energy. There are various mechanisms of dislocation interaction. When two dislocations of opposite Burgers vectors approach each other within a certain critical distance of separation they annihilate. The critical distance for *annihilation* of two screw dislocations is $y_{\mathrm{s}} \approx Gb/2\pi\tau_{\mathrm{g}}$, where τ_{g} is the shear stress required for dislocation glide. The value of y_{s} for metals is about 2 μm. However, their approach can also achieve a stable configuration, known as a *dipole*, if the dislocation pair remains both a distance greater than y_{s} apart and with a relative position angle of around 45°. Typical dipole lengths are on the order of tenths of a micron [6.103]. Dipoles are composed only of edge dislocations since screw dislocations annihilate easily by cross-slip. Furthermore, the attractive forces can lead to the formation of dislocation *junctions* and *walls*. The energy of a dislocation bounded in a stable wall configuration (e.g., a low-angle grain boundary) is about four times lower than the energy of a single dislocation.

Dislocations can multiply by cross-glide and the *Frank–Read mechanism*. In the latter case, multiplication occurs by pinning of the dislocation, bowing out and wrapping around the pinning points. Possible pinning points are precipitates and impurity clusters, but also junction segments produced by two dislocations [6.103]. *Nabarro* [6.104] discussed the bowing out by multistep climb under conditions of vacancy supersaturation, known as the *Bardeen–Herring mechanism*.

The above-mentioned mechanisms of dislocation dynamics are considered to be the basic processes taking place during crystal growth from melt under the action of a thermoelastic stress field.

Dislocation Density Versus Thermomechanical Stress

Dislocations in a growing crystal come first of all from the seed (grown-in dislocations), or in special cases from lattice mismatches, e.g., at the interface between the crystal matrix and foreign inclusions, and are multiplied by viscous–plastic phenomena initiated by the thermoelastic stresses experienced by the crystal. Therefore, the content of dislocations is determined by the time- and space-dependent stress level during growth, which is related to the temperature field in the crystal growing and cooling-down procedure. Firstly, *Billig* [6.105] discovered that the dislocation density correlates with the imposed temperature gradient. *Indenbom* [6.106] specified that thermally induced stresses arise from temperature nonlinearity, i. e., divergence of the isotherm curvature from an idealized linear course, in other words not from the temperature profile but from its second derivative. Theoretically, this implies the simplified but quite useful formula

$$\sigma = \alpha_T E L^2 \left(\frac{\partial^2 T}{\partial z^2} \right) \approx \alpha_T E \delta T_{max} , \qquad (6.24)$$

where σ is the thermal stress, α_T is the coefficient of thermal expansion, E is the Young's modulus, L is the characteristic length (about the crystal diameter), T is temperature, z is the given coordinate (pulling axis), and δT_{max} is the maximum deviation of the isotherm from a linear course. The extremely critical situation in most semiconductor compounds such as GaAs will be obvious by using the material constants near to the melting point ($\alpha_T = 8 \times 10^{-6}\,K^{-1}$, $E = 7.5 \times 10^4$ MPa). As can be seen, only very small isotherm deviations from linearity δT_{max} of 1–2 K are enough to reach the critical-resolved shear stress (CRSS) for dislocation multiplication (0.5 MPa). This is one order of magnitude lower than in silicon, in which much greater isotherm curvatures are tolerated without disturbing dislocation-free growth. Even during pulling of a cylindrical crystal from the melt, steep temperature curvatures can occur due to different temperature gradients in the inner and outer regions of the crystal. From (6.24) follows a direct correlation between stress level and crystal diameter (characteristic length L) which has been observed in reality. The larger the crystal diameter, the higher the mean dislocation density. For instance, in undoped 3, 4, and 6 inch LEC GaAs crystals typical mean dislocation densities are 2×10^4, 5×10^4, and $1 \times 10^5\,cm^{-2}$, respectively.

Modeling of Dislocation Density

From the discussion above it follows that both knowledge and control of the temperature field at all process stages are of essential significance. Due to the difficulties of measurement, numerical simulation is of increasing importance for heat flow, thermomechanical stress, and dislocation density analysis. Two approaches have been used so far [6.7]:

1. Calculation of the thermoelastic stress field (linear theory, isotropic and anisotropic analysis) of the crystal for a given temperature field using available computer packages and comparison of the resolved shear stress (RSS) in the glide systems or the von Mises invariant with the critical resolved shear stress (CRSS), taking into account its temperature dependence known from high-temperature creep experiments. The (local) dislocation density can then be concluded from the total excess shear stress $\sigma_{ex} = \sum_{i=1}^{n} \sigma_i^e$ (where n is the number of effective slip systems in the given crystal structure), where $\sigma_i^e = |\sigma_{RSS,i}| - \sigma_{CRSS}$ for $|\sigma_{RSS,i}| > \sigma_{CRSS}$ and $\sigma_i^e = 0$ for $|\sigma_{RSS,i}| < \sigma_{CRSS}$, i. e., the maximum stress at any time of growth determines the local dislocation density. Examples of this approach can be found in *Jordan* et al. [6.107] and *Miyazaki* et al. [6.108].
2. Estimation of the local dislocation density from the constitutive law of Alexander and Haasen linking the relation between plastic shear rate and movement and density of dislocations $d\varepsilon/dt = \rho_0 v b$ (the Orowan equation) with the applied stress in the course of the cooling-down procedure of the crystal [6.109, 110]. The dislocation multiplication is proportional to the effective stress τ_{eff}, the mobile starting dislocation density ρ_0, and velocity $\boldsymbol{v}$, i. e., $d\rho/dt = K \tau_{eff} \rho_0 \boldsymbol{v}$. By using (6.22) the differential equations of state become

$$\frac{d\varepsilon}{dt} = \rho_0 b \boldsymbol{v}_0 \tau_{eff}^{m} \exp\left(\frac{E_a}{k_B T}\right) , \qquad (6.25)$$

$$\frac{d\rho}{dt} = K \rho_0 b \boldsymbol{v}_0 \tau_{eff}^{m+1} \exp\left(\frac{E_a}{k_B T}\right) , \qquad (6.26)$$

where K is a multiplication constant (all other values are introduced below (6.22)). A detailed review of this approach is given by *Völkl* [6.111]. *Lohonka* et al. [6.112] extended this approach to the case of semiconductor compounds with zincblende structure. They considered the different dynamic characteristics of $\alpha(60°)$, $\beta(60°)$ and screw dislocations by modifying the Orowan equation to $d\varepsilon/dt = b(\rho_{0\alpha}v_\alpha + \rho_{0\beta}v_\beta + \rho_{0S}v_S)$. Note that, although only screw dislocations contribute to the total plastic deformation, their dynamics depends also on the motion of $\alpha(60°)$ and $\beta(60°)$ dislocations formed together with screw dislocation loops [6.113], and which therefore have to be considered. *Grondet* et al. [6.114] included the annihilation of dislocations by pairs, leading to a certain decrease of dislocation density $d\rho^*/dt = d\rho/dt - d\rho^-/dt$ due to its proportionality to the square of the density of dislocation and their velocity: $d\rho^-/dt = A\rho_0^2 v^2$ (A is a constant to be fitted by experimental results). This addition leads to a more realistic dislocation density, which becomes overestimated if this effect is neglected.

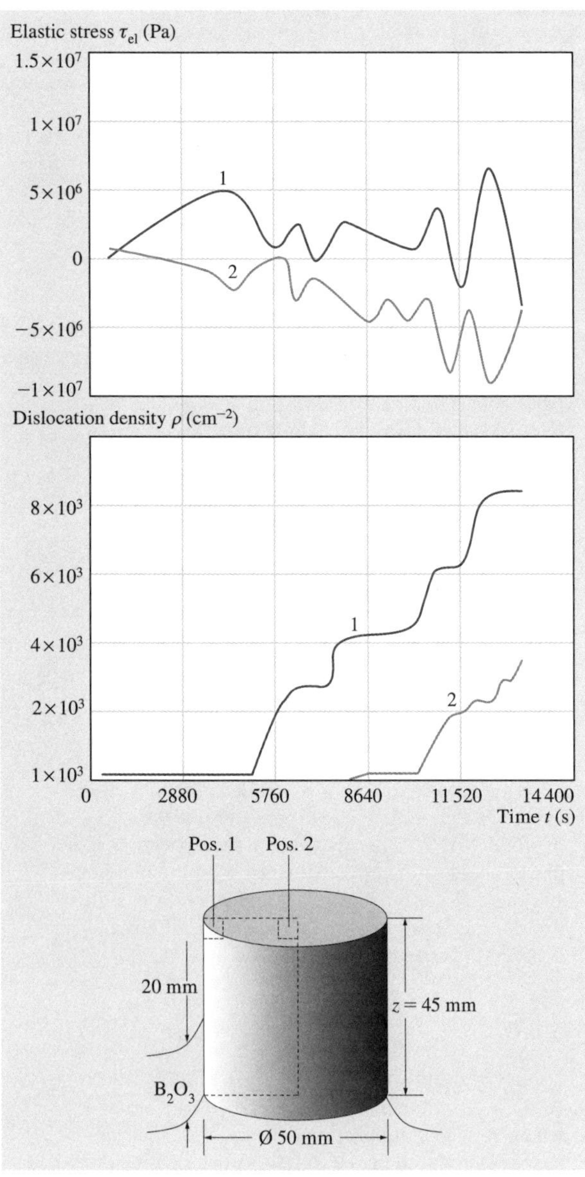

In general, approach 2 is the modern and more realistic way. As emphasized by *Völkl* [6.111], the plastic relaxation varies due to time-dependent experimental conditions and due to continuously acting dislocation dynamics. Therefore, the driving force for dislocation multiplication is given by the actual value of the elastic strain and not by the total deformation. Thus, one has to consider that the often-published *von Mises contours* (as part of approach 1), calculated along growing crystal cuts and cross-sections, reflect snapshots of the unrelaxed thermoelastic stress only. Certainly, they are usable for estimation of the dislocation density at the growing interface but not within the cooling crystal volume where the majority of the elastic strain is already relaxed by plastic deformation. Note that changing stress situations can appear during the growth run, as in the LEC case when the crystal emergences from the liquid encapsulant towards the streaming gas atmosphere, creating a thermoelastic shock, for example.

Recently *Pendurti* et al. [6.115] reported the global numeric modeling of the nonstationary elastic stress and related dislocation development in growing LEC InP crystals by considering the history of the thermal field in the furnace and crystal as well as the convection in the melt and vapor phase. Figure 6.9 shows their calculated elastic stress history and related dislocation density evolution at two selected crystal points.

Fig. 6.9 The history of elastic stress and dislocation density at the edge (1) and in the center (2) of a 2 inch InP crystal growing from the melt under a 20 mm-thick B_2O_3 layer (*sketch at the bottom*) calculated by a broadened Haasen–Alexander model with Grashof numbers of gas and melt of 10^8 and 10^6, respectively, and rotational Reynolds number of 500 (after *Pendurti* et al. [6.115]) ◀

As the elastic stress alternates, the dislocation density increases stepwise accordingly. The authors found that gas convection has a significant effect on the total dislocation density – a quite important fact that had not yet been considered.

The modeling of thermoelastic stress fields and the related dislocation density during a crystal growth process based on constitutive principles of continuum mechanics has made marked progress during the last decade. For both semiconductor compounds [6.114, 115] and oxides [6.116] the comparison between the theoretical results and real structural quality (i. e., mean EPD distributions) along the growth axis and radius show good conformities. However, even today one is still unable to predict exactly the dislocation rearrangements in characteristic microstructures, such as cell patterns and dislocation lineages and bundles, observed in most as-grown crystals under evolving thermomechanical stresses (see below). There are numerous efforts to simulate correctly the three-dimensional dislocation dynamics representing the collective interaction processes [6.117]. However, because of the high computational cost of discrete simulations, present-day modeling is restricted to metals and small system sizes (typically about $10\,\mu m^3$) so that only the very first stages of dislocation cell patterning can be studied. Since the cell size of the incipient dislocation cell patterns is of the same order as the size of the simulation box, not much information about the spatial morphology of the emergent patterns can be obtained. Hence, further theoretical methods such as continuum and stochastic approaches are under current development [6.117].

Dislocation Patterning

During plastic relaxation the interacting dislocation populations tend to rearrange spontaneously into characteristic heterogeneous formations. They can be subdivided into two basic patterning phenomena observed in numerous as-grown crystals, i. e.:

i) Dislocation *cell structuring*
ii) Dislocation *bunching*.

In compounds and alloys, patterning leads to physical and chemical parameter inhomogeneities. Whereas (i) includes formations of three-dimensionally ordered honeycomb-like (often named mosaic) structures, group (ii) compiles local accumulations such as slightly wavey dislocation walls (lineages) and vein-like bundles and gnarls. Today, dislocation theory can explain the properties of individual dislocations reasonably well, but is still unable to solve their collective behavior in all its details [6.118]. Hence, the following sections will concentrate on some typical features and practical relationships only. For more theoretical understanding the recently published reviews [6.117, 119] are recommended.

Cellular Structuring

The self-rearrangement of dislocations present in cellular networks during single-crystal growth is typical of most substances used. Figure 6.10a shows such cells in a 4 inch GaAs wafer, revealed by a standard etching process. The cell structure can be analyzed in more detail by laser scattering tomography (LST), which takes advantage of dislocation decoration by precipitates of the excess component in the case of nonstoichiometric crystals [6.120] (Fig. 6.10b), or by high-resolution x-ray synchrotron topography (ST) [6.121] (Fig. 6.10c).

The cells are of globular-like shape, consisting of walls with high dislocation density separated by interiors of markedly dislocation-reduced or even dislocation-free material. Their size decreases with increasing average dislocation density, yielding diameters of 1–2 mm and 500 μm at dislocation densities of $\rho \leq 10^4$ and $\approx 10^5\,cm^{-2}$, respectively. Note that they are often termed mosaic structures due to their appearance in two-dimensional cuts. Such patterns are also well known from as-grown crystalline metals (e.g., Fe, Al, Ni, Mo), metallic alloys (e.g., Fe-Si, Ti_3Al, Cu-Mn), and dielectric crystals (e.g., LiF, CaF_2, $SrTiO_3$, quartz) [6.122]. However, some differing morphological details are noteworthy. For instance, in Mo, Cu-Mn, and GaAs the cell interiors are nearly free of dislocations and the walls are fuzzy and of certain thickness, consisting of many tangled dislocations (Fig. 6.10c). It is noteworthy that the tilt angle between the cells in high-quality GaAs standard crystals is around 10 arcs. Compared with that in CdTe, PbTe, and CaF_2 the mean disorientation angle is higher (> 20 arcs up to some arcmin) and the cell walls are very thin, of the order of one dislocation row, resembling classical low-angle grain boundaries. In these crystals the matrix shows mostly individual dislocations that can occasionally form a subcell structure. Finally, there are crystals in which cell structures are not well distinguishable in the as-grown state, as in InP, or even missing if special dopants are added, as for Si in GaAs or Se in CdTe [6.122]. After *Rudolph* [6.122] the most probable reasons for missing cell structures are:

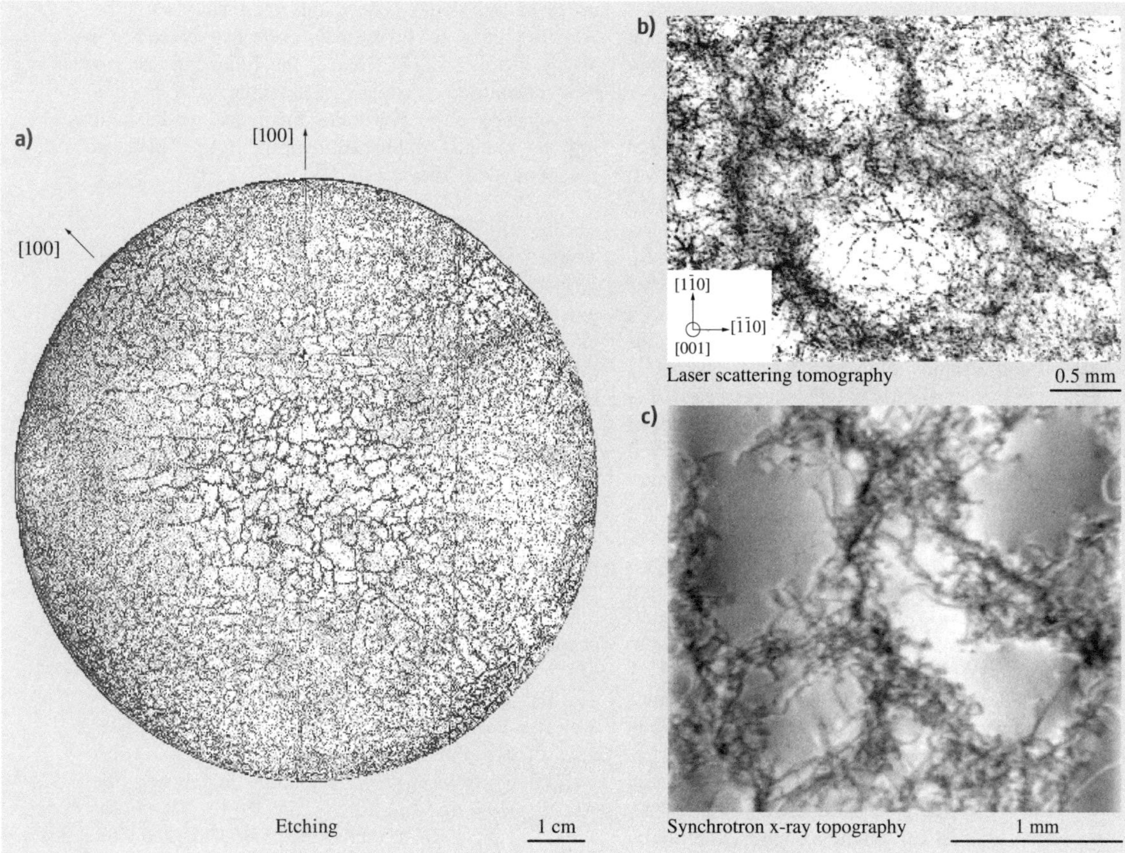

Fig. 6.10a–c Dislocation cell structures in GaAs crystals. **(a)** KOH-etched 4 inch wafer of a VGF crystal (after [6.56]). **(b)** LST analysis with integrated depth 2 mm (after [6.120]). **(c)** X-ray synchrotron topography (after [6.121, 122]) (reproduced by permission of Elsevier **(a,b)** and Wiley-VCH **(c)**)

i) Low stacking fault energy preventing cross-slip
ii) Small compound existence regions or stoichiometric growth conditions delivering only low native point-defect reservoir for climb
iii) Retarded dislocation movement
iv) Solution hardening by doping.

Undoped InP meets (i–iii). A possible explanation for the markedly differing wall morphologies in the cell structures in as-grown crystals is the different ripening levels within the framework of polygonization. Obviously, in Cu and GaAs the high tendency for dislocation screening by pronounced sessile junction formation leads to an entangled dislocation jungle within the walls being even stable against postannealing. Such morphology resembles the so-called incidental dislocation boundaries (IDBs), which are assumed to be a result of statistical mutual trapping of dislocations. On the other hand CdTe, PbTe, and CaF_2 show typical characteristics of well-ripened low-angle grain boundaries, i.e., geometrically necessary boundaries (GNBs) [6.123].

Cell patterning is studied best in metals under an external load, but also in postdeformed elemental and semiconductor compound crystals. Today, there are a large number of papers dealing with cell patternings, especially in the field of metal physics and mechanics. For growing crystals, however, there is not yet detailed knowledge about the genesis of DD at high temperatures. In all probability, cellular substructures are due to the action of the internal thermomechanical stress field.

It can be assumed that cell formation is coupled with dynamic polygonization behind the growing interface where the plastic relaxation by dislocation multiplication takes place. There is a well-confirmed in situ observation on thin crystallizing and remelting Al plates by *Grange* et al. [6.124], who observed by real-time synchrotron x-ray topography that the cellular dislocation structure appears due to the thermally induced strain within the region already some millimeters behind the melt–solid phase boundary. Recently, *Jakobson* et al. [6.125] confirmed such fast dynamics by in situ x-ray reflection analysis on deforming Cu crystals. The observation strongly indicated that subgrain formation is initiated shortly after the onset of plastic deformation. This result is of great importance for understanding cell genesis in growing crystals, whereupon the cell pattern in the cooled crystal can be assumed to be identical to the structure formed under high temperatures, and is therefore generated by the initially acting thermoelastic stress.

Note that not all types of patterning can be attributed to DD that takes place within the crystal volume. As discussed in Sect. 6.2.3, in the case of morphological instability of a fluid–solid phase boundary induced by constitutional supercooling, the former planar shape changes into a characteristic cellular profile [6.90, 96]. As a result, a lamellar-like structure with longitudinally extended walls is formed. However, the distinction from dislocation patterning is sometimes not trivial, even when cross-sectional crystal wafers are investigated. The best way to distinguish between them is the application of analytical methods with three-dimensional (3-D) imaging such as LST to reveal the globular cell morphology typical of dislocation patterning [6.120]. Moreover, as is well known, dislocation cells may disappear completely if certain dopants are added, such as In in GaAs or Se in CdTe, although their presence could promote constitutional supercooling.

A systematic analysis of the origins and genesis of cell formation during growth of semiconductor compound crystals, especially GaAs, was started by *Rudolph* et al. [6.20, 126, 129]. First, the relation between the stored dislocation density ρ and the cell size (diameter) d has been determined. To deduce the 3-D cell diameters from the 2-D etch pit images obtained on cut wafers, a stereological analysis method was used, as described in [6.126]. Figure 6.11a shows this correlation taken from experimental data. The literature data of deformed metals are added [6.127]. For the GaAs samples with EPD $\geq 10^4\ \mathrm{cm}^{-2}$ nearly the same correlation as in deformed metals has been found, i. e.,

$$d \approx K\rho^{-1/2}\,, \qquad (6.27)$$

with the factor of proportionality $K \approx 10$–20. This is surprising if one considers the marked differences between dislocation densities and cell dimensions in as-grown GaAs crystals and those in metals under load. The result shows that *Holt*'s scaling relation [6.127] is

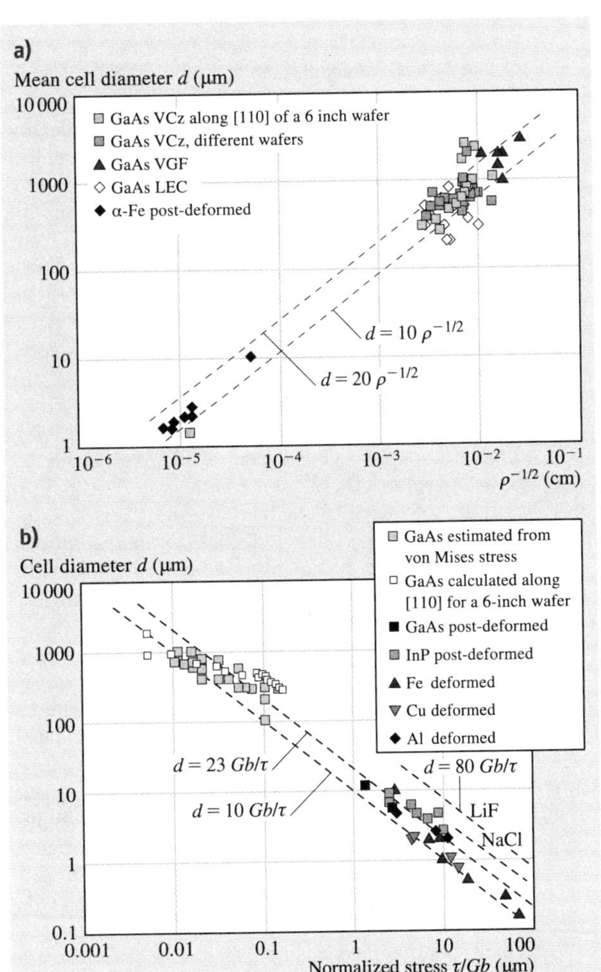

Fig. 6.11a,b Scaling of dislocation cell patterns in growing GaAs crystals in comparison with other crystalline materials post-deformed by mechanical stress (after [6.122, 126–128]). (**a**) Cell size d versus mean dislocation spacing $\rho^{-1/2}$, i. e., dislocation density ρ. (**b**) Cell size d versus flow stress τ (reproduced by permission of Wiley-VCH)

fulfilled over a wide range of materials and dislocation densities. At dislocation densities below $5\times10^3\,cm^{-2}$, however, the cells begin to dissociate.

Furthermore, the relation between acting stress and cell dimension was investigated. From postgrowth deformation experiments on numerous metals and dielectrics a universal relation between cell diameter and shear stress τ was found [6.130]:

$$d = \alpha K G \boldsymbol{b} \tau^{-1}\,, \tag{6.28}$$

where α is another proportionality factor, G is the Young's shear modulus, and $\boldsymbol{b}$ is the Burgers vector. The analysis of this correlation in the case of growing crystals is more difficult due to the impracticality of in situ measurement of the acting thermomechanical stress values. As pointed out above, only the elastic term is responsible for dislocation movement, and thus for the collective dislocation rearrangement in cells. Considering that the subgrain generation is initiated shortly after the onset of plastic deformation [6.125] it can be assumed that the cell formation process takes place immediately behind the growing interface and is completed in the course of plastic relaxation. Thus, the frontal elastic strain acting immediately after the propagating phase boundary can be assumed to be the driving force. This stress value is today readily calculable by global modeling. Therefore, the authors [6.20, 126, 129] used the calculated frontal thermoelastic shear stresses of growth situations, being identical to the real growth positions of each crystal from which the EPD distributions and cell size measurements were taken. The correlation between d and τ^{-1} in the form of (6.28) with $\alpha K = K'$ is shown in Fig. 6.11b. For comparison, the results from deformed metals as well as the slopes for NaCl and LiF are included [6.128, 130]. As can be seen, for cell sizes smaller than 700 μm and calculated stresses larger than about 1 MPa, the functional slope is similar to those of deformed materials. Independent of the growth conditions, it was found that d is inversely proportional to τ^{-1}. Obviously, in this region dislocation glide is the prevailing driving force for cell formation, due to the fact that the stress is larger than the critical resolved shear stress. In the case of larger cell dimensions the trend changes, showing a slope smaller than -1. In these regions a resolved shear stress of $\tau < 1$ MPa was calculated. One can suppose that, for such very low thermomechanical stress, even below the critical resolved shear stress ($\tau_{CRSS} \approx 0.5$ MPa), glide-driven plastic relaxation can no longer be the prevailing driving force for cell formation. Other cell structuring mechanisms must become dominant, such as point-defect-controlled diffusive creep. This could be in accordance with the observation that the residual dislocations in low-EPD GaAs VGF crystals probably no longer lie within the basal glide system [6.131, 132].

A number of theories have been proposed to account for the stress dependence of the cell size [6.117]. It is usually thought that the decreasing cell dimension with increasing stress is due to cell splitting in the course of growing stress, leading to progressive construction of new walls [6.118]. Newer papers favor a stochastic dynamics of the entire dislocation ensemble [6.117, 125] whereby the dislocation-free regions emerge and vanish in a fluctuating fashion in the course of the acting elastic stress. Until now, however, the question is whether the cell patterning is driven energetically or by a self-organizing process, in the former case by equilibrium, or in the latter by nonequilibrium thermodynamics where the entropy production is stopped, leading to dissipative structuring. There are well-known facts to be stated for energy-related processes. In the classical sense, the driving force for network formation is the reduction in strain energy resulting from the clustering (i. e., mutual field screening) of dislocations in cell boundaries. It is important to note that, for the formation of cells with globular morphology, in addition to dislocation glide, even spatial mechanisms such as climb and cross-glide are absolutely required [6.133]. However, the process of plasticity cannot be explained exclusively by equilibrium thermodynamics due to the presence of typical preconditions for irreversibility, such as temperature and stress gradients during the growth process. Hence, a growing crystal can be treated as a thermodynamically open system with continuous import and export of entropy. As a result, a rate of entropy is produced within the crystal, evoking self-ordered patterning of the stored dislocations. Much more fundamental coworking between crystal growers and theoretical physicists is required to clarify this still open problem.

Grain Boundaries

In numerous melt-grown compound crystals, such as in as-grown CdTe, PbTe, and CaF_2, for example, the cell arrangement resembles the classical low-angle grain boundary structure. The grain matrix contains mostly isolated dislocations and the walls are formed very abruptly, consisting of only one row of dislocation pits. *Sabinina* et al. [6.134] investigated the cell wall structure in melt-grown CdTe samples by transmission electron microscopy and observed that the dislocations that constitute the boundaries are nearly all parallel and most have the same Burgers vector. Such behavior is well known from the standard type of polygonized low-angle grain boundaries containing only the excess dislocations of similar Burgers vector after the annihilation process is completed. Obviously, in such crystals the DD contributes much more effectively to substructure ripening than in Cu and GaAs. Also, the larger disorientation angle between the neighboring cells refers to a typical polygonized grain-boundary structure. Tilt angles of 60–120 arcsec were reported for melt-grown CdTe crystals. Higher disorientation angles in the range of 2–30 arcmin, and in some cases even up to 3°, have been ascertained in PbTe crystals. Such a feature is also characteristic of dielectric materials such as CaF_2 and NaCl [6.135, 136]. It is obvious that in these crystals we have to deal with typical well-ripened grain boundaries, i. e., primary subboundaries, which are often superimposed by cellular structure and secondary subboundaries formed previously [6.136]. The scheme in Fig. 6.12 shows the possible stages of dislocation patterning during crystal growth beginning from cell formation towards a ripened small-angle (primary) grain-boundary structure. Depending on the dislocation mobility (highest in CdTe, PbTe), intrinsic point defect content (lowest in InP), and stacking fault energy (lowest in InP, CdTe), the ripening time and frozen-in contents are, however, different in various materials.

Conventionally, a low-angle grain boundary structure, formed by the above-discussed mechanism of dynamic polygonization, still belongs to a single-crystalline state. It is convenient to separate grain boundaries by the extent of the misorientation between two grains $\Theta[\text{rad}] = b/h$ (where h is the dislocation spacing within the grain boundary). Whereas a low-angle grain boundary is composed of an array of dislocations and its properties and structure (i. e., boundary energy) are a function of the misorientation, large-angle grain boundaries are those with a misorientation greater than 10–15° and are normally found to be independent of the tilt angle. A crystal with such grain structure is considered polycrystalline. It can be appear by:

i) Spontaneous nucleation in unseeded melt growth processes
ii) Disturbance of the heat balance between melt and solid phases
iii) Prenucleation in a supercooled melt region before the growing interface.

Also a cellular interface shape, generated under conditions of morphological instability (Sect. 6.2.3), can produce large-angle grain boundaries elongated parallel to the growth direction. Large-angle grain structure is a typical feature of cast solar silicon, affecting the photovoltaic efficiency (Table 6.1).

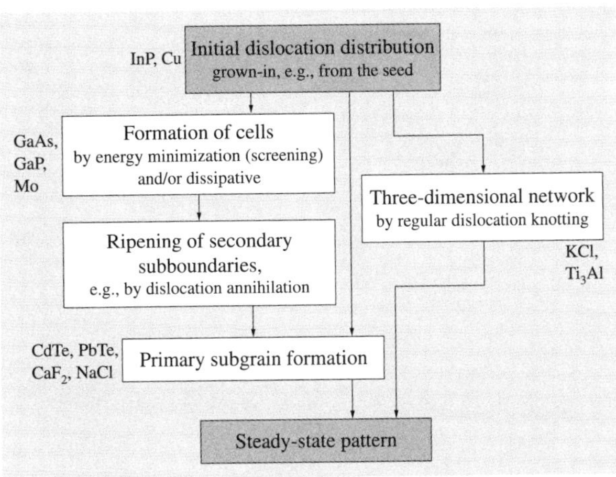

Fig. 6.12 Schematic demonstrating the different types and ripening stages of dislocation substructures that may develop from an initial uniform dislocation distribution. Some materials with typical related patterns, observed after crystallization, are added (after [6.136])

An interesting phenomenon can be observed during the Bridgman growth of semiconductor compounds with high ionicity, e.g., high degree of association in the melt (over 90%), such as CdTe and ZnSe [6.28, 137, 138]. There is a correlation between the number of large-angle grains and superheating of the melt before the crystallization process without an artificial seed is started (Fig. 6.13). Obviously, the high degree of association in slightly overheated melts causes stochastic preformation of structural elements (rings, chains, tetrahedrons), and their docking at the interface markedly affects the single crystallinity. Contrary to that, at a high level of superheating, the melt structure is altered to nearly monomolecular type, promoting nearly grain-free growth.

Dislocation Bunching

Dislocation bundles, often described as gnarls, tangles or clusters, mostly appear sporadically and in isolation. Once nucleated they may propagate through the whole growing crystal, typically parallel to the direction of solidification. Such defects have been detected in GaAs and InP crystals, independently of the growth method. Even in VGF crystals they have been observed. Typi-

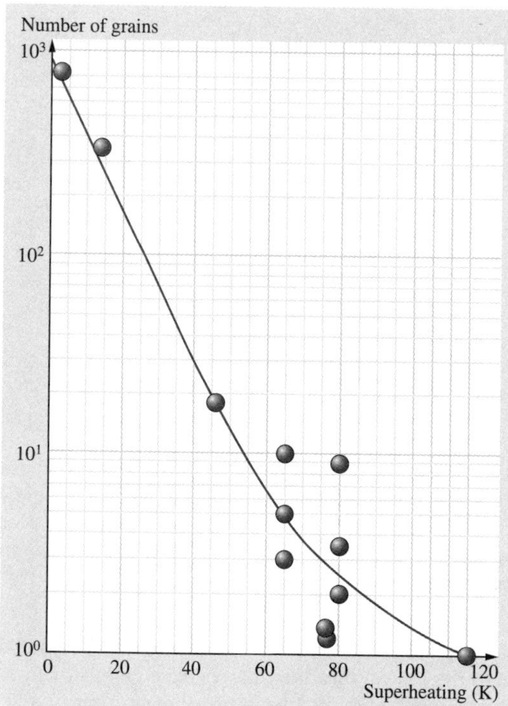

Fig. 6.13 The number of grains with large-angle boundaries in VB-grown ZnSe crystals in dependence on the degree of superheating before crystallization. The growth process was started by spontaneous nucleation within the ampoule tip (after [6.138])

cally, they appear in cast silicon ingots too. They are also known from plastically deformed metals and are often called veins. Figure 6.14 shows such bundles as detected on etched wafers of some VCz GaAs crystals. As was revealed by EPD and LST analysis, such bundles consist of very high-density parallel-arranged sessile dislocations. Principally, one has to differ between two types of *bunching*. They can be originated from inclusions (type 1) and contain a characteristic core of the second phase, or they may consist of a high number of tangled dislocations only (type 2) [6.20]. At first glance the two types are often not distinguishable and a locally good resolving analysis is required to differentiate between them.

Today the genesis of type 1 defects is well understood [6.56, 97, 139]. For instance, in GaAs they appear if Ga-rich inclusions are presented. Possible preconditions are growth from Ga-rich melts and/or the use of nonstoichiometric seed crystals already containing Ga inclusions. They have been also found in VB/VGF-grown CdTe crystals if growth from nonstoichiometric melts was employed [6.140]. In LEC crystals the main origin, however, is an unprotected dissociating crystal surface if it is in contact with too hot a gas ambience. In the case of GaAs the selective As evaporation releases Ga droplets penetrating into the crystal by a traveling solvent mechanism towards increasing temperature, i. e., following the growing interface [6.56, 97]. Usually, behind such defects, a tail of as-generated misfit dislocations are released [6.56, 139]. This mechanism makes LEC growth of semiconductor compounds in low temperature gradients impossible. Its prevention requires the protection of the crystal surface, whether by full encapsulation Czochralski measures (FEC), or VB/VGF or control of the thermodynamic equilibrium with the surrounding gas phase by partial pressure of the volatile component (VCZ mode). Considering these conditions, today type 1 defects are completely preventable.

There are various formation concepts for type 2 dislocation bundles. Mostly they appear at concave parts of the crystallization fronts where favorable conditions of dislocation focusing exist [6.141]. According to etching analysis, dislocation gnarls are mostly localized at the concave-to-convex transition regions on the $\langle 110 \rangle$ axis (Fig. 6.14a). Obviously, this has to do with collision of dislocation glides along the basis glide system $\langle 110 \rangle \{111\}$ according to the Schmidt contour. Once they are formed they follow the propagating crystallization front through the whole crystal, as was theoretically derived by *Klapper* et al. [6.141]. This fact was proved very carefully in GaAs crystals by *Shibata* et al. [6.142]. *Wang* et al. [6.143] attributed such bundles to localized composition variation, i. e., stoichiometry fluctuations, along the interface area, which create dislocation sources by vacancy agglomeration. Such instabilities are conceivable if convection-driven turbulence is present in the melt phase. Even newer concepts on stressed metals couple dislocation bunching with oscillating strain [6.144, 145]. The results of simulations demonstrated that, under cyclic loading regions of low and high dislocation density, a vein structure is formed. From that arises the question: do convective oscillations or even heating temperature fluctuations during a melt growth process play a similar stimulating role in dislocation bunching? Further investigations are required to solve this phenomenon. Generally, it is proved experimentally that the probability of bunching decreases with flattening of the growing interface. In fact, a slightly convex morphol-

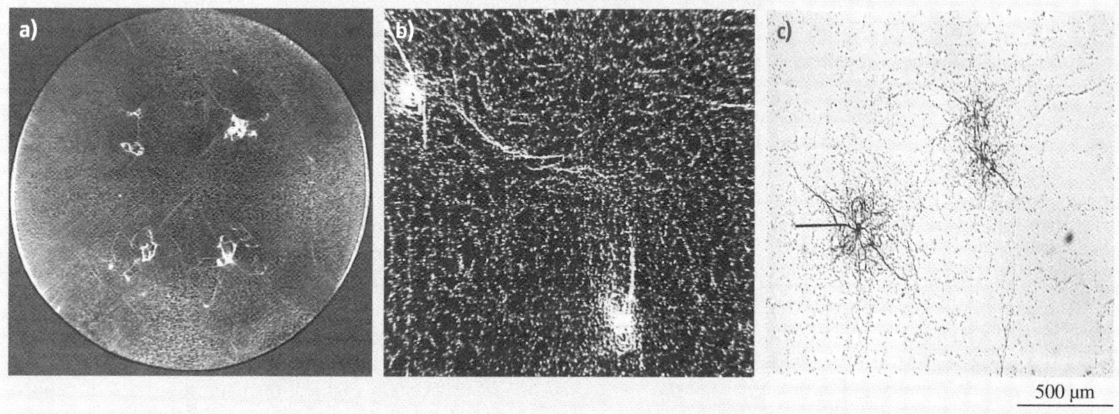

Fig. 6.14a–c Dislocation bunching in melt-grown crystals. (**a**) Dislocation bundles in a GaAs wafer positioned along ⟨110⟩ directions. The comparison with longitudinal striation analysis shows that the bundles are generated at the concave parts of a markedly w-shaped interface (after [6.20]) (reproduced by permission of Elsevier). (**b**) Magnified dislocation bundles in GaAs. (**c**) Dislocation bundles in cast silicon used for solar-cell production passing through the whole ingot as veins (see also Fig. 6.7b; (**c**) courtesy of U. Juda from IKZ Berlin)

ogy, if the ratio between the diameter of the interface curvature r_{IF} and the crystal radius r is > 0.5, proves to be optimum for prevention of dislocation bunching [6.20].

A special case of slip line assembling has been found in *lineages* formed like wavey *dislocation walls*. According to [6.120] in GaAs such lineages are composed of an enormous number of slip lines piled up within thin stripes parallel to {110}. Due to their slope against the crystal edge when the interface was slightly convex they could be related to isotherm curvature. After *Ono* [6.146], lineages are the result of slip interactions leading to sessile dislocations by the Lomer–Cotrell mechanism. The degree of waving is influenced by high-temperature climb processes. Probably, they reflect a morphological feature of the growing interface and are, therefore, coupled with the melt–solid phase boundary. Today the exact origin of lineages is still unclear. From growth experiments it follows that they can be prevented by nearly flat interface shapes when only minimal thermoelastic stresses are present.

6.3.3 Dislocation Engineering

Dislocation engineering deals with practical measures of control of dislocation density and patterns or even their prevention in situ, i. e., during the crystal growth process. Generally, for dislocation-reduced growth of compound and mixed crystals with large diameters, the proper combination of the following conditions are required:

i) Use of a *dislocation-free seed* crystal, in order to prevent grown-in dislocations being the most serious sources of dislocation multiplication
ii) Achievement of a strongly *uniaxial heat flow* with very small temperature gradients, i. e., nearly flat isotherms at all stages of the growth process
iii) *Omission of fluid encapsulants* (boric oxide), the presence of which introduces marked thermomechanical stresses at the crystal periphery, and maybe its replacement by a *detached growth* mode [6.147]
iv) In situ *stoichiometry control* by partial vapor pressure regulation over the melt in order to reduce the intrinsic point defect content which promotes high-temperature dislocation multiplication by climb and also cell structure formation
v) Prevention of *constitutional supercooling* at the interface by proper selection of a noncritical G/R ratio
vi) Minimization of atmospheric *pressure fluctuations* around the growing crystal to prevent heterogeneous dislocation rearrangements in bundles and veins.

The highest-temperature nonlinearities, and hence related thermal stress values, increasing very sensitively with diameter, appear in LEC crystals. Today, in such undoped 4 and 6 inch GaAs crystals, the mean dislocation densities are $(5–7)\times10^4$ and $(1–2)\times10^5\,cm^{-2}$, respectively. The situation can be improved by using modified Czochralski growth with low thermal gradients, for which the vapor-pressure-controlled Czochralski (VCZ) method is available [6.56]. In 6 inch SI (semi-insulating) VCz crystals of more than 20 cm length (25 kg) the average etch pit density (EPD) along the ⟨110⟩ and ⟨100⟩ directions can be reduced to $(1.8–2.6)\times10^4$ and $(2–3)\times10^4\,cm^{-2}$, respectively. Minimum values of $(6–8)\times10^3\,cm^{-2}$ were ascertained near the $r/2$ region (where r is the wafer radius). In 4 inch VCz crystals a somewhat lower average EPD of $(5–10)\times10^3\,cm^{-2}$ was found [6.148].

The best EPD results, however, can be obtained by the VGF method, matured on the industrial scale since the mid-1990s as the most promising growth variant for important semiconductor (InP, GaAs, CdTe), fluoride (CaF_2), and oxide ($Bi_4Ge_3O_{12}$ (BGO), $Pb(Mg, Nb)_{1-x}Ti_xO_3$ (PMNT), $Pb(Zn, Nb)_{1-x}Ti_xO_3$ (PZNT)) compound crystals. The decisive technological measure proved to be the maintenance of a uniaxial heat flow through the growing crystal during the whole growth run by proper control of the cooling rate between a top and bottom heater flanked by a booster heater to avoid radial heat outflow [6.149]. In undoped GaAs VGF crystals with diameters between 3 and 6 inch, dislocation densities in the range from 500 to $5000\,cm^{-2}$ have been reported [6.150, 151]. *Müller* and *Birkmann* [6.152] succeeded in the growth of Si-doped 4 inch GaAs crystals with the lowest EPD of $31\,cm^{-2}$ by optimized VGF. The few residual dislocations in ⟨100⟩-oriented crystals are accumulated cross-like along the ⟨100⟩ directions, obviously connected with the pronounced joint of the {111} facets along the ⟨100⟩ directions in the crystal cone after the seed. This phenomenon favors growth with a flat bottom from a seed of the same diameter [6.153, 154] in order to maintain the rotational symmetry without pronounced faceting.

Another phenomenon to be controlled in situ is dislocation cell patterning. This problem has not yet been solved on an industrial scale. However, there are already some in-principal laboratory experiences usable for future melt growth improvements. First, independently of the materials used, the cell formation can be reduced very effectively by doping. No cell structuring was observed in GaAs doped with In or Si at concentrations $> 10^{18}\,cm^{-3}$. Such an effect is due to the impurity gettering at the dislocation core increasing with temperature because of the increasing diffusion rate. As a consequence, the yield stress is enhanced by dislocation locking. No low-angle grain structure was found in CdTe and PbTe crystals when solution hardening by mixing components Se ($x > 0.4$) and Sn ($x > 0.15$) was provided, respectively. However, there is the drawback of segregation when dopants are added to the melt (Sect. 6.2.2) and the danger of morphological interface instability by constitutional supercooling (Sect. 6.2.3). Obviously, the best way to exclude dislocation patterning is reduction of the dislocation density by minimization of the thermomechanical stress. In undoped GaAs it was observed that at ρ values $< 5\times10^{-3}\,cm^{-2}$ the cell structure began to disappear. However, for compound crystals with larger diameters over 100 mm the attainment of such low dislocation densities is not yet solved empirically when hardening dopants are not added. Hence, current efforts are directed to homogenization of the thermal field in the growing crystals in order to reduce the dislocation multiplication and mobility by minimizing the thermomechanical stress. Another important way to prevent cell patterning is the minimization of the native point defect content by in situ control of stoichiometry during growth. The stoichiometry can be regulated by the partial pressure of the volatile component over the melt, applying an extra heat source within the growth chamber [6.56]. Using such a VCz arrangement without boric oxide encapsulant, the cellular structure could be suppressed in undoped GaAs crystals when the stoichiometry was controlled by growth from Ga-rich melt composition [6.155, 156]. Recently, this result was confirmed theoretically [6.157]. The stored dislocation density can be reduced under stoichiometric growth conditions. This was achieved for GaAs by horizontal Bridgman growth [6.158], hot-wall Czochralski method [6.159], and VCz [6.155]. Tomizawa et al. explained it as a result of the lowest intrinsic point defect concentration taking part in dislocation motion and multiplication.

6.4 Second-Phase Particles

The presence of second-phase particles, often named COPs (crystal-originated particles), markedly affects the optical and electronic bulk quality as well as the surface perfection of epiready substrates (Sect. 6.1.2). COPs are some of the most studied harmful objects in as-grown crystals. They are present even in undoped $A^{II}B^{VI}$ (e.g., CdTe [6.28, 140], ZnSe [6.137]), $A^{III}B^{V}$ (e.g., GaAs [6.160, 161]), and $A^{IV}B^{VI}$ (e.g., PbTe [6.162]) compounds, but also in numerous oxides (e.g., $Gd_3Ga_5O_{12}$ (GGG) [6.163]) and fluorides (e.g., CaF_2 [6.164]). One of the most serious consequences of compound crystal growth under conditions of native point defect formation is their condensation in precipitates and microvoids. This phenomenon is due to the retrograde behavior of the boundary of the compound existence region, and therefore related to nonstoichiometry (Sect. 6.2.1). In addition, foreign particles, melt–solution droplets of the excess component, and gas bubbles can be incorporated at the growing melt–solid interface.

As explained in detail by *Rudolph* et al. [6.28, 165], it is important to distinguish between these two different second-phase particle formation mechanisms:

i) Precipitation
ii) Inclusion incorporation.

A schematic sketch of the origins of both processes is shown in Fig. 6.15, where their relations to the phase diagram are also illustrated.

In the case of a particle with near-spherical geometry it is known that the total concentration of the second-phase component N_i (in cm^{-3}) can be estimated according to the formula [6.28]

$$N_i = \frac{4\pi \rho_i N_A}{3A_i} \sum_{i=1}^{n} r_i^3 \rho_i \, , \tag{6.29}$$

where ρ_i and A_i are the density and the relative atomic mass of the second-phase component, respectively, N_A is Avogadro's constant, r_i is the particle radius, and i is an index for each class of particle diameter.

6.4.1 Precipitates

Fundamentals of Generation

Precipitates are formed due to the retrograde solubility of native point defects in nonstoichiometric solid compositions (Fig. 6.15). As the as-grown crystal is cooling down, the solidus is crossed and nucleation of the second phase takes place. Probably, Ostwald ripening has to be considered. Favored sites of precipitate ripening are dislocations, as has been concluded from IR laser scattering tomography (e.g., [6.120, 161]; Fig. 6.16). Average precipitate densities of about 10^8 and up to 10^{12} cm^{-3} have been found in GaAs [6.160] and CdTe [6.28], respectively. Typical sizes between 10–100 nm have been determined for As precipitates in GaAs [6.166] and Te precipitates in CdTe [6.167]. Half-empty precipitates have been found by transmission electron microscopy (TEM) in GaAs [6.168] and

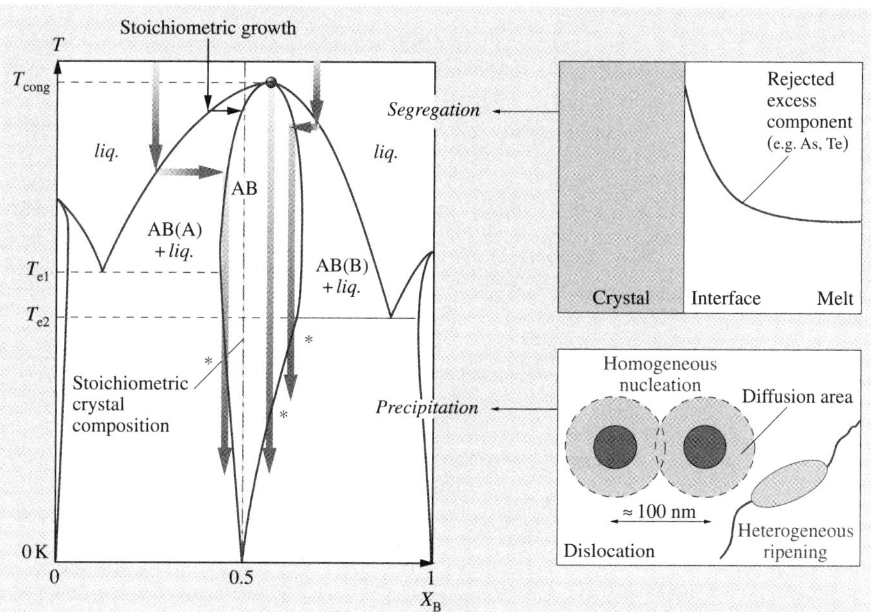

Fig. 6.15 Nonstoichiometry-related effects of second-phase particle formation in a growing compound crystal explained by a sketched phase diagram with elongated phase extent. The segregation evokes the rejection, and hence enrichment, of excess component at the interface that may lead to inclusion incorporation. Homogeneous (matrix) and heterogeneous (decoration) precipitations take place due to second-phase nucleation at the retrograde slope of the *solidus curves*, probably with subsequent Ostwald ripening

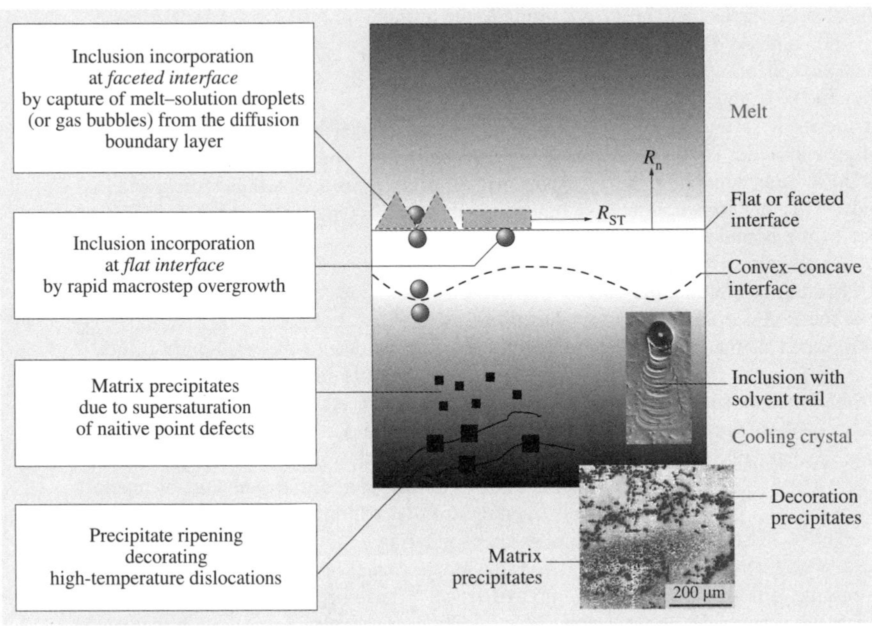

Fig. 6.16 Scheme of different inclusion incorporation and precipitation mechanisms during crystal growth from melt. Two images showing a Ga inclusion with traveling Ga-rich solvent trail and As precipitates in GaAs crystals (after [6.57, 161], reproduced by permission of Elsevier)

CdTe [6.169], probably caused by vacancy condensation in one of the sublattice components accompanied by conglomeration of excess atoms of the opposite sublattice.

In CdTe both tellurium and cadmium precipitates have been found [6.170]. This depends on the melt composition from which the crystal was grown (Te- or Cd-rich, respectively). According to the CdTe phase diagram their density can be effectively minimized by growth under near-stoichiometric conditions [6.28]. Contrary to this, in melt-grown GaAs crystals, only As precipitates were observed. *Fornari* et al. [6.166] found a correlation between As precipitate size and deviation from stoichiometry. Whereas at the stoichiometric melt composition ($x_L = 0.5$) the diameter is about 120 nm, from Ga-rich melts As precipitates of 40 nm diameter form, even when the mole fraction is markedly Ga-enriched ($x_L = 0.425$). This would to be in accordance with a GaAs existence region located completely on the As-rich side, as obtained from current thermochemical calculations and stoichiometry-controlled VCz experiments [6.19, 47, 48].

Control of Precipitate Density

There are two effective ways to minimize the precipitation concentration:

i) In situ control of stoichiometric crystal composition (see also Sect. 6.2.1)
ii) Postgrowth annealing under controlled partial pressure.

As was demonstrated by *Kießling* et al. [6.57] in VCz growth of GaAs without boric oxide encapsulant, melt compositions less than or around a mole fraction of $x_L \approx 0.45$ yield near-stoichiometric crystals essentially without precipitation.

Oda et al. [6.53] developed a multiple postgrowth wafer annealing technology for semi-insulating GaAs. Highly uniform substrates with markedly decreased arsenic precipitate density were obtained. Postgrowth wafer annealing was also successfully used by other authors for InP, GaP, CdTe, and ZnSe wafers.

6.4.2 Inclusions

Incorporation Mechanisms

In contrast to precipitates, inclusions are formed by capture of melt–solution droplets, gas bubbles or foreign particles from the diffusion boundary layer adjacent to the growing interface and enriched by the rejected excess component (Fig. 6.15). Preferred sites are reentrant angles of grain boundaries and twins crossing the interface. *Dinger* and *Fowler* [6.171] found lineages made of tellurium along the growth direction of CdTe crystals grown from Te-rich melt. They attributed this phenomenon to the enhanced cellular growth caused by constitutional supercooling. The inclusions are concentrated in the interlamella notches (Fig. 6.16). Typical inclusion diameters are 1–2 μm, but sizes up to 30 μm have been also observed in HB and VB CdTe crystals grown without Cd pressure control [6.28]. Their axial distribution increases slightly with increasing excess component by segregation. Melt–solution inclusions due to nonstoichiometric melt compositions show a specific crystallization genesis within the already solidified matrix [6.140]. Often they are embedded in a negative polyhedron formed by adjacent zincblende {111} planes. An inclusion could also be captured at a nearly flat interface by an overgrowth or embedding mechanism, as discussed by *Chernov* [6.38].

Prevention of Inclusion Trapping

There are two essential technological measures against inclusion incorporation:

i) Growth from the congruent melting point (mostly located close to stoichiometry)
ii) The choice of undercritical growth velocities.

Furthermore, accelerated crucible rotation techniques [6.85] and control by ultrasonic [6.86] or nonsteady magnetic fields [6.92–94] can be adopted as effective additional steps to disassemble phase boundary layers in melt-growth processes.

Note, even microgravity conditions are not favorable for the prevention of inclusion incorporation. *Salk* et al. [6.172] demonstrated during growth under microgravity that the danger of Te inclusion incorporation in CdTe is markedly increased. Due to the absence of convection in the melt the enrichment of the excess component at the interface, and hence the thickness of the diffusion boundary layer, is increased. This enhances the probability of inclusion capture. The growth of inclusion-reduced crystals in space proved to be successful only when well-controlled melt mixing by a rotating magnetic field was applied [6.172].

Chernov [6.38] estimated a critical interface rate R_{cr} to prevent inclusion or gas bubble incorporation that depends on the particle (bubble) radius r_{in}, the interface energy α, and the dynamical viscosity of the melt η as

$$R_{cr} \leq \left(\frac{0.11B}{\eta r_{in}}\right)\left(\frac{\alpha}{Br_{in}}\right)^{1/3} , \tag{6.30}$$

where B is a constant ($\approx 10^{-17}\,cm^2\,kg\,s^{-2}$). For instance, in the growth of GaAs crystals from the melt with $\eta = 2.8\times10^{-5}\,kg\,cm^{-1}\,s^{-1}$ a solid spherical foreign particle with $r_{in} = 1.5\times10^{-3}\,cm$ and $\alpha = 0.19\,kg\,s^{-2}$ would be rejected from the propagating interface if its normal rate R_n is $< 2\,mm/h$. Chernov showed that the hydrostatic pressing force of gaseous bubbles towards the interface plane is lower than that for solid particles. Hence, prevention of incorporation of microbubbles (e.g., of gaseous arsenic) of the same radius requires consideration of a somewhat enhanced critical velocity of about 8 mm/h.

6.5 Faceting

When *Hulme* and *Mullin* [6.174] looked in the 1950s at the segregation of a number of solutes in InSb they found that the segregation coefficient was different for crystals grown in a ⟨111⟩ direction as compared with growth in any other direction. Using radiotracer techniques, they demonstrated that a nonequilibrium concentration was incorporated into those parts of the crystal that had been grown on a faceted interface [6.2, 3] (Fig. 6.17a).

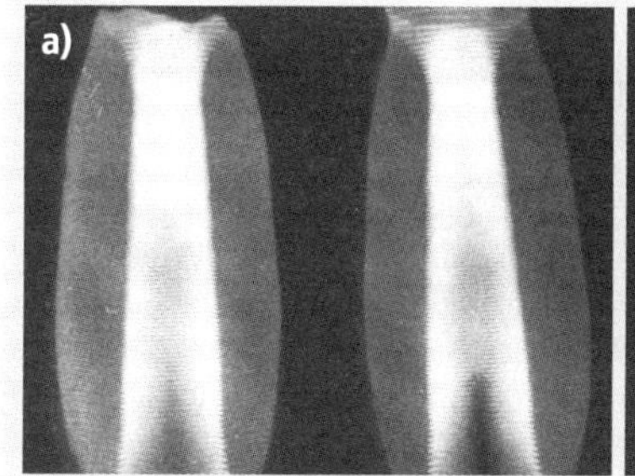

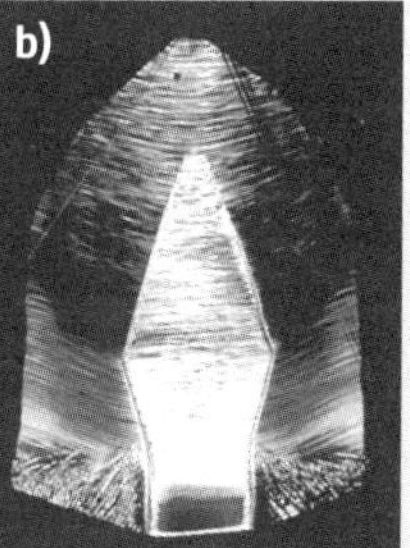

Fig. 6.17a,b Dopant redistribution in crystals grown with core facets. **(a)** Audioradiographs of a longitudinal section of ⟨111⟩-oriented InSb crystals doped with Te. The *bright central column* in each section is a region of enhanced Te concentration where the crystals grew with a faceted interface (after [6.2,3]). **(b)** Optical diffraction along a longitudinal cut of an InP crystal doped with sulfur. The core region corresponding with {111} facet shows enhanced S concentration (after [6.173], reproduced by permission of Elsevier) ▶

Until today faceting plays a problematic role during crystal growth, especially of oxide materials [6.175, 176]. Facets are the macroscopic indication of a given crystallographic structure and express, therefore, the natural tendency of single crystals to form polyhedra. Hence, the higher the crystal quality, the more developed the facets. This means that they are not defects in reality, but rather a serious cause of dopant and solvent redistributions as well as thermal field inhomogeneities due to their enhanced radiation effects. Further, facets at the rim of a pulling crystal may affect meniscus stability. Careful observation of a growing $LiNbO_3$ crystal periphery has revealed sudden repelling of the melt from a facet followed by rewetting, i. e., the meniscus jumps several mm down and then up. Such meniscus instability can lead to spontaneous nucleation and subsequent growth of polycrystals [6.176].

Facets form on crystal planes, for which 2-D nucleation is required in order to initiate the growth of a new layer. On nonfaceted (atomically rough) surface atoms can be added singly without the need for nucleation. At a given growth temperature, all crystals will have some surfaces which are rough. However most crystals will have one or more surfaces that are atomically smooth requiring nucleation, especially dielectric materials. *Jackson* [6.37] proposed a simple relation (Jackson factor α) for the faceting probability

$$\alpha = \frac{\Delta H_{SL}}{k_B T_m} \frac{w}{u} \tag{6.31}$$

(T_m is the melting temperature; w is the number of nearest neighbors of an atom on the growing face; u is the number of nearest neighbors of an atom in the crystal), which indicates that the magnitude of the entropy of fusion ΔH_{SL} of a material is the guide to its likelihood of forming facets during growth, i. e., materials having a low entropy of fusion (such as metals with $\alpha < 2$) have the lowest probability. In contrast, dielectric crystals with large ΔH_{SL} due to their strong ionic bond energy show a high α factor (> 2), and hence the greatest faceting probability. The common semiconductor materials, with their covalent bonding and $\alpha \approx 2$, tend to form facets during melt growth only on their most closely packed $\{111\}$ planes, i. e., where the w/u ratio is nearest to unity, or in other words the surface energy is lowest. For instance, in silicon for the $\{100\}$ and $\{111\}$ planes the Jackson factors for $\alpha_{\{100\}}$ and $\alpha_{\{111\}}$ are 1.75 (atomically rough) and 2.63 (atomically smooth), respectively.

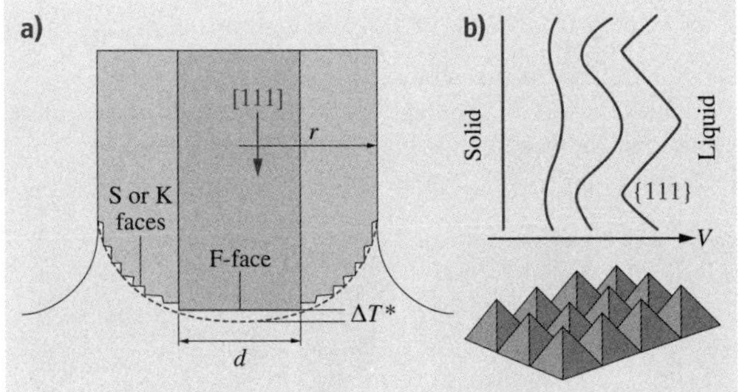

Fig. 6.18a,b Faceting phenomena at the growing melt–solid interface. **(a)** Sketch of Czochralski growth of a crystal with diamond or zincblende structure in $\langle 111 \rangle$ direction with convex interface where a [111] facet is formed. **(b)** Morphological instability of a growing interface where microfacets are formed (after *Hurle* and *Rudolph* [6.2, 3])

A further geometric requirement for facet formation during crystal growth from melts is that the radial temperature gradient be such that the freezing-point isotherm is convex when viewed from the crystal (Fig. 6.18a). This ensures that, if the crystal starts to lag behind the isotherm, it experiences increased supercooling at the facet, which ultimately promotes the nucleation of a new layer (note that, the higher the crystal perfection, i. e., the lower the dislocation density, the larger the supercooling). The lateral extension of the facet d is proportional to the supercooling ΔT [6.176] (Fig. 6.18a) as

$$d = 2\sqrt{\frac{2r_{IF}\Delta T}{G_r}}\,, \tag{6.32}$$

where G_r is the radial temperature gradient, ΔT is the undercooling, and r_{IF} is the radius of curvature of the interface. As can be seen, d increases with decreasing convexity of the interface, i. e., reducing radial temperature gradient. The rapid lateral growth tends to trap in the surface-adsorbed (equilibrium) solute concentration, thereby increasing the effective interface segregation coefficient of solutes that are preferentially adsorbed at the interface. The most dramatic effect occurs with Te-doping of InSb. Here the equilibrium segregation coefficient of Te is ≈ 0.5, whereas the effective segregation coefficient on the $\{111\}$ facets is ≈ 4.0, giving the remarkable ratio $8:1$ [6.177].

For several years the symptom of faceting at growing interfaces has been considered in numeric modeling of crystal growth processes [6.178]. To do this, the transport phenomena must be coupled with interfacial attachment kinetics. Hence, the isotherm condition, typically employed at the melt–crystal interface, is replaced by an equation accounting for undercooling due to interface kinetics. As a result, the interplay between evolving thermal fields and anisotropic interface kinetics is investigated. In particular, the evolution of facets and the dependence of their size on growth conditions, especially of oxide crystals such as $Y_3Al_5O_{12}$ (YAG), are explored. Of course, most realistic modeling of crystal growth systems involving partial faceting will usually require three-dimensional analysis techniques [6.179].

6.6 Twinning

Grown-in twins are one of the most serious macroscopic defects, the presence of which makes a crystal of no commercial use because of the twin-induced growth disorientation over the whole crystal body (Fig. 6.19). To date, there is no absolutely reliable measure to prevent twinning due to the stochastic character of its appearance. However, one can rank the material and growth parameters most responsible for enhancing the twinning probability. *Gotschalk* et al. [6.181] correlated it with the stacking fault energy, whereupon the greatest danger of twinning exists in materials with high ionicity, showing the lowest stacking fault energies. In fact, InP and CdTe with degrees of ionicity of 42% and 72%, i. e., stacking fault energies of 18 and 11×10^{-7} J/cm^2, respectively, show an extremely high twinning statistics among the semiconductor compounds. Growth conditions enhancing twin appearance are:

i) Temperature instabilities, i. e., remelting of the interface
ii) Presence of impurities
iii) Foreign particles swimming on the melt surface
iv) Interface contact with wetting inner container walls
v) Morphological instability of the crystallization front

The recently successful twin-free growth of GaAs crystals from Ga-rich melts without boric oxide encapsulant by the VCz technique [6.156] refutes the former conclusion that twin-free growth was highly improbable for growth at marked deviation from stoichiometry, even of GaAs from Ga-rich melts [6.182].

In the diamond and zincblende structures, twinning is closely related to facet formation (Sect. 6.5) and specified by a rotation of the lattice by 60° about a ⟨111⟩ axis, the twin lying on the orthogonal {111} plane. As a result, a former [001]-oriented crystal after twinning becomes completely disoriented with the [221] axis (Fig. 6.19a). It was recognized early on by *Billig* [6.183] that such twinning occurred principally on {111} facets, which form adjacent to the three-phase boundary of melt, crystal, and ambient. Billig studied Ge crystals where the problem is not serious and totally avoidable with carefully controlled growth. The problem is more serious in the III–V compounds, notably In-containing ones such as InSb, InAs, and InP.

The mechanism by which such twins form during growth defied explanation for many years, but in 1995, *Hurle* [6.182] provided a possible thermodynamic description based on ideas due to *Voronkov* [6.184], which can explain the key features of the process. The model demonstrates that, because of the orientation dependence of interfacial energies in the presence of facets, there is a configuration of the three-phase boundary for which, for sufficiently large supercooling, the free

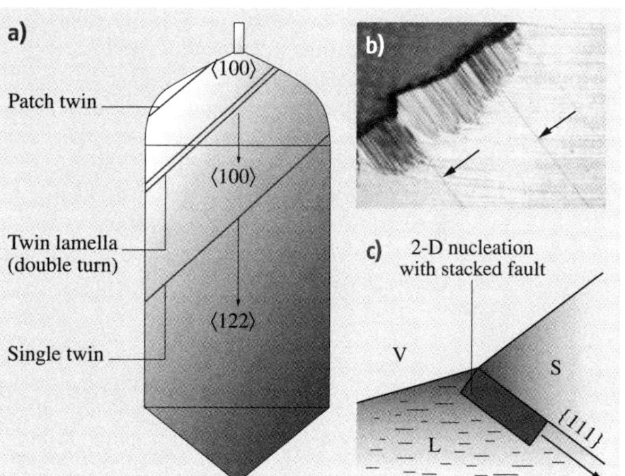

Fig. 6.19a–c Formation of grown-in twins in Czochralski crystals with diamond and zincblende structures. (**a**) Sketch showing different kinds of twins and the misorientation of the whole crystal from the initial [001] into [122] direction after generation of a single twin. (**b**) Twinning on {111} facets (*white arrows*) at the shoulder region of a InP crystal revealed by *Inada* [6.180] (reproduced by permission of Elsevier). (**c**) Scheme demonstrating the stacking fault (i. e., twinned) 2-D growth mode along the {111} facets when a misoriented nucleus is generated at the vapor–liquid–solid (VLS) interface

energy of formation of a critical nucleus is actually lowered by forming that nucleus at the three-phase boundary in twinned orientation. This will occur only if a critical angle of conical growth presenting a portion of crystal surface normal to $\langle 111 \rangle$ is sampled during the growth.

Such a twinned nucleus is thermodynamically favored if the supercooling exceeds the critical value

$$\delta T^* = \left(\frac{\sigma T_m}{h \Delta H_{SL}} \right) A^* , \quad (6.33)$$

where σ is the twin plane energy, T_m is the melting temperature, h is the nucleus height, ΔH_{SL} is the latent heat of fusion, and A^* is the reduced work of formation of a nucleus intersecting the three-phase boundary. Experimental test of this model on InP had been provided by the groups of *Müller* [6.185] and *Dudley* [6.186, 187]. Especially *Dudley* et al. [6.187] found, by synchrotron x-ray anomalous scattering analysis of as-grown InP crystals, that not 60° twinning but 180° rotation between the matrix and the twins takes place. They observed that twinning immediately follows the formation of a $(1\bar{1}\bar{5})$ external shoulder facet which, upon twinning was converted to a $(\bar{1}11)$ one. This can only occur when the edge facets exist in a region where the shoulder angle is close to 74.21°. The result demonstrates the importance of shoulder geometry during growth. Additionally, it was ascertained that twin nucleation occurs preferentially on $\{111\}_P$ faces due to the 30% lower surface energy than those of $\{111\}_{In}$. In GaAs preferential twinning on $\{111\}_{As}$ planes has been reported, showing a 12% lower surface energy than that of the Ga-terminated one.

Many LEC and Bridgman experiments have demonstrated that the twin probability is markedly reduced if the temperature oscillations of the growth system, and therefore excursions of the angle of the contacting meniscus, are minimized. This is due to the reduced probability of encountering the critical angle described above when the meniscus angle fluctuations are reduced. In fact, *Hosokawa* et al. [6.188] succeeded in growing twin-free InP crystal with diameters up to 100 mm by careful maintenance of thermal stability during growth, which was achieved by VCz and applying damping magnetic fields around the melt.

An interesting correlation has been observed by *Rudolph* during the vertical Bridgman growth of CdTe [6.28] and ZnSe crystals [6.137] by self-seeding. The number of twins was reduced and even prevented when the melt was markedly overheated and held for a longer time before the nucleation process in the ampoule tip was started. Similar results were described by *Khattak* and *Schmid* [6.189], with reduced twin formation in CdTe by overheating of the melt to above 110 K. Such a phenomenon can be explained when the well-known high degree of melt association of these materials at low superheating is taken into account. Probably, at a few Kelvin above the melting point, there are still enough preserved melt associates, such as tetrahedra, which can be incorporated into the crystalline phase by a false stacking sequence (e.g., 60° rotated in relation to a correct stacked {111}-plane). *Witt* [6.190] proposed a clustering model of the melt, whereupon the smallest cluster of relative stability capable of nucleating oblique twins comprises eight atoms. In general, more investigations of the melt structure and its influence on the growth kinetics are necessary.

6.7 Summary

The most important defect types and possible origins during crystal growth from the melt have been discussed. Today, most of the defect-forming mechanisms are well understood. However, some important questions still remain to be solved. For instance, the influence of the degree of association of the melt on the growth kinetics, which probably plays an essential role in II–VI systems, requires further consideration. Further, the dislocation patterning and bunching mechanisms in correlation with the growth conditions are not yet completely decoded. Also, the main origin of twinning has not yet been revealed.

Eminent success has been achieved in technological developments. During the last decade, computer-controlled vertical gradient freezing has matured to become the leading industrial production method for semiconductor compound crystals. However, that is not to say that all defects can be avoided. For example, the relatively poor thermal conductivity and low yield stresses of III–V and II–VI compounds as compared with Ge and Si mean that it is not possible to reduce the thermal stresses to a sufficiently low level to avoid dislocations.

Scaling up to achieve cost reduction is an ever-present pressure. Increasing crystal diameter increases the thermal stresses experienced during cooling. Avoiding increased dislocation density requires continual refinement of furnace design, and here computer modeling plays a key role. An additional problem posed by scaling-up is the increased turbulence which occurs in the melt. The use of magnetic fields is being exploited to achieve damping of this turbulence. Recently, in the author's laboratory, the concept of simultaneous generation of heat and a traveling magnetic field by a heater placed within the growth chamber was successfully tested and its first industrial application started [6.191].

References

6.1 H. Föll: *Defects in Crystals*, Hyperscript, http://www.tf.uni-kiel.de/matwis/amat/

6.2 D.T.J. Hurle, P. Rudolph: A brief history of defect formation, segregation, faceting, and twinning in melt-grown semiconductors, J. Cryst. Growth **264**, 550–564 (2003)

6.3 R.S. Feigelson (Ed.): *50 Years Progress in Crystal Growth* (Elsevier, Amsterdam 2004) p.109

6.4 H. Pick: Festkörperphysik, Naturwissenschaft **41**, 346–354 (1954), in German

6.5 S. Mahajan: Defects in semiconductors and their effects on devices, Acta Mater. **48**, 137–149 (2000)

6.6 I. Baumann, P. Rudolph, D. Krabe, R. Schalge: Orthoscopic investigations of the axial optical and compositional homogeneity of Czochralski grown $LiNbO_3$ crystals, J. Cryst. Growth **128**, 903–908 (1993)

6.7 P. Rudolph, M. Jurisch: Bulk growth of GaAs – An overview, J. Cryst. Growth **198/199**, 325–335 (1999)

6.8 N. Mainzer, E. Lakin, E. Zolotoyabko: Point-defect influence on $1/f$ noise in HgCdTe photodiodes, Appl. Phys. Lett. **81**, 763–765 (2002)

6.9 T. Fukuda, P. Rudolph, S. Uda: *Fiber Crystal Growth from the Melt* (Springer, Berlin 2004)

6.10 V. Swaminathan, A.T. Macrander: *Materials Aspects of GaAs and InP Based Structures* (Prentice Hall, Upper Saddle River 1991)

6.11 H. Chen, B. Raghotharmachar, W. Vetter, M. Dudley, Y. Wang, B.J. Skromme: Effects of different defect types on the performance of devices fabricated on a 4H-SiC homoepitaxial layer, Mater. Res. Soc. Symp. Proc. **911**, 1–6 (2006)

6.12 S. Miyazawa: Effect of dislocations on GaAs-MESFET Threshold voltage, and growth of dislocation-free, semi-insulating GaAs, Prog. Cryst. Growth Charact. Mater. **23**, 23–71 (1991)

6.13 J.R. Niklas, W. Siegel, M. Jurisch, U. Kretzer: GaAs wafer mapping by microwave-detected photoconductivity, Mater. Sci. Eng. B **80**, 206–209 (2001)

6.14 T.E. Schlesinger, J.E. Toney, H. Yoon, E.Y. Lee, B.A. Brunett, L. Franko, R.B. James: Cadmium zinc telluride and its use as a nuclear radiation detector material, Mater. Sci. Eng. R **32**, 103–189 (2001)

6.15 B.G. Ivanov, M.T. Kogan, V.M. Reiterov: Small-angle disorientation in Bridgman-Stockbarger-grown lithium fluoride crystals, J. Opt. Technol. **68**, 32–34 (2001)

6.16 P. Sadrabadi, P. Eisenlohr, G. Wehrhan, J. Stablein, L. Parthier, W. Blum: Evolution of dislocation structure and deformation resistance in creep exemplified on single crystals of CaF_2, Mater. Sci. Eng. A **510**, 46–50 (2009)

6.17 H. Halloin, P. von Ballmoos, J. Evrard, G.K. Skinner, N. Abrosimov, P. Bastie, G. Di Cocco, M. George, B. Hamelin, P. Jean, J. Knödleseder, P. Laporte, C. Badenes, P. Laurent, R.K. Smither: Performance of CLAIRE, the first balloon-borne γ-ray lens telescope, Nucl. Instrum. Methods Phys. Res. A **504**, 120–125 (2003)

6.18 J.R. Weertman, D. Farkas, K. Hemker, H. Kung, M. Mayo, R. Mitra, H. Van Swygenhoven: Structure and mechanical behavior of bulk nanocrystalline materials, MRS Bull. **24**(2), 44–50 (1999)

6.19 P. Rudolph: Non-stoichiometry related defects at the melt growth of semiconductor compound crystals – A review, Cryst. Res. Technol. **38**, 542–554 (2003)

6.20 P. Rudolph, C. Frank-Rotsch, U. Juda, M. Naumann, M. Neubert: Studies on dislocation patterning and bunching in semiconductor compounds (GaAs), J. Cryst. Growth **265**, 331–340 (2004)

6.21 F.A. Kröger: *The Chemistry of Imperfect Crystals* (North-Holland Publ., Amsterdam 1973)

6.22 D.I. Takamura: Point defects. In: *Physical Metallurgy*, ed. by R.W. Cahn (North-Holland Publ., Amsterdam 1965), Chap. XIII–XX

6.23 E. Dornberger: Prediction of OFS ring dynamics and grown-in voids in Czochralski silicon crystals. Ph.D. Thesis (Universite Catholique de Louvain, Louvain-la-Neuve 1997)

6.24 K.M. Luken, R.A. Morrow: Formation energies and charge states of native defects in GaAs: A selected compilation from the literature, Semicond. Sci. Technol. **11**, 1156–1158 (1996)

6.25 J.L. Rouviere, Y. Kim, J. Cunningham, J.A. Rentschler, A. Bourret, A. Ourmazd: Measuring properties of point derects by electron microscopy: The Ga vacancy in GaAs, Phys. Rev. Lett. **68**, 2798–2801 (1992)

6.26 L. Yujie, M. Guoli, J. Wanqi: Point defects in CdTe, J. Cryst. Growth **256**, 266–275 (2003)

6.27 M.A. Berding, M. van Schilfgaarde, A.T. Paxton, A. Sher: Defects in ZnTe, CdTe, and HgTe: Total energy calculations, J. Vac. Sci. Technol. A **8**, 1103–1107 (1990)

6.28 P. Rudolph: Fundamental studies on Bridgman growth of CdTe, Prog. Cryst. Growth Charact. Mater. **29**, 275–381 (1995)

6.29 R. Grill, J. Franc, P. Hoeschl, E. Belas, I. Turkevych, L. Turjanska, P. Moravec: Semiinsulating CdTe, Nucl. Instrum. Methods Phys. Res. A **487**, 40–46 (2002)

6.30 D.T.J. Hurle: Point defects in compound semiconductors. In: *Crystal Growth – From Fundamentals to Technology*, ed. by G. Müller, J.-J. Metois, P. Rudolph (Elsevier, Amsterdam 2004) pp. 323–343

6.31 P. Rudolph: Elements of thermodynamics for the understanding and design of crystal growth processes. In: *Theoretical and Technological Aspects of Crystal Growth*, ed. by R. Fornari, C. Paorici (Trans Tech Publications, Switzerland 1998) pp. 1–26

6.32 V.V. Voronkov, R. Falster, F. Quast: On the properties of the intrinsic point defects in silicon: A perspective from crystal growth and wafer processing, Phys. Status Solidi (b) **222**, 219–244 (2000)

6.33 D.T.J. Hurle: A comprehensive thermodynamic analysis of native point defect and dopant solubilities in gallium arsenide, J. Appl. Phys. **85**, 6957–7022 (1999)

6.34 R. Brown, D. Maroudas, T. Sinno: Modelling point defect dynamics in the crystal growth of silicon, J. Cryst. Growth **137**, 12–25 (1994)

6.35 E. Dornberger, J. Virbulis, B. Hanna, R. Hoelzl, E. Daub, W. von Ammon: Silicon crystals for future requirements of 300 mm wafers, J. Cryst. Growth **229**, 11–16 (2001)

6.36 M.A. Berding: Native defects in CdTe, Phys. Rev. **60**, 8943–8950 (1999)

6.37 K.A. Jackson: Liquid metals and solidification, Am. Soc. Met. (Cleveland, Ohio 1958) 174–180

6.38 A.A. Chernov: *Modern Crystallography III* (Springer, Berlin 1984)

6.39 I.V. Markov: *Crystal Growth for Beginners* (World Scientific, Singapore 1995)

6.40 A.A. Chernov: Notes on interface growth kinetics 50 years after Burton, Cabrera and Frank, J. Cryst. Growth **264**, 499–518 (2004)

6.41 K. Fujiwara, K. Nakajima, T. Ujihara, N. Usami, G. Sazaki, H. Hasegawa, S. Mizoguchi, K. Nakajima: In situ observations of crystal growth behavior of silicon melt, J. Cryst. Growth **243**, 275–282 (2002)

6.42 S. Arai, S. Tsukimoto, S. Muto, H. Saka: Direct observation of the atomic structure in a solid-liquid interface, Microsc. Microanal. **6**, 358–361 (2000)

6.43 T. Motooka, K. Nishihira, R. Oshima, H. Nishizawa, F. Hori: Atomic diffusion at solid/liquid interface of silicon: transition layer and defect formation, Phys. Rev. B **65**, 813041–813044 (2002)

6.44 R. Falster, V. Voronkov: Engineering of intrinsic point defects in silicon wafers and crystals, Mater. Sci. Eng. B **73**, 87–94 (2000)

6.45 F. Rosenberger: *Fundamentals of Crystal Growth I* (Springer, Berlin 1979)

6.46 S. Erdei, F.W. Ainger: Trends in the growth of stoichiometric single crystals, J. Cryst. Growth **174**, 293–300 (1997)

6.47 M. Jurisch, H. Wenzl: *Workshop on Simulations in Crystal Growth* (DGKK, Memmelsdorf 2002)

6.48 W. Dreyer, F. Duderstadt: *On the modelling of semi-insulating GaAs including surface tension and bulk stresses (EMS)* (Weierstraß-Institut, Berlin 2004), Treatise No. 995

6.49 P. Rudolph, F.-M. Kießling: Growth and characterization of GaAs crystals produced by the VCz method without boric oxide encapsulation, J. Cryst. Growth **292**, 532–537 (2006)

6.50 P. Rudolph, F.-M. Kießling: The horizontal Bridgman method, Cryst. Res. Technol. **23**, 1207–1224 (1988)

6.51 E. Monberg: Bridgman and related growth techniques. In: *Handbook of Crystal Growth*, Vol. 2a, ed. by D.T.J. Hurle (Elsevier, North-Holland 1994), Chap. 2

6.52 P. Rudolph, S. Kawasaki, S. Yamashita, S. Yamamoto, Y. Usuki, Y. Konagaya, S. Matada, T. Fukuda: Attempts to growth of undoped CdTe single crystals with high electrical resistivity, J. Cryst. Growth **161**, 28–33 (1996)

6.53 S. Oda, M. Yamamoto, M. Seiwa, G. Kano, T. Inoue, M. Mori, R. Shimakura, M. Oyake: Defects in and device properties of semi-insulating GaAs, Semicond. Sci. Technol. **7**, 215–223 (1992)

6.54 J. Nishizawa: Stoichometry control for growth of III–V crystals, J. Cryst. Growth **99**, 1–8 (1990)

6.55 J. Nishizawa, Y. Oyama: Stoichiometry of III–V compounds, Mater. Sci. Eng. R **12**, 273–426 (1994)

6.56 M. Neubert, P. Rudolph: Growth of semi-insulating GaAs crystals in low-temperature gradients by using the vapour pressure controlled Czochralski method (VCz), Prog. Cryst. Growth Charact. Mater. **43**, 119–185 (2001)

6.57 F.-M. Kießling, P. Rudolph, M. Neubert, U. Juda, M. Naumann, W. Ulrici: Growth of GaAs crystals from Ga-rich melts by the VCz method without liquid encapsulation, J. Cryst. Growth **269**, 218–228 (2004)

6.58 U.A. Borovlev, N.V. Ivannikova, V.N. Shlegel, Y.V. Vasiliev, V.A. Gusev: Progress in growth of large sized BGO crystals by the low-thermal-gradient Cz technique, J. Cryst. Growth **229**, 305–311 (2001)

6.59 E. Pfeiffer: Untersuchungen zur Optimierung der Züchtungstechnologie von $PbMoO_3$-Einkristallen nach der Czochralski-Methode. Ph.D. Thesis (Humboldt-University, Berlin 1990), in German

6.60 E. Pfeiffer, P. Rudolph: German patent DD 290-226 (1989)

6.61 K. Hein, E. Buhrig (Eds.): *Kristallisation aus Schmelzen* (Verlag für Grundstoffindustrie, Leipzig 1983), in German

6.62 P. Rudolph, U. Rinas, K. Jacobs: Systematic steps towards exactly stoichiometric and uncompensated CdTe Bridgman crystals, J. Cryst. Growth **138**, 249–254 (1994)

6.63 E. Northrup, S.B. Zhang: Dopant and defect energetics: Si in GaAs, Phys. Rev. B **47**, 6791–6794 (1993)

6.64 H. Zimmermann, R. Boyn, C. Michel, P. Rudolph: Absorption-calibrated determination of impurity concentrations in CdTe from excitonic photoluminescence, Phys. Status Solidi (a) **118**, 225–234 (1990)

6.65 J.A. Burton, R.C. Prim, W.P. Slichter: The distribution of solute in crystals grown from the melt, J. Chem. Phys. **21**, 1987–1991 (1953)

6.66 V.G. Levich: *Physicochemical Hydrodynamics* (Prentice-Hall, Englewood Cliffs 1961)

6.67 A. Ostrogorsky, G. Müller: A model of effective segregation coefficient, accounting for convection in the solute layer at the growth interface, J. Cryst. Growth **121**, 587–598 (1992)

6.68 W.A. Tiller: Principles of solidification. In: *The Art and Science of Growing Crystals*, ed. by J.J. Gilman (Wiley, New York 1963), Chap. 15

6.69 N. Sato, M. Kakimoto, Y. Kadota: The carbon and boron concentration control in GaAs crystals grown by liquid encapsulated Czochralski method. In: *Semi-Insulating III–V Materials*, ed. by A. Milnes, C. Miner (Hilger, Bristol 1990)

6.70 M. Jurisch, F. Börner, T. Bünger, S. Eichler, T. Flade, U. Kretzer, A. Köhler, J. Stenzenberger, B. Weinert: LEC- and VGF-growth of SI GaAs single crystals – Recent developments and current issues, J. Cryst. Growth **275**, 283–291 (2005)

6.71 K. Jacob, C. Frank, M. Neubert, P. Rudolph, W. Ulrici, M. Jurisch, J. Korb: A study on carbon incorporation in semi-insulating GaAs crystals grown by the vapor pressure controlled Czochralski technique (VCz), Cryst. Res. Technol. **35**, 1163–1171 (2000)

6.72 S. Eichler, A. Seidl, F. Börner, U. Kretzer, B. Weinert: A combined carbon and oxygen segregation model for the LEC growth of SI GaAs, J. Cryst. Growth **247**, 69–76 (2003)

6.73 J. Bohm, A. Lüdge, W. Schröder: Crystal growth by floating zone melting. In: *Handbook of Crystal Growth*, Vol. 2a, ed. by D.T.J. Hurle (Elsevier, North-Holland 1994), Chap. 4

6.74 R. Triboulet: The travelling heater method (THM) for $Hg_{1-x}Cd_xTe$ and related materials, Prog. Cryst. Growth Charact. Mater. **28**, 85–144 (1994)

6.75 C. Genzel, P. Gille, I. Hähnert, F.-M. Kießling, P. Rudolph: Structural perfection of (Hg,Cd)Te grown by THM, J. Cryst. Growth **101**, 232–236 (1990)

6.76 H.J. Koh, Y. Furukawa, P. Rudolph, T. Fukuda: Oxide mixed crystals grown by heater-immersed zone melting method with multi-capillary holes, J. Cryst. Growth **149**, 236–240 (1995)

6.77 R.S. Feigelson: Pulling optical fibers, J. Cryst. Growth **79**, 669–680 (1986)

6.78 T. Fukuda, P. Rudolph, S. Uda (Eds.): *Fiber Crystal Growth from the Melt* (Springer, Berlin 2004)

6.79 K.M. Kim, A.F. Witt, H.C. Gatos: Crystal growth from the melt under destabilizing thermal gradients, J. Electrochem. Soc. **119**, 1218–1222 (1972)

6.80 H.J. Scheel: Theoretical and technological solutions of the striation problem, J. Cryst. Growth **287**, 214–223 (2006)

6.81 E. Bauser: Atomic mechanisms of LPE. In: *Handbook of Crystal Growth*, Vol. 3b, ed. by D.T.J. Hurle (Elsevier, North-Holland 1994), Chap. 20

6.82 J. Barthel, M. Jurisch: Oszillation der Erstarrungsgeschwindigkeit beim Kristallwachstum aus der Schmelze mit rotierendem Keimkristall, Kristall und Technik **8**, 199–206 (1973), in German

6.83 D.T.J. Hurle, E. Jakeman: Effects of fluctuations on measurement of distribution coefficient by directional solidification, J. Cryst. Growth **5**, 227–232 (1969)

6.84 D.T.J. Hurle, R.W. Series: Use of magnetic field in melt growth. In: *Handbook of Crystal Growth*, Vol. 2a, ed. by D.T.J. Hurle (Elsevier, North-Holland 1994), Chap. 5

6.85 H.J. Scheel: Accelerated crucible rotation: A novel stirring technique in high-temperature solution growth, J. Cryst. Growth **13/14**, 560–565 (1972)

6.86 G.N. Kozhemyakin: Imaging of convection in a Czochralski crucible under ultrasound waves, J. Cryst. Growth **257**, 237–244 (2003)

6.87 E. Gilioli, J.L. Weyher, L. Zanotti, C. Mucchino: Growth striations in GaAs as revealed by DSL photoetching, Mater. Sci. Forum **203**, 13–17 (1996)

6.88 J.L. Weyher, P.J. van der Wel, G. Frigerio, C. Mucchino: DSL photoetching and high spatial resolution PL study of growth striations in undoped semi-insulating LEC-grown GaAs, Proceedings of the 6th Conference on Semi-Insulating III–V (1990) pp. 161–166

6.89 R.T. Gray, M.F. Larrousse, W.R. Wilcox: Diffusional decay of striations, J. Cryst. Growth **92**, 530–542 (1988)

6.90 B. Billia, R. Trivedi: Pattern formation in crystal growth. In: *Handbook of Crystal Growth*, Vol. 1b, ed. by D.T.J. Hurle (Elsevier, North-Holland 1994), Chap. 14

6.91 W.A. Tiller, K.A. Jackson, J.W. Rutter, B. Chalmers: The redistribution of solute atoms during the solidification of metals, Acta Metallurg. **1**, 428–437 (1953)

6.92 P. Dold, K.W. Benz: Rotating magnetic fields: Fluid flow and ctrystal growth applications, Prog. Cryst. Growth Charact. Mater. **38**, 7–38 (1999)

6.93 C. Stelian, Y. Delannoy, Y. Fautrelle, T. Duffar: Solute segregation in directional solidification of GaInSb concentrated alloys under alternating magnetic fields, J. Cryst. Growth **266**, 207–215 (2004)

6.94 V. Socoliuc, D. Vizman, B. Fischer, J. Friedrich, G. Müller: 3D numerical simulation of Rayleigh-Bénard convection in an electrically conducting melt acted on by a travelling magnetic field, Magnetohydrodynamics **39**, 187–200 (2003)

6.95 W.W. Mullins, R.F. Sekerka: Stability of planar interface during solidification of a dilute alloy, J. Appl. Phys. **35**, 444–451 (1964)

6.96 S.R. Coriell, G.B. McFadden: Morphological stability. In: *Handbook of Crystal Growth*, Vol. 1b, ed. by D.T.J. Hurle (Elsevier, North-Holland 1994), Chap. 12

6.97 H. Wenzl, W.A. Oates, K. Mika: Defect thermodynamics and phase diagrams in compound crystals. In: *Handbook of Crystal Growth*, Vol. 1a, ed. by D.T.J. Hurle (Elsevier, North-Holland 1994), Chap. 3

6.98 P. Schlossmacher, K. Urban: Dislocations and precipitates in gallium arsenide, J. Appl. Phys. **71**, 620–629 (1992)

6.99 S. Brochard, J. Rabier, J. Grilhé: Nucleation of partial dislocations from a surface-step in semiconductors: a first approach of the mobility effect, Eur. Phys. J. Appl. Phys. **2**, 99–105 (1998)

6.100 G. Grange, C. Jourdan, A.L. Coulet, J. Gastaldi: Observation of the melting-solidification process of an Al crystal by synchrotron x-ray topography, J. Cryst. Growth **72**, 748–752 (1985)

6.101 K. Sumino, I. Yonenaga: Interactions of impurities with dislocations: Mechanical effects, Solid State Phenom. **85/86**, 145–176 (2002)

6.102 E. Nadgorny: Dislocation dynamics and mechanical properties of crystals. In: *Progress in Materials Science*, Vol. 31, ed. by J.W. Christian, P. Haasen, T.B. Massalski (Pergamon, Oxford 1988)

6.103 R.J. Amodeo, N.M. Ghoniem: Dislocation dynamics. I. A proposed methodology for deformation micromechanics; Dislocation dynamics. II. Applications to the formation of persistent slip bands, planar arrays, and dislocation cells, Phys. Rev. B **41**, 6958–6976 (1990)

6.104 R.N. Nabarro: Steady-state diffusional creep, Philos. Mag. A **16**, 231–238 (1967)

6.105 E. Billig: Some defects in crystals grown from the melt, Proc. R. Soc. Lond. Ser. A **235**, 37–55 (1956)

6.106 V.L. Indenbom: Ein Beitrag zur Entstehung von Spannungen und Versetzungen beim Kristallwachstum, Kristall und Technik **14**, 493–507 (1979), in German

6.107 A.S. Jordan, A.R. von Neida, R. Caruso: The theory and practice of dislocation reduction in GaAs and InP, J. Cryst. Growth **70**, 555–573 (1984)

6.108 N. Miyazaki, H. Uchida, S. Hagihara, T. Munakata, T. Fukuda: Thermal stress analysis of bulk single crystal during Czochralski growth (comparison between anisotropic analysis and isotropic analysis), J. Cryst. Growth **113**, 227–241 (1991)

6.109 S. Motakef, A.F. Witt: Thermoelastic analysis of GaAs in LEC growth configuration: I. Effect of liquid encapsulation on thermal stresses, J. Cryst. Growth **80**, 37–50 (1987)

6.110 C.T. Tsai, A.N. Gulluoglu, C.S. Hartley: A crystallographic methodology for modeling dislocation dynamics in GaAs crystals grown from melt, J. Appl. Phys. **73**, 1650–1656 (1993)

6.111 J. Völkl: Stress in cooling crystals. In: *Handbook of Crystal Growth*, Vol. 2b, ed. by D.T.J. Hurle (Elsevier, North-Holland 1994), Chap. 14

6.112 R. Lohonka, G. Vanderschaeve, J. Kratochvil: Modelling of the plastic behaviour of III-V compound semiconductors during compressive tests, Mater. Sci. Eng. A **337**, 50–58 (2002)

6.113 K. Sumino, I. Yonenaga: Interactions of impurities with dislocations: mechanical effects, Solid State Phenom. **85/86**, 145–176 (2002)

6.114 S. Grondet, T. Duffar, F. Louchet, F. Theodore, N. Van Den Bogaert, J.L. Santailler: A visco-plastic model of the deformation of InP during LEC growth taking into accound dislocation annihilation, J. Cryst. Growth **252**, 92–101 (2003)

6.115 S. Pendurti, V. Prasad, H. Zhang: Modelling dislocation generation in high pressure Czochralski growth of InP single crystals: parts I and II, Modelling Simul. Mater. Sci. Eng. **13**, 249–297 (2005)

6.116 N. Miyazaki, Y. Matsuura, D. Imahase: Thermal stress analysis of lead molybdate single crystal during growth process: Discussion on relation between thermal stress and crystal quality, J. Cryst. Growth **289**, 659–662 (2006)

6.117 M. Zaiser: Dislocation patterns in crystalline solids – phenomenology. In: *Crystal Growth – From Theory to Technology*, ed. by G. Müller, J.-J. Metois, P. Rudolph (Elsevier, Amsterdam 2004) pp. 215–238

6.118 L. Kubin: Collective defect behavior under stress, Science **312**, 864–865 (2006)

6.119 F.R.N. Nabarro, M.S. Duesbery (Eds.): *Dislocations in Solids*, Vol. 11 (North-Holland, Amsterdam 2002)

6.120 M. Naumann, P. Rudolph, M. Neubert, J. Donecker: Dislocation studies in VCz GaAs by laser scattering tomography, J. Cryst. Growth **231**, 22–33 (2001)

6.121 T. Tuomi, L. Knuuttila, J. Riikonen, P.J. McNally, W.-M. Chen, J. Kanatharana, M. Neubert, P. Rudolph: Synchrotron x-ray topography of undoped VCz GaAs crystals, J. Cryst. Growth **237**, 350–355 (2002)

6.122 P. Rudolph: Dislocation patterning in semiconductor compounds, Cryst. Res. Technol. **40**, 7–20 (2005)

6.123 W. Pantleon: The evolution of disorientations for several types of boundaries, Mater. Sci. Eng. A **319–321**, 211–215 (2001)

6.124 G. Grange, C. Jourdan, A.L. Coulet, J. Gastaldi: Observation of the melting-solidification process of an Al crystal by synchrotron x-ray topography, J. Cryst. Growth **72**, 748–752 (1985)

6.125 B. Jakobson, H.F. Poulsen, U. Lienert, J. Almer, S.D. Shastri, H.O. Sørensen, C. Gundlach, W. Pantleon: Formation and subdivision of deformation structures during plastic deformation, Science **312**, 889–892 (2006)

6.126 C. Frank-Rotsch, U. Juda, F.-M. Kießling, P. Rudolph: Dislocation patterning during crystal growth of semiconductor compounds (GaAs), Mater. Sci. Technol. **21**, 1450–1454 (2005)

6.127 D.L. Holt: Dislocation cell formation in metals, J. Appl. Phys. **41**, 3197–3201 (1970)
6.128 J. P. Poirier: *Creep of Crystals – High-Temperature Deformation Processes in Metals, Ceramics*, Cambridge Earth Science Series (Cambridge Univ. Press, Cambridge 1985)
6.129 P. Rudolph, C. Frank-Rotsch, U. Juda, F.-M. Kießling: Scaling of dislocation cells in GaAs crystals by global numeric simulation and their restraints by in situ control of stoichiometry, Mater. Sci. Eng. A **400/401**, 170–174 (2005)
6.130 S.V. Raj, G.M. Pharr: A compilation and analysis of data for the stress dependence of the subgrain size, Mater. Sci. Eng. **81**, 217–237 (1986)
6.131 B. Birkmann, J. Stenzenberger, M. Jurisch, J. Härtwig, V. Alex, G. Müller: Investigations of residual dislocations in VGF-grown Si-doped GaAs, J. Cryst. Growth **276**, 335–346 (2005)
6.132 G. Müller, P. Schwesig, B. Birkmann, J. Härtwig, S. Eichler: Types and origin of dislocations in large GaAs and InP bulk crystals with very low dislocation densities, Phys. Status Solidi (a) **202**, 2870–2879 (2005)
6.133 B. Devincre, L.P. Kubin: Mesoscopic simulations of dislocations and plasticity, Mater. Sci. Eng. A **234–236**, 8–14 (1997)
6.134 I.V. Sabinina, A.K. Gutakovski, T.I. Milenov, N.N. Lykakh, Y.G. Sidorov, M.M. Gospodinov: Melt growth of CdTe crystals and transmission electron microscopic, Cryst. Res. Technol. **26**, 967–972 (1991)
6.135 L. Parthier, C. Poetsch, K. Pöhl, J. Stäblein, G. Wehrhan: About the influence of lattice-defects on the optical homogeneity of CaF_2 crystals for use in high performance microlithography, Gemeinsame Jahrestagung der DGK und DGKK, Jena, Referate (Oldenburg, München 2004) p. 5
6.136 S.V. Raj, I.S. Iskovitz, A.D. Freed: Modeling the role of dislocation substructure during class M and exponential creep, NASA Technical Memorandum 106986, 1–77 (1995)
6.137 P. Rudolph, N. Schäfer, T. Fukuda: Crystal growth of ZnSe from the melt, Mater. Sci. Eng. R **15**, 85–133 (1995)
6.138 P. Rudolph, K. Umetsu, H.J. Koh, T. Fukuda: Correlation between ZnSe crystal growth conditions from melt and generation of large-angle grain boundaries and twins, Jpn. J. Appl. Phys. **33**, 1991–1994 (1994)
6.139 J.P. Tower, R. Tobin, P.J. Perah, R.M. Ware: Interface shape and crystallinity in LEC GaAs, J. Cryst. Growth **114**, 665–675 (1991)
6.140 P. Rudolph, A. Engel, I. Schentke, A. Grochocki: Distribution and genesis of inclusions in CdTe and (Cd,Zn)Te single crystals grown by the Bridgman method and by the travelling heater method, J. Cryst. Growth **147**, 297–304 (1995)
6.141 H. Klapper: Generation and propagation of dislocations during crystal growth, Mater. Chem. Phys. **66**, 101–109 (2000)
6.142 M. Shibata, T. Suzuki, S. Kuma, T. Inada: LEC growth of large GaAs single crystals, J. Cryst. Growth **128**, 439–443 (1993)
6.143 F.-C. Wang, M.F. Rau, J. Kurz, M.F. Ehman, D.D. Liao, R. Carter: Correlation of growth phenomena to electrical properties of gnarl defects in GaAs. In: *Defect Recognition and Image Processing in III–V Compounds II*, ed. by E.R. Weber (Elsevier, Amsterdam 1987) p. 117
6.144 J. Kratochvil: Self-organization model of localization of cyclic strain into PSBs and formation of dislocation wall structure, Mater. Sci. Eng. A **309/310**, 331–335 (2001)
6.145 O. Politano, J.M. Salazar: A 3D mesoscopic approach for discrete dislocation dynamics, Mater. Sci. Eng. A **309/310**, 261–264 (2001)
6.146 H. Ono: Dislocation reactions and lineage formation in liquid encapsulated Czochralski grown GaAs crystals, J. Cryst. Growth **89**, 209–219 (1988)
6.147 T. Duffar, P. Dusserre, F. Picca, S. Lacroix, N. Giacometti: Bridgman growth without crucible contact using the dewetting phenomenon, J. Cryst. Growth **211**, 434–439 (2000)
6.148 P. Rudolph, M. Czupalla, C. Frank-Rotsch, U. Juda, F.-M. Kießling, M. Neubert, M. Pietsch: Semi-insulating 4-6-inch GaAs crystals grown in low temperature gradients by the VCz method, J. Ceram. Proc. Res. **4**, 1–8 (2003)
6.149 M. Althaus, K. Sonnenberg, E. Küssel, R. Naeven: Some new design features for vertical Bridgman furnaces and the investigation of small angle grain boundaries developed during VB growth of GaAs, J. Cryst. Growth **166**, 566–571 (1996)
6.150 T. Kawase, Y. Hagi, M. Tasumi, K. Fujita, R. Nakai: Low-dislocation-density and low-residual-strain semi-insulating GaAs grown by vertical boat method. In: *1996 IEEE Semiconducting and Semi-insulating Materials Conference, IEEE SIMC-9, Toulouse 1996*, ed. by C. Fontaine (IEEE, Piscataway 1996) pp. 275–278
6.151 T. Bünger, D. Behr, S. Eichler, T. Flade, W. Fliegel, M. Jurisch, A. Kleinwechter, U. Kretzer, T. Steinegger, B. Weinert: Development of a vertical gradient freeze process for low EPD GaAs substrates, Mater. Sci. Eng. B **80**, 5–9 (2001)
6.152 G. Müller, B. Birkmann: Optimization of VGF-growth of GaAs crystals by the aid of numerical modelling, J. Cryst. Growth **237–239**, 1745–1751 (2002)
6.153 P. Rudolph, F. Matsumoto, T. Fukuda: Studies on interface curvature during vertical Bridgman growth of InP in a flat-bottom container, J. Cryst. Growth **158**, 43–48 (1996)
6.154 U. Sahr, I. Grant, G. Müller: Growth of S-doped 2″ InP-crystals by the vertical gradient freeze technique. In: *Indium Phosphide and Related Materials*,

2001. IPRM. IEEE International Conference on 14–18 May 2001 in Nara, Japan, pp. 533–536

6.155 F.-M. Kießling, P. Rudolph, M. Neubert, U. Juda, M. Naumann, W. Ulrici: Growth of GaAs crystals from Ga-rich melts by the VCz method without liquid encapsulation, J. Cryst. Growth **269**, 218–228 (2004)

6.156 P. Rudolph, F.-M. Kießling: Growth and characterization of GaAs crystals produced by the VCz method without boric oxide encapsulation, J. Cryst. Growth **292**, 532–537 (2006)

6.157 B. Bakó, I. Groma, G. Györgyi, G. Zimányi: Dislocation patterning: The role of climb in meso-scale simulations, Comput. Mater. Sci. **38**, 22–28 (2006)

6.158 J.M. Parsey, F.A. Thiel: A new apparatus for the controlled growth of single crystals by horizontal Bridgman techniques, J. Cryst. Growth **73**, 211–220 (1985)

6.159 K. Tomizawa, K. Sassa, Y. Shimanuki: J. Nishizawa, Growth of low dislocation density GaAs by as pressure-controlled Czochralski method, J. Electrochem. Soc. **131**, 2394–2397 (1984)

6.160 P. Schlossmacher, K. Urban, H. Rüfer: Dislocations and precipitates in gallium arsenide, J. Appl. Phys. **71**, 620–629 (1992)

6.161 T. Steinegger, M. Naumann, M. Jurisch, J. Donecker: Precipitate engineering in GaAs studied by laser scattering tomography, Mater. Sci. Eng. B **80**, 215–219 (2001)

6.162 M. Mühlberg, D. Hesse: TEM precipitation studies in Te-rich as-grown PbTe single crystals, Phys. Status Solidi (a) **76**, 513–524 (1983)

6.163 K.-T. Wilke, J. Bohm: *Kristallzüchtung* (H. Deutsch, Thun, Frankfurt 1988) p. 356, in German

6.164 L. Su, Y. Dong, W. Yang, T. Sun, Q. Wang, J. Xu, G. Zhao: Growth, characterization and optical quality of CaF_2 single crystals grown by the temperature gradient technique, Mater. Res. Bull. **40**, 619–628 (2005)

6.165 P. Rudolph, M. Neubert, M. Mühlberg: Defects in CdTe Bridgman monocrystals caused by nonstoichiometric growth conditions, J. Cryst. Growth **128**, 582–587 (1993)

6.166 R. Fornari, C. Frigeri, R. Gleichmann: Structural and electrical properties of n-type bulk gallium arsenide grown from non-stoichiometric melts, J. Electron. Mater. **18**, 185–189 (1989)

6.167 R.S. Rai, S. Mahajan, S. McDevitt, D.J. Johnson: Characterisation of CdTe, (Cd,Zn)Te, and Cd(Te,Se) single crystals by transmission electron microscopy, J. Vac. Sci. Technol. B **9**, 1892–1896 (1991)

6.168 K. Sonnenberg: Defect studies in GaAs by NIR-microscopy with different contrast techniques, IFF Bull. **51**, 14–55 (1997)

6.169 J. Shen, D.K. Aidun, L. Regel, W.R. Wilcox: Characterization of precipitates in CdTe and $Cd_{1-x}Zn_xTe$ grown by vertical Bridgman-Stockbarger technique, J. Cryst. Growth **132**, 250–260 (1993)

6.170 H.G. Brion, C. Mewes, I. Hahn, U. Schäufele: Infrared contrast of inclusions in CdTe, J. Cryst. Growth **134**, 281–286 (1993)

6.171 R.J. Dinger, I.L. Fowler: Te inclusions in CdTe grown from a slowly cooled Te solution and by traveling solvent method, Rev. Phys. Appl. **12**, 135–139 (1977)

6.172 M. Salk, M. Fiederle, K.W. Benz, A.S. Senchenkov, A.V. Egorov, D.G. Matioukhin: CdTe and $CdTe_{0.9}Se_{0.1}$ crystals grown by the travelling heater method using a rotating magnetic field, J. Cryst. Growth **138**, 161–167 (1994)

6.173 J. Donecker, B. Lux, P. Reiche: Use of optical diffraction effects in crystals for growth characterization, J. Cryst. Growth **166**, 303–308 (1996)

6.174 K.F. Hulme, J.B. Mullin: Facets and anomalous solute distributions in InSb crystals, Philos. Mag. **41**, 1286–1288 (1959)

6.175 M.T. Santos, C. Marin, E. Dieguez: Morphology of $Bi_{12}GeO_{20}$ crystals grown along the ⟨111⟩ directions by the Czochralski method, J. Cryst. Growth **160**, 283–288 (1996)

6.176 P. Reiche, J. Bohm, H. Hermoneit, D. Schultze, P. Rudolph: Effect of an electrical field on the growth of lithium niobate single crystals, J. Cryst. Growth **108**, 759–764 (1991)

6.177 J.B. Mullin, K.F. Hulme: Orientation-dependent distribution coefficients in melt-grown InSb crystals, J. Phys. Chem. Solids **17**, 1–6 (1960)

6.178 Y. Liu, A. Virozub, S. Brandon: Facetting during directional growth of oxides from the melt: coupling between thermal fields, kinetics and melt/crystal interface shapes, J. Cryst. Growth **205**, 333–353 (1999)

6.179 O. Weinstein, S. Brandon: Dynamics of partially faceted melt-crystal interfaces III: Three-dimensional computational approach and calculations, J. Cryst. Growth **284**, 235–253 (2005)

6.180 M. Shibata, Y. Sasaki, T. Inada, S. Kuma: Observation of edge-facets in ⟨100⟩ InP crystals grown by LEC method, J. Cryst. Growth **102**, 557–561 (1990)

6.181 H. Gottschalk, G. Patzer, H. Alexander: Stacking fault energy and ionicity of cubic III–V compounds, Phys. Status Solidi (a) **45**, 207–217 (1978)

6.182 D.T.J. Hurle: A mechanism for twin formation during Czochralski and encapsulated vertical Bridgman growth of III–V compound semiconductors, J. Cryst. Growth **147**, 239–250 (1995)

6.183 E. Billig: Some defects in crystals grown from the melt, Proc. R. Soc. Lond. Ser. A **235**, 37–55 (1956)

6.184 V.V. Voronkov: Structure of crystal surfaces, Sov. Phys. Cryst. **19**, 573 (1975), (see also [6.179])

6.185 J. Amon, F. Dumke, G. Müller: Influence of the crucible shape on the formation of facets and twins in the growth of GaAs by the vertical gradient freeze technique, J. Cryst. Growth **187**, 1–8 (1998)

6.186 H. Chung, M. Dudley, D.J. Larson, D.T.J. Hurle, D.F. Bliss, V. Prassad: The mechanism of growth-twin formation in zincblende crystals: New insights from a study of magnetic liquid encapsulated

Czochralski-grown InP single crystals, J. Cryst. Growth **187**, 9–17 (1998)

6.187 M. Dudley, B. Raghothamachar, Y. Guo, X.R. Huang, H. Chung, D.T.J. Hurle, D.F. Bliss: The influence of polarity on twinning in zincblende structure crystals: new insights from a study of magnetic liquid encapsulated, Czochralski grown InP single crystals, J. Cryst. Growth **192**, 1–10 (1998)

6.188 Y. Hosokawa, Y. Yabuhara, R. Nakai, K. Fujita: Development of 4-inch diameter InP single crystal with low dislocation density using VCZ method, *Indium Phosphide and Related Materials*, 1998, IPRM IEEE International Conference on 11–15 May 1998 in Tsukuba, Japan, pp. 34–37

6.189 C.P. Khattak, F. Schmid: Growth of CdTe crystals by the heat exchanger method (HEM), SPIE **1106**, 47–55 (1989)

6.190 A.F. Witt: Growth of CdTe under controlled heat transfer conditions, *Final Report*, DAAG No. 29-82-K-0119 (Mater. Proc. Center M.I.T., Cambridge 1986)

6.191 P. Rudolph: Travelling magnetic fields applied to bulk crystal growth from the melt: The step from basic research to industrial scale, J. Cryst. Growth **310**, 1298–1306 (2008)